Toxin-Antitoxin Systems in Pathogenic Bacteria

Toxin-Antitoxin Systems in Pathogenic Bacteria

Editor

Juan Carlos Alonso

MDPI • Basel • Beijing • Wuhan • Barcelona • Belgrade • Manchester • Tokyo • Cluj • Tianjin

Editor
Juan Carlos Alonso
National Centre for Biotechnology (CNB-CSIC)
Spain

Editorial Office
MDPI
St. Alban-Anlage 66
4052 Basel, Switzerland

This is a reprint of articles from the Special Issue published online in the open access journal *Toxins* (ISSN 2072-6651) (available at: https://www.mdpi.com/journal/toxins/special_issues/bacterial_toxin_antitoxin_systems).

For citation purposes, cite each article independently as indicated on the article page online and as indicated below:

LastName, A.A.; LastName, B.B.; LastName, C.C. Article Title. *Journal Name* **Year**, *Volume Number*, Page Range.

ISBN 978-3-0365-0674-6 (Hbk)
ISBN 978-3-0365-0675-3 (PDF)

Contents

About the Editor

Juan Carlos Alonso is a Research Professor in the Department of Microbial Biotechnology at the Centro Nacional de Biotecnología (CNB-CSIC) of Madrid, Spain. He received his M.V., B.S, and Ph.D. degrees in Veterinary Science, and in Bacteriology, all from the National University of La Plata, La Plata, Argentina. He has been in his current position since 1994. His research is focused on understanding factors in genetic recombination, repair-by-recombination, and segregation of extrachromosomal elements (via toxin–antitoxin and genuine partition systems). Dr. Alonso has authored and co-authored more than 220 peer-reviewed journal articles.

Editorial

Toxin–Antitoxin Systems in Pathogenic Bacteria

Juan C. Alonso

Department of Microbial Biotechnology, Centro Nacional de Biotecnología, CNB-CSIC, 28049 Madrid, Spain; jcalonso@cnb.csic.es

Received: 15 January 2021; Accepted: 18 January 2021; Published: 20 January 2021

Toxin–antitoxin (TA) systems, which are ubiquitously present in plasmids, bacterial and archaeal genomes, are classified as types I to VI, according to the nature of the antitoxin and to the mode of toxin inhibition [1–5]. TA systems are not essential for normal cell proliferation, but they control a diverse repertoire of cell transition states in response to various environmental stresses [1–5]. In order to survive such stress, cells slow down their growth rate and redirect their metabolic resources until conditions improve and growth can increase [6]. During unperturbed growth, the effect of a type II toxin is neutralized by its binding to a cognate antitoxin, but under certain stress conditions, the unstable antitoxin is rapidly degraded, enabling the stable toxin to reversibly block cell proliferation, without cell lysis [1,2]. The free toxin triggers a dormant state, thus protecting cells from deleterious environments [1,2]. The toxins are bacteriostatic unless neutralized by their cognate antitoxin. Indeed, when the stress is overcome, the levels of the antitoxin rise, the toxin is neutralized and the cell returns to unperturbed growth [1–5]. Toxins are implicated in the fine tuning of multiple cellular processes, such as transcription, translation, DNA replication, the regulation of nucleotide pool, cell-wall synthesis, biofilm formation, phage predation, etc. [1–5].

This Special Issue is focused on understanding the unique response of TAs to stress, the contribution for the maintenance of drug-resistant strains, and their contribution to therapy failure and the development of chronic and recurrent infections. Understanding how TAs contribute to the mechanisms of phenotypic heterogeneity and pathogenesis may enable the rational development of new treatment for infections caused by pathogens. A review paper provides a good overview of how the widespread family of membrane active peptides, Fst/Ldr, is regulated by small RNAs in the TA type I system [7]. Weaver's review shows that the regulation of the Fst and Ldr toxins is distinct in their respective Gram-positive and Gram-negative hosts, but the effects of ectopic over-expression are similar. Limited toxin expression could conceivably function to slow bacterial growth and halt cell proliferation, playing its canonical role in plasmid stabilization [7].

A manuscript and a review describe specific aspects of the tight control that the TA interaction requires to ensure protection of the cell, and potentially to limit cross-talk between TA pairs of the same family. While each toxin interacts with its cognate antitoxin, TA systems from the same family might present non-cognate interactions and regulate the expression of non-cognate systems. Tandon et al. performed a comprehensive computational analysis on the available 3D structures and generated structural models of paralogous of VapBC and MazEF TA systems [8]. They concluded that for a majority of the systems, the non-cognate TA interactions are structurally incompatible, except for complexes such as VapBC15 and VapBC11, which show similar interfaces and a potential for cross-reactivity. This work contributes to the understanding of TA interfaces and it offers a structure-based explanation for non-cognate toxin–antitoxin interactions [8]. Tu et al. revised the concept that a toxin paralogue may provide a "cure" against the acquisition of highly similar TA systems, such as those found on plasmids or invading genetic elements that frequently carry virulence and resistance genes [9]. Only limited cross-reactions have been observed between chromosomal and mobile genetic elements systems, perhaps due to bias in the type of experiments and functions of TA systems pursued. The Ariyachaokun et al. paper examined the expression of the *mbcAT* operon and its regulation [10]. The *Mycobacterium tuberculosis* type II MbcT toxin halts cell proliferation through the phosphorolysis of the essential metabolite NAD^+, and its effect is neutralized by physical interaction with its cognate antitoxin MbcA. The authors developed a dual fluorescent reporter system, which was used to dissect the operon promoter/operator region at the genetic level. Using this system, it was demonstrated that transcription from the *PmbcA* promoter is induced by a range of stress conditions, reflecting those encountered inside the infected host and uncovering that this TA system could be exploited to treat tuberculosis [10].

Two papers within this Special Issue are focused on different aspects of toxins of the MazF/PemK superfamily that cleaves mRNA in a sequence-specific and ribosome-independent manner [11,12]. Kang et al. conducted an in-depth structural and functional analysis on the mRNA interferase (RNase) of the MazF/PemK family in *Bacillus cereus* [11]. Bleriot et al. studied the molecular mechanisms associated with chlorhexidine (CHLX) adaptation in two clinical strains of *Klebsiella pneumoniae* by phenotypic and transcriptomic analyses and their association with a new PemK/PemI TA system [12]. Klimkaitė et al. have found by bioinformatic analysis 49 putative TA systems in *Stenotrophomonas maltophilia*, and asked whether clinical and environmental isolates contain a different set of type II TA systems [13]. The authors observed that RelBE, HicAB, and the previously undescribed COG3832-ArsR operon were present solely in clinical *S. maltophilia* isolates collected in Lithuania, while HipBA was more frequent in the environmental ones. The paper by Moreno-Del Alamo et al. explores toxin ζ, which reduces the ATP and GTP levels, increases the (p)ppGpp and c-di-AMP pool and inactivates a fraction of uridine diphosphate-N-acetylglucosamine (UNAG), transiently inducing reversible dormancy. The authors, using a genetic orthogonal control of toxin ζ and antitoxin ε levels in *B. subtilis* cells, have shown that transient toxin ζ expression causes a metabolic heterogeneity that induces toxin and Amp dormancy over a long window of time rather than cell persistence. Antitoxin ε expression, by reversing ζ activities, facilitates the exit of Amp-induced dormancy both in *rec*$^+$ and *recA* cells. It has been proposed that an unexploited target to fight against antibiotic persistence is to disrupt toxin–antitoxin interactions [14]. Lastly, the paper by Tuchscherr et al. examined how the TA systems could be developed as targets for novel antimicrobials, and discussed possible undesirable effects of such therapeutic intervention, such as the induction of persister cells, biofilm formation and toxicity in eukaryotic cells [15].

Funding: Our laboratory was funded by Ministerio de Ciencia e Innovación/Agencia Estatal de Investigación (MCI/AEI)/FEDER, EU, grant number PGC2018-097054-B-I00.

Acknowledgments: The editor is grateful to all the authors who contributed their work and to the expert peer reviewers that contribute with their rigorous evaluations of all the manuscripts submitted to this Special Issue. The valuable contribution, organization, and editorial support of the MDPI management team and staff is greatly acknowledged.

Conflicts of Interest: The author declares no conflict of interest.

References

1. Yamaguchi, Y.; Inouye, M. Regulation of Growth and Death in *Escherichia coli* by Toxin-Antitoxin Systems. *Nat. Rev. Microbiol.* **2011**, *9*, 779–790. [CrossRef] [PubMed]
2. Gerdes, K.; Maisonneuve, E. Bacterial Persistence and Toxin-Antitoxin Loci. *Annu. Rev. Microbiol.* **2012**, *66*, 103–123. [CrossRef] [PubMed]
3. Page, R.; Peti, W. Toxin-Antitoxin Systems in Bacterial Growth Arrest and Persistence. *Nat. Chem. Biol.* **2016**, *12*, 208–214. [CrossRef] [PubMed]

4. Diaz-Orejas, R.; Espinosa, M.; Yeo, C.C. The Importance of the Expendable: Toxin-Antitoxin Genes in Plasmids and Chromosomes. *Front. Microbiol.* **2017**, *8*, 1479. [CrossRef] [PubMed]
5. Fraikin, N.; Goormaghtigh, F.; Van Melderen, L. Type II Toxin-Antitoxin Systems: Evolution and Revolutions. *J. Bacteriol.* **2020**, *202*. [CrossRef] [PubMed]
6. Balaban, N.Q.; Helaine, S.; Lewis, K.; Ackermann, M.; Aldridge, B.; Andersson, D.I.; Brynildsen, M.P.; Bumann, D.; Camilli, A.; Collins, J.J.; et al. Definitions and Guidelines for Research on Antibiotic Persistence. *Nat. Rev. Microbiol.* **2019**, *17*, 441–448. [CrossRef] [PubMed]
7. Weaver, K. The Fst/Ldr Family of Type I TA System Toxins: Potential Roles in Stress Response, Metabolism and Pathogenesis. *Toxins* **2020**, *12*, 474. [CrossRef] [PubMed]
8. Tandon, H.; Melarkode Vattekatte, A.; Srinivasan, N.; Sandhya, S. Molecular and Structural Basis of Cross-Reactivity in *M. tuberculosis* Toxin-Antitoxin Systems. *Toxins* **2020**, *12*, 481. [CrossRef] [PubMed]
9. Tu, C.H.; Holt, M.; Ruan, S.; Bourne, C. Evaluating the Potential for Cross-Interactions of Antitoxins in Type II TA Systems. *Toxins* **2020**, *12*, 422. [CrossRef] [PubMed]
10. Ariyachaokun, K.; Grabowska, A.D.; Gutierrez, C.; Neyrolles, O. Multi-Stress Induction of the *Mycobacterium tuberculosis* MbcTA Bactericidal Toxin-Antitoxin System. *Toxins* **2020**, *12*, 329. [CrossRef] [PubMed]
11. Kang, S.M.; Koo, J.S.; Kim, C.M.; Kim, D.H.; Lee, B.J. mRNA Interferase *Bacillus cereus* BC0266 Shows MazF-Like Characteristics Through Structural and Functional Study. *Toxins* **2020**, *12*, 380. [CrossRef] [PubMed]
12. Bleriot, I.; Blasco, L.; Delgado-Valverde, M.; Gual de Torella, A.; Ambroa, A.; Fernandez-Garcia, L.; Lopez, M.; Oteo-Iglesias, J.; Wood, T.K.; Pascual, A.; et al. Mechanisms of Tolerance and Resistance to Chlorhexidine in Clinical Strains of Klebsiella pneumoniae Producers of Carbapenemase: Role of New Type II Toxin-Antitoxin System, PemIK. *Toxins* **2020**, *12*, 566. [CrossRef] [PubMed]
13. Klimkaite, L.; Armalyte, J.; Skerniskyte, J.; Suziedeliene, E. The Toxin-Antitoxin Systems of the Opportunistic Pathogen Stenotrophomonas maltophilia of Environmental and Clinical Origin. *Toxins* **2020**, *12*, 635. [CrossRef] [PubMed]
14. Moreno-Del Alamo, M.; Marchisone, C.; Alonso, J.C. Antitoxin ε Reverses Toxin ζ-Facilitated Ampicillin Dormants. *Toxins* **2020**, *12*, 801. [CrossRef] [PubMed]
15. Rownicki, M.; Lasek, R.; Trylska, J.; Bartosik, D. Targeting Type II Toxin-Antitoxin Systems as Antibacterial Strategies. *Toxins* **2020**, *12*, 568. [CrossRef] [PubMed]

Review

The Fst/Ldr Family of Type I TA System Toxins: Potential Roles in Stress Response, Metabolism and Pathogenesis

Keith Weaver

Division of Basic Biomedical Sciences, Sanford School of Medicine, University of South Dakota, Vermillion, SD 57069, USA; kweaver@usd.edu

Received: 30 June 2020; Accepted: 23 July 2020; Published: 25 July 2020

Abstract: The par_{pAD1} locus was the first type I toxin–antitoxin (TA) system described in Gram-positive bacteria and was later determined to be the founding member of a widely distributed family of plasmid- and chromosomally encoded TA systems. Indeed, homology searches revealed that the toxin component, Fst_{pAD1}, is a member of the Fst/Ldr superfamily of peptide toxins found in both Gram-positive and Gram-negative bacteria. Regulation of the Fst and Ldr toxins is distinct in their respective Gram-positive and Gram-negative hosts, but the effects of ectopic over-expression are similar. While, the plasmid versions of these systems appear to play the canonical role of post-segregational killing stability mechanisms, the function of the chromosomal systems remains largely obscure. At least one member of the family has been suggested to play a role in pathogenesis in *Staphylococcus aureus*, while the regulation of several others appear to be tightly integrated with genes involved in sugar metabolism. After a brief discussion of the regulation and function of the foundational par_{pAD1} locus, this review will focus on the current information available on potential roles of the chromosomal homologs.

Keywords: type I toxin–antitoxin system; small protein toxin structure; Fst/Ldr family

Key Contribution: This is an in-depth comparison of the sequence, structure, regulation and function of a widespread family of type I toxin-antitoxin system small-membrane protein toxins and their potential role in stress response, metabolism and pathogenesis.

1. Introduction

The par_{pAD1} locus of the *Enterococcus faecalis* pheromone-responsive conjugative plasmid pAD1 was the first type I toxin-antitoxin (TA-1) system identified and characterized in Gram-positive bacteria [1–3]. It consists of a ~230 nucleotide mRNA, designated RNA I_{pAD1} and encoding the 33 amino acid toxin of the system, Fst_{pAD1}, and the ~40 nucleotide sRNA antitoxin, RNA II_{pAD1}. The toxin and antitoxin RNAs are transcribed from convergent promoters and share a bidirectional factor-independent transcriptional terminator [3] (Figure 1A). The resultant complementary terminator stem-loops provide the site of initiation of complex formation between toxin and antitoxin RNAs with a U-turn motif in the toxin RNA playing a key role in forming the initial reversible kissing complex [4] (Figure 1B). In addition, the toxin and antitoxin RNAs are transcribed in opposite directions across a pair of DNA direct repeats, DRa and DRb, which provide a second region of complementarity between the RNAs. Interaction in this region suppresses translation of the toxin by sequestering the Fst_{pAD1} translation initiation site [4–6]. An intramolecular stem-loop within RNA I_{pAD1}, the 5′-SL in Figure 1B, sequesters the Fst_{pAD1} Shine–Dalgarno (SD) sequence until interaction with RNA II_{pAD1} can be completed [7]. Once irreversible interaction is established at all of the complementary regions, RNA I_{pAD1} and RNA II_{pAD1} form a stable complex that facilitates the accumulation of a translationally inactive pool of the toxin message [8]. RNA II_{pAD1} is preferentially degraded from the pool by an unknown mechanism, although an intramolecular "upstream helix" (5′-UH in Figure 1B) sequestering the extreme 5′ end of RNA I_{pAD1} contributes to its differential stability [5,9]. At steady state, the two RNAs are maintained at an approximately equimolar ratio and as long as the plasmid is maintained in the cell, the antitoxin is continually replenished by ongoing transcription [8]. If the plasmid is lost, RNA II_{pAD1} is degraded from the complex pool, Fst_{pAD1} is translated from RNA I_{pAD1}, and host cell growth is inhibited [2,10]. Thus, the par_{pAD1} locus functions as a typical post-segregational killing (PSK) or addiction module [11]. The salient features of the par_{pAD1} locus and the important structural elements of RNA I_{pAD1} and RNA II_{pAD1} are shown in Figure 1.

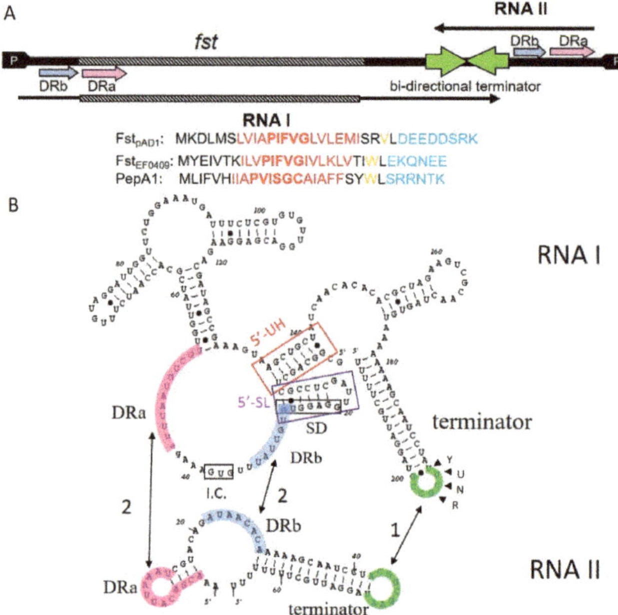

Figure 1. General organization of *par*-like operons and RNA–RNA regulatory interactions as defined in the prototypical par_{pAD1} system. (**A**) Genetic organization of *par* and conserved domains within Fst-like proteins. Convergent promoters (arrows labeled P on each end of the DNA map) produce two transcripts (arrows above and below the DNA map): the toxin message, RNA I, and the regulatory RNA, RNA II. RNAs are transcribed in opposite directions across a pair of direct repeats, DRa and DRb, and a bidirectional terminator providing regions of complementarity for interaction. RNA–RNA interactions at DRa and DRb prevent ribosome binding at the toxin coding sequence, *fst*. Shown below the map are the sequences of the two *E. faecalis* Fst paralogs, the plasmid-encoded Fst_{pAD1} and the chromosomally encoded Fst_{EF0409}, and the *S. aureus* PepA1 sequence. The conserved hydrophobic domain is shown in red with the PXXXG(C) motif bolded, the charged C-terminal tail is shown in blue and the location of the conserved tryptophan residue is show in orange. Fst_{pAD1} has a valine substitution at this position. (**B**) Map of RNA I-RNA II interaction sites. Interaction is initiated at the loop of the intrinsic terminators (green) stimulated by the YUNR U-turn motif in RNA I. Interaction then spreads to the DRa and DRb sequences that sequester the GUG initiation codon (I.C.) and the Shine–Dalgarno sequence (SD) for Fst. The interaction sites are color coordinated with those shown in the genetic map in Figure 1A. Intramolecular structures within RNA I sequester the ribosome binding site to further delay translation of *fst* (5′-SL) and stabilize the RNA I transcript (5′-UH). These figures were modified from [12].

Exhaustive searches later revealed that *par*-like loci are ubiquitous in the genomes of the Firmicutes (Gram-positive bacteria with a low G + C content) with hundreds of homologs of Fst_{pAD1} identified [12–15]. While many of these homologs were present on apparent mobile genetic elements (MGE), including phage and pathogenicity islands as well as plasmids, several loci were also identified on core genomes. Expanded examination of the surrounding sequences revealed that all of the structural features present in par_{pAD1} were highly conserved, including the convergent promoters, the bidirectional intrinsic terminator, the direct repeats and the regions of intramolecular complementarity in RNA I sequestering the 5′ end and the SD sequence, suggesting that the mechanisms of regulation of toxin expression were broadly conserved. However, not all homologs contained a U-turn motif in any of the predicted loop structures and examination of some of the regulatory details of the chromosomal loci suggest that there is variability in the mechanism of regulation (see below). Since the functions of the non-MGE-located loci must be distinct from par_{pAD1}, variability in regulatory details are to be expected. In spite of efforts made to characterize other par_{pAD1}-related loci, the only experimentally verified function of any member of the par_{pAD1} family remains the plasmid stabilization function of its founding member.

The purpose of this review is to provide a detailed comparison of the sequence, structure, regulation and effects of the toxins related to Fst_{pAD1}. This information will be used to speculate on the possible functions of the toxins in the stress response and pathogenesis of the host organisms.

2. Sequence and Structural Comparisons of the Fst/Ldr Family of Small Protein Toxins

Alignment of the Fst_{pAD1} homologs revealed a superfamily signature consisting of a highly conserved tryptophan located between a hydrophobic predicted transmembrane (TM) helix and a highly charged C-terminus [13] (Figure 1A). Interestingly, the conserved tryptophan is replaced by a valine in the Fst_{pAD1} prototype, indicating that it is not essential for function. Surprisingly, this signature was also present in the Ldr family of peptide toxins present in the Gram-negative enterobacterial TA-1 *ldr/rdl* loci [16]. In addition to the defined signature, the majority of the toxins contain a P/D/S/TXXXG(C) motif within the putative transmembrane domain (unpublished observation). The N-terminal residue of the motif varies by clade with the most common PXXXG motif present in *E. faecalis*, *Lactobacillus*, one *Streptococcus pneumoniae* and several *Staphylococcus* clades and the Ldr toxins. The DXXXG motif is present in one *S. pneumoniae* clade, the SXXXG motif in one *Staphylococcus aureus* clade, and the TXXXG motif in several *Staphylococcus* species. The *Listeria* clade is the most divergent of the Fst/Ldr family and has three glycines, one within a PKN(L/I)GF motif which fits in the PXXXG motif clade. Interestingly, all of the *Staphylococcus* isolates, but none of the isolates from other species, contain a cysteine residue at the C-terminal end of the motif and this residue has been postulated to play an important role in the function of this class (see below). While the Ldr and Fst toxins are clearly related, their mechanisms of regulation are distinct; with *ldr/rdl* loci regulated more like the *hok/sok* system of *Escherichia coli* [17].

NMR structures of two Fst family members in membrane mimetics have been determined: the Fst_{pAD1} prototype [18] and the PepA1 toxin of the *S. aureus* SprA1-$SprA1_{AS}$ locus [19] (Figure 2). Fst_{pAD1} forms a TM α-helix with the first ~two and last ~seven amino acids protruding. The α-helix structure is maintained at the proline residue in spite of its potential destabilizing effect, but is slightly bent at the glycine residue. The highly charged C-terminal seven amino acids are intrinsically unstructured. Molecular dynamic simulations suggested that the TM α-helix extends from and includes aspartic acid residues at positions 3 and 26. Since the C-terminus contains a stretch of five negatively charged amino acids at the membrane exit point and the external side of Gram-positive membranes is known to be negatively charged, the authors postulated that the C-terminus is located within the cytosol. The TM helix prediction tool TMpred [20] also predicted this orientation. A similar orientation was also proposed for LdrA by molecular modeling [21]. NMR analysis of PepA1 revealed a discontinuous α-helical TM domain with a flexible hinge near the cysteine residue that is unique to the staphylococcal toxins. Molecular dynamic simulations predicted that this α-helix would condense into an extended straight helix when inserted into the cell membrane. Unlike Fst_{pAD1}, it was predicted that the C-terminal domain of PepA1 was folded while the N-terminal domain was unstructured. It was also proposed that the arginine-rich C-terminus interacts with the anionic head groups of membrane phospholipids to lock the TM domain in place. PepA1 lacks the multiple acidic amino acids present in the Fst_{pAD1} C-terminus, so the possibility that the two toxins insert into the membrane in opposite orientations must be considered. These structural differences might account for the differences in the behavior of the two toxins. Fst_{pAD1}, and also the closely related *fst-Sm* from *Streptococcus mutans*, functions strictly from the inside of the cell and neither lyses bacterial cells nor red blood cells when added extracellularly [18,22,23]. It was proposed that, rather than forming oligomeric pores, the unstructured C-terminus of Fst_{pAD1} might interact with a specific intracellular target [18]. Conversely, PepA1 is capable of lysing both bacterial cells and red blood cells when added extracellularly [24]. It was proposed that the cysteine residue might promote intermolecular disulfide bonds facilitating oligomerization and pore formation [19]. Curiously, PepA2, a *S. aureus* TA-1 toxin 50% identical to PepA1 and containing the same PXXXGC motif, does not lyse bacterial cells extracellularly but is approximately 10X more effective in lysing red blood cells [25]. Thus far, only PepA1 and LdrA have been experimentally determined to be located in the cytoplasmic membrane of their native hosts [19,21].

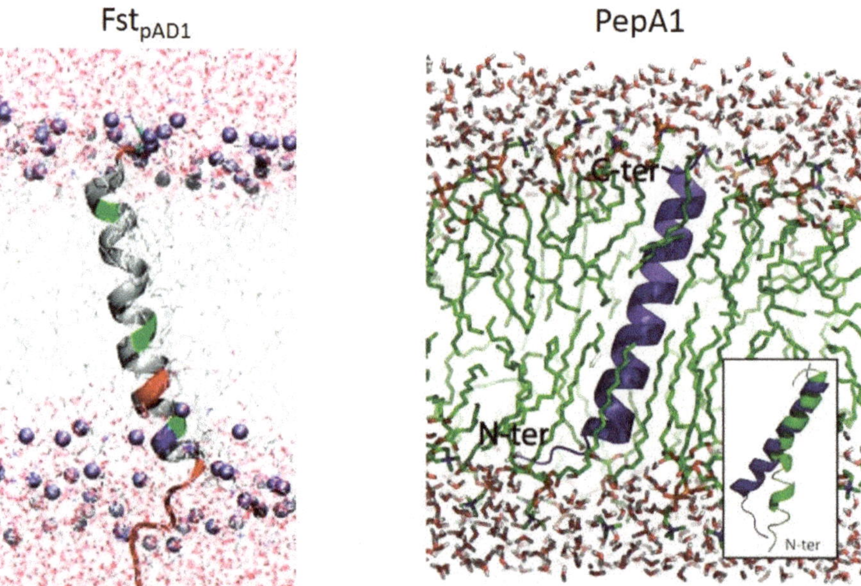

Figure 2. NMR-derived structure of two Fst-family proteins in membrane mimetics. Fst_{pAD1} presents a linear helix postulated to have the N-terminus outside the cell (top) and the C-terminus inside the cell (bottom). Polar amino acids are shown in green, negatively charged residues in red and positively charged residues in blue. Reprinted with permission from [18]. PepA1 forms a similar linear helix spanning the membrane. Membrane interaction appears to straighten a kink produced at the cysteine residue in all staphylococcal Fst-family toxins (inset). Reprinted with permission from [19].

An extensive mutagenesis survey of Fst_{pAD1} revealed several sequence features affecting toxicity in *E. faecalis* [15]. (1) Maintaining the hydrophobicity of the putative TM domain was essential for toxin function. In most cases isoleucine, leucine and valine could be interchanged for one another but alanine substitutions were not tolerated. (2) The conserved proline and glycine residues in the PXXXG motif were essential for toxicity. (3) The two charged amino acids at the N-terminus (lysine and aspartic acid) could be substituted with either negatively or positively charged, but not uncharged, amino acids without disrupting function. (4) Truncation of the charged C-terminus retained toxicity, although recent data suggests that toxicity is reduced (K. Weaver, unpublished results). In the only other mutagenic analysis of an Fst/Ldr family member published to date, the proline residue of the PXXXG motif of the *Lactobacillus rhamnosus* Lpt toxin was shown to be essential for toxicity when expressed in *E. coli* as for Fst_{pAD1} [26]. Unlike Fst_{pAD1}, however, the charged C-terminus was also required.

3. Effects of Toxin Over-Expression

As with most TA toxins, the effects of Fst/Ldr superfamily toxins have been determined primarily in toxin over-producing strains. Thus far, overproduction of Fst_{pAD1} [1] and Fst_{EF0409} [27] of *E. faecalis*, PepA1 [19] and PepA2 [25] of *S. aureus*, *fst-Sm* of *S. mutans* [22], and LdrA [21] and LdrD [16] of *E. coli* have all been shown to cause cell death of their native hosts and, in the case of Fst_{pAD1} [15,28], PepA1 [24] and PepA2 [25], several heterologous hosts as well. In addition, the Lpt toxin from a *par* homolog on a *L. rhamnosus* plasmid was demonstrated to be toxic when over-produced in *E. coli* [26]. Disruption of the cell membrane was demonstrated directly for Fst_{pAD1} [23,28], PepA1 [19] and Lpt [26] by DNA staining with membrane impermeant dyes and inferred for LdrA by simultaneous inhibition of macromolecular synthesis (which was also observed for Fst_{pAD1} [23]) and inhibition of ATP production [21]. Lpt was shown to form pores in the outer membrane of *E. coli* cells by atomic force microscopy, but the relevance of this observation to function in a native Gram-positive host without an outer membrane is uncertain [26]. It was also demonstrated that Fst_{pAD1} and nisin, a lantibiotic that disrupts both peptidoglycan biosynthesis and membrane integrity, had synergistic effects on *E. faecalis* cells, strongly supporting a cell envelope effect of Fst_{pAD1} [23]. However, membrane disruption is most likely a secondary effect as time course experiments revealed that nucleoid condensation along with aberrant chromosomal segregation and cell division occurred prior to permeation to DNA staining dyes in both *E. faecalis* and heterologous hosts [28]. Nucleoid condensation was also observed with both LdrD and Lpt [16,26]. A primary role for membrane perturbation has yet to be demonstrated for PepA1 as well, with possible effects on membrane associated functions and/or nucleoid condensation still under consideration [29].

Microarray and RNA-seq analyses revealed the induction of a preponderance of transporters, particularly ATP-utilizing transporters, in response to Fst_{pAD1} expression in *E. faecalis* cells [27,30]. A spontaneous Fst_{pAD1}-resistant mutant with a single base change in the *rpoC* gene, encoding the β' subunit of RNA polymerase, showed reduced transporter induction, suggesting that over-expression of one or more transporters might be responsible for growth inhibition, perhaps by depleting ATP pools. In support of this hypothesis, the inhibition of transporter function with the broad-spectrum translocase inhibitor reserpine had a protective effect against Fst_{pAD1} over-expression [30]. A microarray analysis of cells overproducing LdrD showed the upregulation of genes involved in purine metabolism [16].

Other than the loss of plasmid stability due to the deletion of plasmid-encoded systems, no phenotype has been associated with the deletion of any Fst/Ldr TA-1 system. Plasmid-encoded ectopic over-expression of the complete *S. mutans* Fst-Sm/srSm system led to a dramatic decrease in the number of oxacillin, cefotaxime and vancomycin tolerant persister cells [22]. Since the plasmid-encoded expression of the locus would be expected to proportionately increase both toxin mRNA and antitoxin regulatory RNA, it is not clear how this effect is established and no molecular mechanism has been suggested. Somewhat surprisingly, Michaux et al. were able to construct a mutant in the antitoxin gene of the chromosomal par_{EF0409} locus in *E. faecalis* strain V583 [31]. Such a deletion would be expected to be lethal as a result of the loss of the repression of Fst_{Ef0409} expression. However, the transcription of the toxin message from this locus is quite low and, given the intramolecular translational inhibitory structures, expression of the toxin might be below that required to induce cell death even in the absence of the antitoxin. The mutant strain showed increased virulence in a *G. mellonella* larval model and a mouse urinary tract infection model and showed increased resistance to oxidative stress, bile salts, and acidity. The authors suggested that these results implicated the par_{EF0409} locus in colonization rather than virulence. Proteomic analysis was also performed to determine the response of the mutant to the absence of the antitoxin RNA. Numerous changes potentially related to the observed increased resistance to stress were identified, but curiously the gene expression changes were distinct from those observed by RNA-seq during mild over-expression of Fst_{EF0409} in *E. faecalis* strain OG1RF [27,31]. Indeed, in one case, the expression of the *arc* operon, the effects were opposite. The reason for these differences is currently unknown, but could relate to differences in toxin expression levels, strain differences, differences between effects on transcription and translation, or the presence of second site suppressor mutations.

Given the various effects of over-production of Fst/Ldr family members, it seems unlikely that they have a generally similar mechanism of action, e.g., forming membrane pores. To examine the degree of variability directly, the transcriptomic response of *E. faecalis* OG1RF cells to Fst_{pAD1} and Fst_{EF0409}, plasmid- and chromosomally encoded toxins that presumably have distinct functions, was determined [27]. Results showed substantial differences in response to the two toxins with 113 genes showing higher expression when exposed to Fst_{pAD1} and 90 showing higher expression when exposed to Fst_{EF0409}. For example, OG1RF_RS02610, annotated as a copper-translocating P-type ATPase, is induced greater than 100-fold in response to Fst_{pAD1} but not induced significantly by Fst_{EF0409}. Conversely, OG1RF_RS01655, encoding an ABC-transporter closely linked genetically to the par_{EF0409} locus, is induced 16-fold by Fst_{EF0409} but only eightfold by Fst_{pAD1}. These results suggest that, rather than simply poking holes in the membrane, the toxin sequences may be optimized for specific functions. Interestingly, TA-1 toxins in *B. subtilis* unrelated to the Fst/Ldr family have been observed to have subtle effects that similarly bring into question a role in pore formation [32]. Given the small size of the toxins, it should be possible to identify specific amino acids responsible for distinct responses. Such results will pave the way for more detailed structure/function analyses.

At this time, it is unclear how small, membrane-localized proteins could have such disparate effects on nucleoid structure, cell division, stress response and gene expression. It seems unlikely that the proteins act directly with the transcription apparatus, so effects on gene expression are probably an indirect effect of the perturbation of the cell envelope structure or interference with membrane protein function. Discerning the mechanistic aspects of toxin function is a focus of ongoing research.

4. Regulation of Toxin Expression

The performance of a post-segregational killing (PSK) function by plasmid-encoded TA-1 systems imposes certain requirements for proper regulation. First, and most obvious, the toxin mRNA must be more stable than the antitoxin so that it can persist in cells that lose the plasmid. Second, the antitoxin must be transcribed at a level high enough to prevent translation from all available active toxin mRNA, but not so high that it allows several generations to pass without killing after plasmid loss. Ideally, the level of antitoxin to translatable toxin mRNA should be close to 1:1. Third, the antitoxin cannot immediately bind and degrade the toxin mRNA, as occurs in most negatively regulated antisense systems, because then no mRNA would remain for translation upon plasmid loss. In the case of the prototypical TA-1 system *E. coli hok/sok*, this is accomplished by the initial adoption of a conformation of *hok* toxin mRNA that can neither be translated nor interact with the *sok* antitoxin, allowing a pool of inactive mRNA to accumulate in the cell. The *hok* mRNA is then slowly degraded from the 3′ end, triggering a conformational change that allows *sok* binding and degradation if the plasmid is retained, or translation if it has been lost (for a review, see [33]). Regulation of par_{pAD1} adopts a different strategy with the formation of a mRNA:antitoxin complex that is more stable than either RNA is alone [8]. A pool of the complex is then maintained in plasmid-containing cells, but with continual replacement of the antitoxin RNA removed from the complex. The 5′-SL sequesters the Fst_{pAD1} ribosome binding site to prevent translation of transiently antitoxin-free mRNA [7]. Only after the plasmid is lost is enough antitoxin removed from the complex to free sufficient toxin mRNA for translation. It seems likely that other plasmid-encoded *par* homologs are regulated in a similar manner. One possible exception is the *L. rhamnosus* plasmid-encoded Lpt TA system which shows elevated levels of both toxin and antitoxin RNA under conditions mimicking those that occur during cheese ripening [34]

The regulation of chromosomally encoded TA systems is most likely tailored to their specific functions, so examining the regulatory features of these systems might shed light on their physiological roles. Unlike its plasmid-encoded paralog, the *E. faecalis* antitoxin RNA of chromosomally encoded par_{EF0409} is transcribed in large excess over the Fst_{EF0409} toxin mRNA. Indeed, toxin mRNA is barely detectable on Northern blots and is barely above background in qRT-PCR [27,31]. Furthermore, induced expression of the toxin mRNA ectopically reduced levels of the antitoxin RNA suggesting that, unlike in par_{pAD1}, complex formation destabilizes the antitoxin [27]. Michaux et al. reported that the antitoxin promoter showed threefold decreased activity in the presence of bile salts, a 13-fold decrease in the presence of H_2O_2 and a twofold decrease during growth in glycerol as compared to glucose [31]. However, because the antitoxin is produced in such molar excess over toxin mRNA, it seems unlikely that such small stressor-induced decreases in antitoxin RNA would liberate sufficient toxin mRNA for translation. Levels of the antitoxin RNA have also been shown to decrease in the stationary phase, but remain in excess over the toxin message [27]. The plasmid-encoded par_{pAD1} and the chromosomally encoded par_{EF0409} systems can co-exist in the same cell, apparently without interfering with each other's function. Thus, RNA II_{pAD1} protects cells from the over-expression of RNA I_{pAD1} but not RNA I_{EF0409} [27]. This is presumably due to variations in the sequence of the DRa and DRb RNA–RNA interaction sites which alignments of *par*-family systems show are poorly conserved.

The regulation of expression of two Fst homologs in *S. aureus*, PepA1 and PepA2, have been examined in detail. As in par_{EF0409}, SprA1 mRNA, encoding PepA1, is constitutively expressed at low levels and expression of the antitoxin, $SprA_{AS}$, is in large molar excess (35-90-fold) over the toxin mRNA and peaks in mid-exponential phase [24]. PepA1 production was increased by acid stress (twofold) and oxidative stress (threefold) and levels of the antitoxin were decreased by 25% and 50% by those two stresses, respectively. This is consistent with the reported decrease in par_{EF0409} antitoxin promoter activity under similar stress conditions [31]. As with Fst_{EF0409}, however, it is not clear how a fractional drop in antitoxin production would lead to toxin expression. Nonetheless, the similar expression levels and responses to stress in the two systems from divergent hosts is intriguing.

The second *S. aureus par*-homolog, SprA2/$SprA2_{AS}$, produces both toxin mRNA and regulatory antisense RNA at easily detectable levels, though the relative quantities of the toxin mRNA and antitoxin RNA were not determined [25]. Unlike par_{EF0409} but similarly to par_{pAD1}, overexpression of the toxin mRNA stabilized the antitoxin RNA. This is curious given that the SprA2 system is part of the core genome and therefore likely does not play a PSK role. Antitoxin expression levels were reduced under osmotic shock and stringent conditions, suggesting that it may play a role in stress response, but to different stresses than the SprA1/$SprA1_{AS}$ system.

In summary, while the mechanism and rationale for the regulation of plasmid-encoded PSK systems is well understood, that of the chromosomal systems, as with their functions, remains obscure. Some appear to produce toxin mRNA and antitoxin sRNA at relatively equal levels, while others produce the antitoxin sRNA in high molar excess. In the latter cases, it remains unclear how or when antitoxin levels are reduced sufficiently to allow toxin expression. Further studies on promoter structure and function would certainly be helpful, as would a broader examination of related systems in different species. An examination of this expression under a greater variety of growth conditions, including biofilm and in vivo infection models, would also be helpful.

5. Speculations on Function

Except for the well-established role of plasmid-encoded TA systems in maintaining stable inheritance, TA system functions remain largely mysterious and the case of Fst/Ldr systems is no exception. Indeed, it is even tempting to question the canonical PSK function of the plasmid-encoded systems. For example, although the par_{pAD1} locus clearly stabilizes heterologous, artificially destabilized plasmids at the expense of host cell growth rate [10], it is reasonable to ask whether this is the most efficient way to stabilize the native pAD1 plasmid. Since the plasmid has a highly efficient conjugation system and enterococci grow in chains, would it not be more efficient to just slow the growth of the plasmid-free segregant and reacquire the plasmid from a neighboring cell in the chain? This possibility certainly raises many questions about how the pheromone-response system required for conjugation would function within an enterococcal chain, but, to the best of our knowledge, no efforts have been directed toward answering these questions.

The functioning of the chromosomally encoded Fst/Ldr systems is even more speculative. As is common for TA systems in general, the deletion of the Fst/Ldr TA systems has not been associated with any phenotypic effect on host cells; so either their functions are redundant with other bacterial genes or conditions have not yet been found under which their function can be observed. As described above, overexpression of the toxins has been demonstrated to have numerous effects on the host cells, but it is unclear whether such high expression levels are ever obtained under normal physiological conditions. A variety of stress conditions have been shown to modestly reduce the expression of antitoxin RNA, which could conceivably increase toxin levels and affect growth, but this observation comes with a number of caveats. First, in the SprA1 and $\mathrm{Fst}_{\mathrm{EF0409}}$ cases, the antitoxin is produced in such excess over the toxin mRNA levels that it is hard to see how modest changes in antitoxin levels could lead to sufficient toxin expression to have an effect [24,31]. Increased toxin expression was observed in the SprA1 system, but only when the entire locus was present ectopically on a multicopy plasmid [19]. With $\mathrm{Fst}_{\mathrm{EF0409}}$, improved growth was observed in an antitoxin knockout under the same conditions that showed reduced antitoxin expression in the wild type, which provides some circumstantial evidence that the system may be involved in stress survival [31]. However, genomic sequencing should be performed to ensure that this mutant does not have compensatory mutations that allow the cell to grow in the presence of the toxin. The SprA2 antitoxin RNA also appears to be regulated by stress conditions [25], but whether this affects toxin expression has not been addressed. Overall, the evidence supporting a role for these TA loci in stress response is not strong.

The *S. aureus* SprA1-$\mathrm{SprA1}_{\mathrm{AS}}$ system is located on a pathogenicity island [24] and therefore could conceivably function as a PSK to ensure maintenance of that MGE. However, the vast excess of antitoxin RNA under the conditions examined thus far makes such a role unlikely. Both PepA1 and PepA2 are cytolytic for human cells and therefore could be considered as virulence factors during *S. aureus* infection [24,25]. However, no evidence from animal models has been presented supporting such a role. Furthermore, it is not clear how the peptide toxin would be released from the membranes of the producing bacterium to attack red blood cells. Synthetic SprA1 also lyses Gram-positive and Gram-negative bacterial cells [24], but, again, problems with release from the producing cell's membrane would limit its usefulness in microbial competition, though it does provide potential as an antibacterial agent [35].

Circumstantial evidence based on gene location suggests that some of the chromosomal *par* homologs may be integrated with core metabolic functions. For example, par_{EF0409} is located in all sequenced *E. faecalis* strains between two paralogous mannitol class phosphotransferase systems [15] (see Figure 3). Our unpublished results suggest that the downstream operon plays an essential role in mannitol transport, while the upstream operon plays a regulatory role, particularly the *mtlR* gene, which is homologous to mannitol regulators in other Gram-positive bacteria [36]. Interestingly, there is no transcriptional terminator between the *mtlF* gene and the antitoxin RNA, suggesting that readthrough transcription could modulate expression of the toxin by altering antitoxin levels. We have also observed that mannitol-grown cells are approximately 10-fold more sensitive to $\mathrm{Fst}_{\mathrm{EF0409}}$ than glucose-grown cells. Finally, a mutation of a genetically linked ABC transporter increased toxicity in both glucose and mannitol, suggesting that this transporter was important for recovery from toxin expression. How exactly Fst_{EF0409} might fit into the regulation of mannitol transport and/or metabolism is not clear, but it is worth noting that small peptides have been shown to directly modulate the transport of other sugars [37]. Alternatively, we also note that there are significant regions of DNA sequence homology between the *mtlA* and *mtlA2* genes that could result in the recombination and excision of the intervening DNA. The par_{EF0409} locus could conceivably perform a PSK role to keep the excised circle of DNA from being lost, although the ratio of antitoxin to toxin mRNA would not be ideal for this function. Several other chromosomally encoded *par* homologs show a similarly integrative association with core metabolism genes, but little is known about how these systems are regulated and function [14,15].

Figure 3. Genetic context of par_{EF0409}. The "T" marks the position of the bidirectional terminator shown in Figure 1A with the shorter white arrow on the left depicting the position of RNA II and the longer white arrow on the right depicting RNA I. The locus is situated between two paralogous mannitol-family phosphotransferase systems. The upstream operon includes *mtlR* encoding a putative positive regulator of the downstream operon.

6. Conclusions

The Fst/Ldr proteins are a widespread family of apparently membrane active peptides regulated by small RNAs in TA-1 systems. The Fst subfamily is present in the Firmicutes, including in pathogenic species of *Staphylococcus*, *Streptococcus*, *Enterococcus* and *Listeria* as well as important commensal species of *Clostridia* and *Lactobacillus*. The Ldr subfamily is present in the enterobacteria, including pathogenic species of *Escherichia*, *Shigella*, and *Salmonella*. While the toxins of the two subfamilies clearly share conserved motifs, their regulatory mechanisms are distinct, and it is unclear whether they share a common ancestor or evolved convergently. Their taxonomic coherence with the phylogenetic trees of their host bacterial species suggest that they have not been spread broadly by horizontal genetic transfer despite their presence on MGE [13].

Although widespread in numerous bacterial pathogens, the demonstration of a specific role in pathogenesis for the Fst/Ldr family has been elusive. The prototype Fst-encoding TA-1 system, par_{pAD1}, has been demonstrated to perform a stabilization function for its host plasmid and related systems are likely to perform similar functions for other MGE. The locus is highly conserved on pheromone-responsive conjugative plasmids of *E. faecalis*, which are known to encode both antibiotic resistance and virulence factors [38–40], and therefore likely plays a critical role in the maintenance and dissemination of these determinants in this important pathogen. Other putative roles in pathogenesis are more speculative. Limited toxin expression could conceivably function to slow bacterial growth and lead to the accumulation of persisters, but no evidence of such an effect has been observed and some evidence actually contradicts such a role [22]. Expression analysis suggests that several of the Fst-related toxins may be involved in stress response [24,25,31], but the mechanisms have yet to be elucidated. An *E. faecalis* strain over-producing the Fst_{EF0409} was found to increase virulence in two model systems [31], but the conditions that would lead to such levels of toxin production have not been defined. Synthetic derivatives of two *S. aureus* Fst-related toxins have been demonstrated to be cytolytic for both human and bacterial cells [24,25], but whether these toxins are actually secreted from natural producers has yet to be demonstrated. Circumstantial evidence suggests that some related loci are integrated with basic sugar metabolism [14,15], but evidence for a specific role is lacking.

Thus, like most other TA system toxins, the role of the Fst/Ldr toxins in pathogenesis, and indeed in metabolism in general, remains frustratingly obscure. A clear demonstration of function remains tantalizingly just beyond our reach. Once a function has been defined for a specific system, it will be important not to jump to the conclusion that all of the related systems perform the same function. The possibility that their functions may be tailored to the needs of specific species must always be considered. Clearly, more work is justified and required to identify the role(s) of these ubiquitous systems.

Funding: This research was funded by PHS grant AI140037.

Conflicts of Interest: The author declares no conflict of interest.

References

1. Weaver, K.E.; Clewell, D.B. Construction of Enterococcus faecalis pAD1 miniplasmids: Identification of a minimal pheromone response regulatory region and evaluation of a novel pheromone-dependent growth inhibition. *Plasmid* **1989**, *22*, 106–119. [CrossRef]
2. Weaver, K.E.; Jensen, K.D.; Colwell, A.; Sriram, S. Functional analysis of the *Enterococcus faecalis* plasmid pAD1-encoded stability determinant *par*. *Mol. Microbiol.* **1996**, *20*, 53–63. [CrossRef] [PubMed]
3. Weaver, K.E.; Tritle, D.J. Identification and characterization of an *Enterococcus faecalis* plasmid pAD1-encoded stability determinant which produces two small RNA molecules necessary for its function. *Plasmid* **1994**, *32*, 168–181. [CrossRef]
4. Greenfield, T.J.; Franch, T.; Gerdes, K.; Weaver, K.E. Antisense RNA regulation of the *par* post-segregational killing system: Structural analysis and mechanism of binding of the antisense RNA, RNAII and its target, RNAI. *Mol. Microbiol.* **2001**, *42*, 527–537. [CrossRef] [PubMed]
5. Greenfield, T.J.; Ehli, E.; Kirshenmann, T.; Franch, T.; Gerdes, K.; Weaver, K.E. The antisense RNA of the *par* locus of pAD1 regulates the expression of a 33-amino-acid toxic peptide by an unusual mechanism. *Mol. Microbiol.* **2000**, *37*, 652–660. [CrossRef] [PubMed]
6. Greenfield, T.J.; Weaver, K.E. Antisense RNA regulation of the pAD1 *par* post-segregational killing system requires interaction at the 5′ and 3′ ends of the RNAs. *Mol. Microbiol.* **2000**, *37*, 661–670. [CrossRef]
7. Shokeen, S.; Patel, S.; Greenfield, T.J.; Brinkman, C.; Weaver, K.E. Translational regulation by an intramolecular stem-loop is required for intermolecular RNA regulation of the *par* addiction module. *J. Bacteriol.* **2008**, *190*, 6076–6083. [CrossRef]
8. Weaver, K.E.; Ehli, E.A.; Nelson, J.S.; Patel, S. Antisense RNA regulation by stable complex formation in the *Enterococcus faecalis* plasmid pAD1 *par* addiction system. *J. Bacteriol.* **2004**, *186*, 6400–6408. [CrossRef]
9. Shokeen, S.; Greenfield, T.J.; Ehli, E.A.; Rasmussen, J.; Perrault, B.E.; Weaver, K.E. An intramolecular upstream helix ensures the stability of a toxin-encoding RNA in *Enterococcus faecalis*. *J. Bacteriol.* **2009**, *191*, 1528–1536. [CrossRef]
10. Weaver, K.E.; Walz, K.D.; Heine, M.S. Isolation of a derivative of *Escherichia coli–Enterococcus faecalis* shuttle vector pAM401 temperature sensitive for maintenance in *E. faecalis* and its use in evaluating the mechanism of pAD1 *par*-dependent plasmid stabilization. *Plasmid* **1998**, *40*, 225–232. [CrossRef]
11. Hayes, F. Toxins-antitoxins: Plasmid maintenance, programmed cell death, and cell cycle arrest. *Science* **2003**, *301*, 1496–1499. [CrossRef] [PubMed]
12. Weaver, K.E. The *par* toxin-antitoxin system from *Enterococcus faecalis* plasmid pAD1 and its chromosomal homologs. *RNA Biol.* **2012**, *9*, 1498–1503. [CrossRef] [PubMed]
13. Fozo, E.M.; Makarova, K.S.; Shabalina, S.A.; Yutin, N.; Koonin, E.V.; Storz, G. Abundance of type I toxin–antitoxin systems in bacteria: Searches for new candidates and discovery of novel families. *Nucleic Acids Res.* **2010**, *38*, 3743–3759. [CrossRef]
14. Kwong, S.M.; Jensen, S.O.; Firth, N. Prevalence of Fst-like toxin–antitoxin systems. *Microbiology* **2010**, *156*, 975–977. [CrossRef] [PubMed]
15. Weaver, K.E.; Reddy, S.G.; Brinkman, C.L.; Patel, S.; Bayles, K.W.; Endres, J.L. Identification and characterization of a family of toxin–antitoxin systems related to the *Enterococcus faecalis* plasmid pAD1 *par* addiction module. *Microbiology* **2009**, *155*, 2930–2940. [CrossRef]

16. Kawano, M.; Oshima, T.; Kasai, H.; Mori, H. Molecular characterization of long direct repeat (LDR) sequences expressing a stable mRNA encoding for a 35-amino-acid cell-killing peptide and a cis-encoded small antisense RNA in *Escherichia coli*. *Mol. Microbiol.* **2002**, *45*, 333–349. [CrossRef] [PubMed]
17. Kawano, M. Divergently overlapping *cis*-encoded antisense RNA regulating toxin-antitoxin systems from *E. coli*. *RNA Biol.* **2012**, *9*, 1520–1527. [CrossRef]
18. Göbl, C.; Kosol, S.; Stockner, T.; Rückert, H.M.; Zangger, K. Solution structure and membrane binding of the toxin Fst of the *par* addiction module. *Biochemistry* **2010**, *49*, 6567–6575. [CrossRef]
19. Sayed, N.; Nonin-Lecomte, S.; Réty, S.; Felden, B. Functional and structural insights of a *Staphylococcus aureus* apoptotic-like membrane peptide from a toxin-antitoxin module. *J. Biol. Chem.* **2012**, *287*, 43454–43463. [CrossRef]
20. Hofmann, K.; Stoffl, W. TMbase—A database of membrane spanning proteins segments. *Biol. Chem. Hoppe-Seyler* **1993**, *374*, 166.
21. Yamaguchi, Y.; Tokunaga, N.; Inouye, M.; Phadtare, S. Characterization of LdrA (Long direct repeat A) protein of *Escherichia coli*. *J. Mol. Microbiol. Biotechnol.* **2014**, *24*, 91–97. [CrossRef] [PubMed]
22. Koyanagi, S.; Lévesque, C.M. Characterization of a *Streptococcus mutans* intergenic region containing a small toxic peptide and its *cis*-encoded antisense small RNA antitoxin. *PLoS ONE* **2013**, *8*, e54291. [CrossRef] [PubMed]
23. Weaver, K.E.; Weaver, D.M.; Wells, C.L.; Waters, C.M.; Gardner, M.E.; Ehli, E.A. *Enterococcus faecalis* plasmid pAD1-encoded Fst toxin affects membrane permeability and alters cellular responses to lantibiotics. *J. Bacteriol.* **2003**, *185*, 2169–2177. [CrossRef]
24. Sayed, N.; Jousselin, A.; Felden, B. A *cis*-antisense RNA acts in *trans* in *Staphylococcus aureus* to control translation of a human cytolytic peptide. *Nat. Struct. Mol. Biol.* **2012**, *19*, 105–112. [CrossRef] [PubMed]
25. Germain-Amiot, N.; Augagneur, Y.; Camberlein, E.; Nicolas, I.; Lecureur, V.; Rouillon, A.; Felden, B. A novel *Staphylococcus aureus cis–trans* type I toxin–antitoxin module with dual effects on bacteria and host cells. *Nucleic Acids Res.* **2018**, *47*, 1759–1773. [CrossRef]
26. Maggi, S.; Yabre, K.; Ferrari, A.; Lazzi, C.; Kawano, M.; Rivetti, C.; Folli, C. Functional characterization of the type I toxin Lpt from *Lactobacillus rhamnosus* by fluorescence and atomic force microscopy. *Sci. Rep.* **2019**, *9*, 15208. [CrossRef]
27. Weaver, K.E.; Chen, Y.; Miiller, E.M.; Johnson, J.N.; Dangler, A.A.; Manias, D.A.; Clem, A.M.; Schjodt, D.J.; Dunny, G.M. Examination of *Enterococcus faecalis* toxin-antitoxin system toxin Fst function utilizing a pheromone-inucible expression vector with tight repression and broad dynamic range. *J. Bacteriol.* **2017**, *199*, e00065-17. [CrossRef]
28. Patel, S.; Weaver, K.E. Addiction toxin Fst has unique effects on chromosome segregation and cell division in *Enterococcus faecalis* and *Bacillus subtilis*. *J. Bacteriol.* **2006**, *188*, 5374–5384. [CrossRef]
29. Brielle, R.; Pinel-Marie, M.L.; Felden, B. Linking bacterial type I toxins with their actions. *Curr. Opin. Microbiol.* **2016**, *30*, 114–121. [CrossRef]
30. Brinkman, C.L.; Bumgarner, R.; Kittichotirat, W.; Dunman, P.M.; Kuechenmeister, L.J.; Weaver, K.E. Characterization of the effects of an *rpoC* mutation that confers resistance to the Fst peptide toxin-antitoxin system toxin. *J. Bacteriol.* **2013**, *195*, 156–166. [CrossRef]
31. Michaux, C.; Hartke, A.; Martini, C.; Reiss, S.; Albrecht, D.; Budin-Verneuil, A.; Sanguinetti, M.; Engelmann, S.; Hain, T.; Verneuil, N.; et al. Involvement of *Enterococcus faecalis* small RNAs in stress response and virulence. *Infect. Immun.* **2014**, *82*, 3599–3611. [CrossRef] [PubMed]
32. Brantl, S.; Muller, P. Toxin-antitoxin systems in *Bacillus subtilis*. *Toxins* **2019**, *11*, 262. [CrossRef] [PubMed]
33. Gerdes, K.; Wagner, E.G.H. RNA antitoxins. *Curr. Opin. Microbiol.* **2007**, *10*, 117–124. [CrossRef] [PubMed]
34. Folli, C.; Levante, A.; Percudani, R.; Amidani, D.; Bottazzi, S.; Ferrari, A.; Rivetti, C.; Neviani, E.; Lazzi, C. Toward the identification of a type I toxin-antitoxin system in the plasmid DNA of dairy *Lactobacillus rhamnosus*. *Sci. Rep.* **2017**, *7*, 12051. [CrossRef] [PubMed]
35. Solecki, O.; Mosbah, A.; Baudy Floc'h, M.; Felden, B. Converting a *Staphylococcus aureus* toxin into effective cyclic pseudopeptide antibiotics. *Chem. Biol.* **2015**, *22*, 329–335. [CrossRef]
36. Deutscher, J.; Aké, F.M.D.; Derkaoui, M.; Zébré, A.C.; Cao, T.N.; Bouraoui, H.; Kentache, T.; Mokhtari, A.; Milohanic, E.; Joyet, P. The bacterial phosphoenolpyruvate:carbohydrate phosphotransferase system: Regulation by protein phosphorylation and phosphorylation-dependent protein-protein interactions. *Microbiol. Mol. Biol. Rev.* **2014**, *78*, 231–256. [CrossRef]

37. Lloyd, C.R.; Park, S.; Fei, J.; Vanderpool, C.K. The small protein SgrT controls transport activity of the glucose-specific phosphotransferase system. *J. Bacteriol.* **2017**, *199*, e00869-16. [CrossRef]
38. Clewell, D.B. Properties of *Enterococcus faecalis* plasmid pAD1, a member of a widely disseminated family of pheromone-responding, conjugative, virulence elements encoding cytolysin. *Plasmid* **2007**, *58*, 205–227. [CrossRef]
39. Weaver, K.E. Enterococcal genetics. *Microbiol. Spectr.* **2019**, *7*. [CrossRef]
40. Zheng, B.; Tomita, H.; Inoue, T.; Ike, Y. Isolation of VanB-type *Enterococcus faecalis* strains from nosocomial infections: First report of the isolation and identification of the pheromone-responsive plasmids pMG2200, encoding VanB-type vancomycin resistance and a Bac41-type bacteriocin, and pMG2201, encoding erythromycin resistance and cytolysin (Hly/Bac). *Antimicrob. Agents Chemother.* **2009**, *53*, 735–747. [CrossRef]

Article

mRNA Interferase *Bacillus cereus* BC0266 Shows MazF-Like Characteristics Through Structural and Functional Study

Sung-Min Kang [1,†], Ji Sung Koo [1,†], Chang-Min Kim [1], Do-Hee Kim [2] and Bong-Jin Lee [1,*]

1 The Research Institute of Pharmaceutical Sciences, College of Pharmacy, Seoul National University, Gwanakgu, Seoul 08826, Korea; men0528@snu.ac.kr (S.-M.K.); koojisung@snu.ac.kr (J.S.K.); kt912@snu.ac.kr (C.-M.K.)
2 College of Pharmacy, Jeju National University, Jeju 63243, Korea; doheekim@jejunu.ac.kr
* Correspondence: lbj@nmr.snu.ac.kr; Tel.: +82-2-880-7869
† These authors contributed equally to this work.

Received: 15 May 2020; Accepted: 6 June 2020; Published: 8 June 2020

Abstract: Toxin–antitoxin (TA) systems are prevalent in bacteria and are known to regulate cellular growth in response to stress. As various functions related to TA systems have been revealed, the importance of TA systems are rapidly emerging. Here, we present the crystal structure of putative mRNA interferase BC0266 and report it as a type II toxin MazF. The MazF toxin is a ribonuclease activated upon and during stressful conditions, in which it cleaves mRNA in a sequence-specific, ribosome-independent manner. Its prolonged activity causes toxic consequences to the bacteria which, in turn, may lead to bacterial death. In this study, we conducted structural and functional investigations of *Bacillus cereus* MazF and present the first toxin structure in the TA system of *B. cereus*. Specifically, *B. cereus* MazF adopts a PemK-like fold and also has an RNA substrate-recognizing loop, which is clearly observed in the high-resolution structure. Key residues of *B. cereus* MazF involved in the catalytic activity are also proposed, and in vitro assay together with mutational studies affirm the ribonucleic activity and the active sites essential for its cellular toxicity.

Keywords: toxin–antitoxin system; mazF; type II; toxin; mRNA interferase; X-ray crystallography

Key Contribution: Structural and functional study on putative mRNA interferase BC0266 revealed that BC0266 is a *B. cereus* MazF toxin in type II TA system. This is the first reported toxin structure in type II TA system of *B. cereus*.

1. Introduction

Toxin–antitoxin (TA) systems are prevalent in prokaryotes and function to regulate cellular growth in response to stress such as antibiotic exposure, nutrient starvation, heat shock, and DNA damage. Initially, TA systems were discovered as a part of the plasmid maintenance system, in which only the daughter cells harboring the vertically transferred TA operon can survive [1–5]. In addition to the primary maintenance function, TA systems are important for bacterial survival. The TA systems are involved in many cellular processes, including cell growth, cell persistence, cell dormancy, biofilm formation, antibiotic resistance, DNA replication, translation, cell division, cell wall synthesis, and cell apoptosis [6–11].

TA systems are a two-component system, composed of a toxin and an antitoxin usually sharing the same operon. Currently, TA systems can be categorized into six phenotypes (I-VI) by the nature of each component and the regulatory mechanisms of the antitoxins to its cognate toxins. In the case of RNA antitoxins, direct inhibition of mRNA encoding toxins and protein toxins constitute the type I and type III systems, respectively. Similarly, protein toxins are directly inhibited by their counterpart protein antitoxins in type II TA system. As for indirect regulatory mechanisms, protein antitoxins counteract the activity of protein toxins in type IV TA system and cleave mRNA encoding protein toxins in type V TA system. Lastly, in type VI TA system, protein toxins are degraded by specific proteases and lose their toxicity as a result of complex formation with protein antitoxins [12–16].

Typically, in type II TA system, the rather stable toxins have much more invariant characteristics than their labile and flexible cognate antitoxins. Upon degradation of antitoxins, free toxins can function in postsegregational cell killing, abortive infection, and even bacterial persistence. Thus, it has been regarded that bacterial persistence would indeed be closely related to type II TA systems due to the cellular process-specific nature of type II toxins [17–21]. One of the best characterized type II toxins is mRNA interferase MazF toxin, belonging to the MazEF family. Comprehensive bioinformatic analyses have shown that MazF is widely distributed in both Gram-positive and Gram-negative bacteria, allowing an in-depth understanding of the consequences of stress and MazF activation on the physiological effects of the bacteria [6,22,23]. In addition, understanding of the cutting specificity of MazF on its target RNA is well established. Specifically, the MazF toxin acts as a ribonuclease and cleaves mRNA in a sequence-specific and ribosome-independent manner, but in some cases also cleaves rRNA at specific sequences, exerting its toxicity [17,24,25].

Bacillus cereus is a Gram-positive, facultatively anaerobic, and rod-shaped bacterium that is well known for its association with foodborne illness [26]. Recently, the pathogenicity potential of the bacteria gained attention due to its ability in causing serious and fatal nongastrointestinal infections [27]. As TA systems are involved in several cellular responses such as stress response, bacterial persistence, and biofilm formation, there is a clear linkage between TA systems and the virulence of the bacteria [28]. For this reason, research was conducted on the putative MazF toxin, mRNA interferase BC0266.

Here, we report 2.0 Å X-ray crystal structure of mRNA interferase MazF as the first toxin structure in the TA system of *B. cereus*. Similar to that of other MazF toxins, *B. cereus* MazF adopts a PemK-like fold and also has an elongated loop between the β1 and β2 strands. Further sequence alignment with several other MazFs supported the key residues in ribonucleic catalysis. Finally, in vitro ribonuclease activity test together with site-directed mutational studies demonstrated that *B. cereus* MazF exhibits a clear ribonuclease activity and its key residues play a significant role in exerting its role as a ribonuclease. In conclusion, we show that mRNA interferase BC0266 belongs to the MazEF family and is the first reported MazF toxin structure in the TA system of *B. cereus*. This is an important study to investigate the diversity of the TA system among many different bacterial strains.

2. Results and Discussion

2.1. *B. cereus MazF Adopts a PemK-Like Fold*

Crystals of *B. cereus* MazF were diffracted to high resolution (2.0 Å) and the final structure was determined using the molecular replacement method with the MazF structure from *Mycobacterium tuberculosis* (PDB code 5XE2) [29]. The structure was refined to R_{work} factor 18.9% and R_{free} factor 22.5%. All 116 amino acids were present and lie within the Ramachandran favored region. Crystals had $P3_121$ space group with the unit cell dimensions of $a = b = 60.648$ Å, $c = 76.247$ Å, and $\alpha = \beta = 90.0°$, $\gamma = 120°$. Detailed crystallographic data collection and refinement statistics are summarized in Table 1.

There are three α-helices and seven β-strands with the order of β-barrel arrangement in a monomer of *B. cereus* MazF (Figure 1A). Among those β-strands, five β-strands (β1, β2, β3, β6, β7) and two β-strands (β4, β5) form a β-sheet antiparallel to each other. Two loops between the β1 and β2 strands, and the β3 and β4 strands are also clearly observed. The total solvent-accessible surface area of the monomeric structure is 6740 Å^2, and the contact area between two monomers is 1440 Å^2 (21.4% per monomeric subunit). The crystal structure shows that the dimeric interface displays a concave structure covered by the neighboring loop (Figure 1B). Calculations of surface and interface areas were conducted using the PISA server [30]. The oligomeric state of *B. cereus* MazF was predicted through size-exclusion chromatography with reference proteins in the Gel Filtration Calibration Kits (GE Healthcare, Chicago, IL, USA) (Figure 1C). Because theoretical molecular weight of the *B. cereus* MazF monomer containing N-terminal His-tag is 15.1 kDa, it is reasonable that oligomeric state of *B. cereus* MazF is homodimer (30.2 kDa) as it was eluted at almost the same time as that of carbonic anhydrase (29 kDa).

2.2. Comparison of B. cereus MazF with MazF Homologs

To relate the structural characteristics of *B. cereus* MazF with mRNA interferase MazF, comparative analysis was performed with previously reported MazF structures. Illustrations of sequence alignment were displayed with *Staphylococcus aureus* MazF (PDB code 4MZM) [31], *M. tuberculosis* MazF3 (PDB code 5UCT), [32]; MazF4 (PDB code 5XE2) [29]; and MazF7 (PDB code 5WYG) [33] (Figure 2A). The results showed that two key residues of Arg25 and Thr48 involved in catalytic activity are well conserved among multiple MazFs. Interestingly, Lys19 is conserved in *M. tuberculosis* MazF4 instead of arginine [29], and Ser51 is conserved in *M. tuberculosis* MazF7 instead of threonine [33]. Yet, mutations of these two residues significantly impaired the activities of corresponding toxins, implying their role in ribonucleic catalysis [29,33]. Although considerable structural similarities are shared between the compared MazFs, there are notable differences in flexibilities and structural variations in their β1–β2 and β3–β4 loops (Figure 2B). These loops are reported as a major binding site with its cognate antitoxin MazE and, therefore, it is supposed that the flexibilities of these loops have relations to interactions with its target RNA substrate depending on MazE binding [32,34,35].

Table 1. Structure data collection and refinement statistics.

(a) Data Collection Details	
X-ray source	BL44XU beamline of Spring-8, Japan
X-ray wavelength (Å)	0.899995
Space group	$P3_121$
Unit cell parameters: *a*, *b*, *c* (Å)	60.648, 60.648, 76.247
Unit cell parameters: α, β, γ (°)	90.0, 90.0, 120.0
Resolution range (Å)	50.0-2.00
Observed reflections (>1σ)	231077
Unique reflections	21852
<I/σ(I)>	10.74 (2.97) [e]
Completeness (%)	99.3 (95.5) [e]
Multiplicity [a]	10.57 (9.76) [e]
R_{merge} (%) [b]	12.1 (47.9) [e]
$CC_{1/2}$	0.997 (0.918) [e]
(b) Refinement statistics	
R_{work} [c] (%)	18.9
R_{free} [d] (%)	22.5
No. of atoms/average *B* factor (Å^2)	997/46.2
RMSD [f] from ideal geometry: Bond distance (Å)	0.008
RMSD [f] from ideal geometry: Bond angle (°)	1.108
Ramachandran statistics: Most favored regions (%)	96.49
Ramachandran statistics: Additional allowed regions (%)	3.51
PDB accession code	7BXY

[a] N_{obs}/N_{unique}, [b] $R_{merge} = \Sigma (I - \langle I \rangle)/\Sigma \langle I \rangle$, [c] $R_{work} = \Sigma_{hkl} ||F_{obs}| - k |F_{calc}||/\Sigma_{hkl} |F_{obs}|$, [d] R_{free} was calculated in the same manner as R_{work} with 5% of the reflections excluded from the refinement. [e] Values in parentheses indicate the highest-resolution shell. [f] Root mean square deviation (RMSD) was calculated using REFMAC.

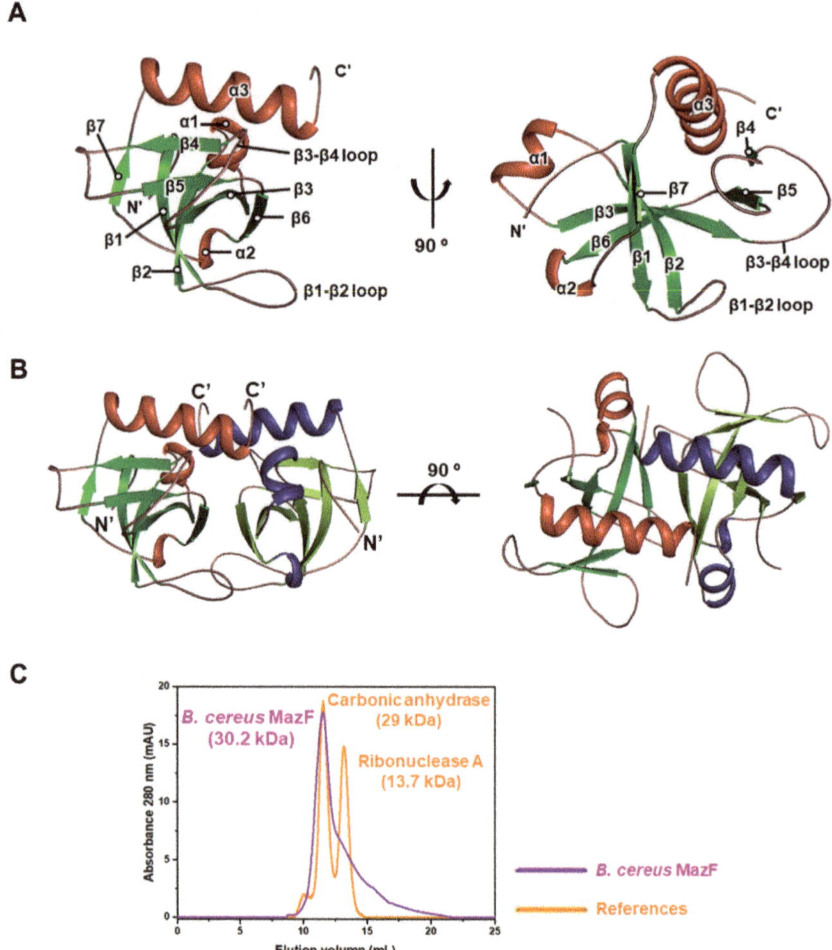

Figure 1. Overall structure of *B. cereus* MazF and gel filtration data with reference proteins. α helices are colored in red and blue. β strands are colored in green. (**A**) 90° rotational views on *B. cereus* MazF monomer. β1–β2 and β3–β4 loops are denoted. (**B**) 90° rotational views on *B. cereus* MazF homodimer. (**C**) Molecular weight estimates of *B. cereus* MazF obtained by size-exclusion chromatography with Superdex 75 10/300 gl column. Overlaid chromatograms were denoted with their names and molecular weights.

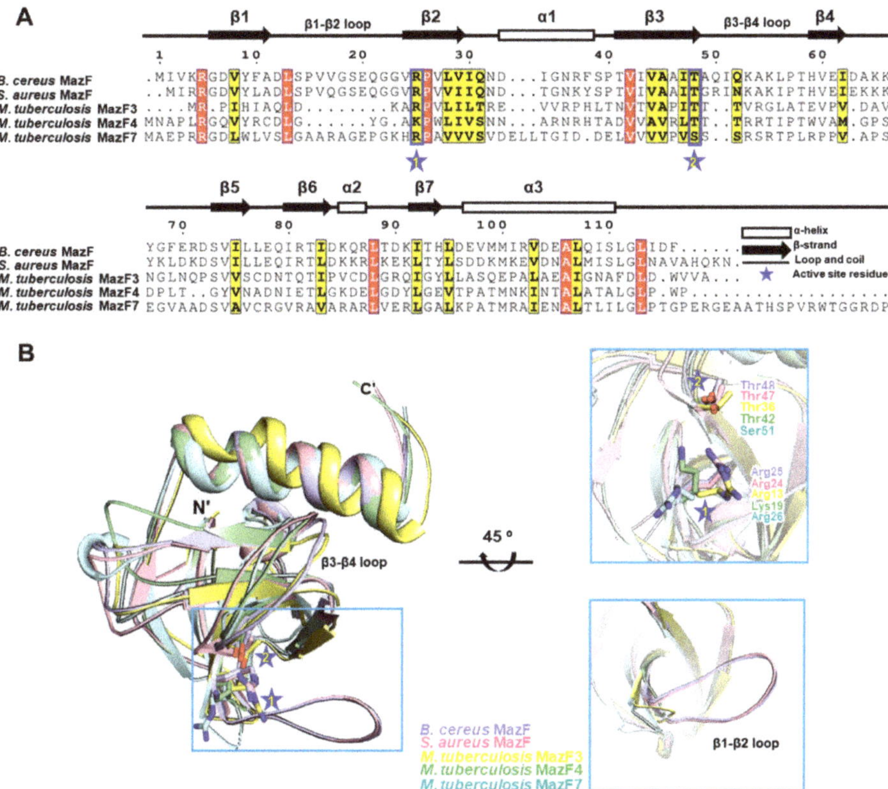

Figure 2. Comparative analysis of MazF toxins. (**A**) Sequence alignment of *B. cereus* MazF with other MazFs. Secondary structural elements are displayed above the alignment. Residues showing similarity are highlighted in red and yellow. Conserved active site residues are emphasized with star symbol. (**B**) Structural comparison of *B. cereus* MazF with other MazFs. Cartoon representations are employed to draw each structures. Conserved active site residues are shown in sticks. Conserved active site residues and variations in β1–β2 loop are illustrated by enlarged view in blue square.

Through early studies on *E. coli* MazEF, it was suggested that the β1–β2 loop in the MazF toxin undergoes conformational transition from disordered to ordered state upon binding to the MazE antitoxin [36]. During this transition, it is assumed that these loops are modulated by MazE, affecting the target RNA accessibility of MazF [36]. For instance, MazF4 and MazF7 toxins from *M. tuberculosis* with truncated β1–β2 loops had significant steric clashes within the β1–β2 loop region due to the high flexibility arising from the structurally unbound MazE antitoxin [29,33]. As for MazF3 toxin in *M. tuberculosis,* the length of the short β1–β2 loop may be insufficient to reach and recognize the downstream of the target RNA substrate [31,32]. Furthermore, it is predicted that absence of an elongated β1–β2 loop would cause even greater conformational transition of the β1–β2 loop, leaving the toxin structurally and functionally impractical. Therefore, it is suggested that target sequence specificity of truncated or short β1–β2 loops would be significantly reduced compared with MazFs having a long β1–β2 loop, such as that of *B. cereus* MazF. Therefore, it can be presumed that the long length of β1–β2 loop may be the most powerful factor that determines the target RNA specificity.

In contrast to the β1–β2 loop, β3–β4 loop shows an opposite conformational transition from ordered to disordered state upon binding to MazE antitoxin [29]. Furthermore, previous research has shown that aromatic residues located on the antitoxin MazEs are involved in essential MazEF interactions [29]. Specifically, His68 of *E. coli* MazE, Tyr61 of *B. subtilis* MazE, and Tyr76 of *M. tuberculosis* MazE4 interacted with the β3–β4 loops via van der Walls interactions [29,34,35].

Thus, considering the *E. coli* and *M. tuberculosis* model proteins, the *B. cereus* MazE antitoxin, BC0265, likely contains C-terminal aromatic residues that might interact with the β3–β4 loop in *B. cereus* MazF. Because of high structural unpredictability and lack of structural information of *B. cereus* MazE and other MazE antitoxins in general, an accurate prediction of which aromatic residue will interact with its cognate toxin is difficult. However, assuming the model is correct, one of the aromatic residues in *B. cereus* MazE, Tyr63, Phe78, Tyr82 or His86, likely engages in a hydrophobic interaction during complex formation.

2.3. Arg25 and Thr48 Act as Key Residues in Catalytic Activity

To assure that two key residues Arg25 and Thr48 are critical in catalytic activity, comparative analysis with *B. subtilis* MazF complexed with RNA (PDB code 4MDX) [34] (Figure 3A) and *E. coli* MazF complexed with DNA substrate analog (PDB code 5CR2) [35] (Figure 3B) was performed. Subsequently, in silico model of *B. cereus* MazF complexed with RNA and DNA was generated by superimposition of *B. cereus* MazF structure onto the 4MDX (left) and 5CR2 (right), respectively (Figure 3C).

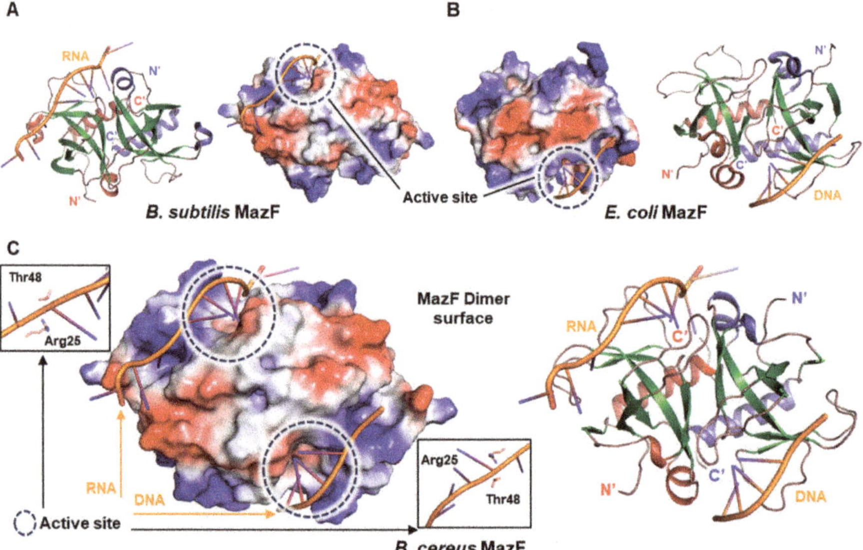

Figure 3. In silico model of previously reported MazFs and *B. cereus* MazF in complex with RNA and DNA. (**A–C**) Active sites for each proteins are illustrated in circles. (**A**) Ribbon representations and electrostatic surface potentials of *B. subtilis* MazF complexed with RNA. RNA is handled to be placed at left chain. (**B**) Ribbon representations and electrostatic surface potentials of *E. coli* MazF complexed with DNA. DNA is handled to be placed at right chain. (**C**) Superimposition of *B. cereus* MazF on *B. subtilis* MazF complexed with RNA (left chain) and *E. coli* MazF complexed with DNA (right chain). Close-up views of the interactions between active site residues and RNA/DNA are displayed.

In the in silico model of *B. cereus* MazF in complex with RNA and DNA substrate, Arg25 and Thr48 residues were well organized with RNA and DNA substrate (Figure 3C). Moreover, mutational study on *B. subtilis* MazF Arg25 and Thr48 resulted in complete inactivation of toxicity of MazF [34]. From these results, it can be assumed that Arg25 and Thr48 from *B. cereus* would also be involved in RNA catalysis. In addition, according to earlier work performed on MazF toxins, Arg25 is predicted to act as a general acid/base during RNA catalysis and Thr48 may serve as a charge stabilizer in transition state during RNA substrate binding [35,37,38].

Interestingly, charge distributions observed in the front surfaces of three proteins (Figure 3A–C) are similar but slightly different from each other, mainly due to the differences in the charged amino acid content. In *E. coli* MazF, negative-charged front surface is mostly formed by Asp16, Asp18, and Glu24 in the β1–β2 loop (residues 16 to 27). However, corresponding residues are substituted with different amino acids in both *B. subtilis* and *B. cereus* MazF. Because these front surfaces are known as the binding region of cognate antitoxin MazE, each MazE is thought to have a surface potential specific to its cognate MazF toxin to effectively bind and neutralize toxin surfaces. This seems to be the one of strong reasons why a specific toxin is neutralized only by its cognate antitoxin, although considerable structural similarities are observed within the same family of TA system [29,39,40].

2.4. *B. cereus MazF Shows Ribonuclease Activity by Two Key Residues*

MazF is a ribonuclease that is known to cleave RNA substrate [41]. Toxins such as VapC in the Type II TA system require metal ions as a cofactor to exhibit ribonuclease activity, but in the case of MazF, cofactor metal is not necessarily needed [42,43]. To demonstrate that *B. cereus* MazF shows ribonuclease activity like other MazF toxins, in vitro ribonuclease activity test was performed (Figure 4A). In this assay, a synthetic RNA strand of unknown sequence is used as substrate. A fluorophore is covalently attached to one end of the RNA strand while the quencher is located at the other end. If this RNA is cleaved by a ribonuclease, the quencher is detached and as a result fluorescence is detected depending on the amount of cleaved RNA. The result from in vitro ribonuclease assay showed an elevation of RFU (resulting fluorescence unit) as a function of time. *B. cereus* MazF exhibits a ribonuclease activity showing time and dose dependency, and hence shares the same functional activity with MazF toxins.

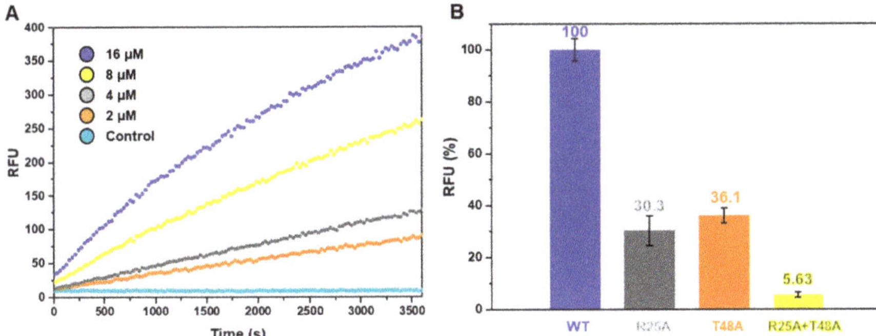

Figure 4. Ribonuclease activity test of *B. cereus* MazF toxin. (**A**) The in vitro ribonuclease activity test of wild-type *B. cereus* MazF. Fluorescence was measured as a function of time (s) during 1 h. Each curve is colored separately according to different *B. cereus* MazF concentrations. Concentrations of *B. cereus* MazF were increased by doubling from 2 to 16 µM. An equal aliquot of fluorescent RNA substrate was incubated with *B. cereus* MazF in different concentrations. Fluorescence was measured upon cleavage of RNA substrate. The control as well as different protein concentrations contained 20 mM Tris-HCl (pH 8.0), 500 mM NaCl, and 40 units of RiboLock™ (Thermo Scientific, Waltham, MA, USA) RNase inhibitor. (**B**) Ribonuclease activities of 10 µM wild-type *B. cereus* MazF and its mutants represented as bar graph with standard deviation derived from triplicate tests. Magnitudes of each reaction were assessed by subtracting the RFU (resulting fluorescence unit) of the starting point from RFU of the end point. The RFU obtained from wild-type *B. cereus* MazF was taken as 100%.

Furthermore, two single and one double site-directed mutagenesis on the two key catalytic residues of Arg25 and Thr48 were performed (Figure 4B). The results of the mutational studies showed that single mutations of either Arg25 or Thr48 resulted in severe reductions in the ribonuclease activity of toxin while double mutations of both Arg25 and Thr48 resulted in close to none ribonuclease activity. Therefore, in agreement with the results of *B. subtilis* MazF, Arg25 and Thr48 are key residues in *B. cereus* MazF, which is a characteristic observed consistently in MazF toxins.

3. Conclusions

We conducted an in-depth characterization and investigation on putative mRNA interferase MazF in *B. cereus*. High-resolution (2.0 Å) structure was obtained by X-ray crystallography, and overall structural folding of *B. cereus* MazF indicated β-barrel arrangement containing two independent β-sheets identical to that of MazF toxins. Through comparison analysis with previously reported structures of MazFs, amino acid residues Arg25 and Thr48 in *B. cereus* MazF were well conserved with other MazFs and were also confirmed as key residues in the catalytic activity of *B. cereus* MazF. Also, β1–β2 and β3–β4 loops were clearly observed in *B. cereus* MazF structure and their role in the recognition of target RNA substrate depending on the binding of its cognate antitoxin was discussed. From previously conducted studies, it can be presumed that the β1–β2 loop acts as a gateway to binding of its target RNA substrate and cognate antitoxin and the β3–β4 loop also acts as a major interaction site in antitoxin-binding via van der Walls interactions. Lastly, ribonuclease activity test via in vitro assay together with site-directed mutational studies on Arg25 and Thr48 showed a decrease in its ribonuclease activity, which was consistent in other MazF toxins. Altogether, our study provides a unified structural and functional basis that mRNA interferase BC0266 is indeed a MazF toxin. This is the first determined MazF and toxin structure in type II TA system of *B. cereus*. Our work may be a rational basis for understanding the TA systems regarding the stress-responsive toxin MazF and the general regulatory mechanisms of TA systems.

4. Materials and Methods

4.1. Cloning and Transformation

The *BC0266* gene was amplified by polymerase chain reaction (PCR). The following primers were used in PCR: forward, 5′- GGAATTCCATATGATGATTGTAAAACGCGGC-3′; reverse, 5′-CCGCTCGAGTTAAAAATCTATTAGTCCTAAAC-3′. *NdeI* and *XhoI* were used as restriction enzymes, and the cutting sites are underlined. The PCR products and pET28a vector were double-cleaved by the same restriction enzymes and ligated to each other. For purification, residual N-terminal His tag (MGSSHHHHHHSSGLVPRGSH) was attached to the gene of *B. cereus* MazF and was transformed into *E.coli* DH5α competent cells (Novagen, Madison, WI, USA). As the next steps, these cloned plasmids were re-transformed into *E. coli* Rosetta2 (DE3) pLysS competent cells (Novagen, Madison, WI, USA).

4.2. Protein Expression and Purification

Transformed cells were grown in Luria broth (LB) at 37 °C until the OD_{600} reached to 0.5. Overexpression of target protein was induced by 0.5 mM isopropyl 1-thio-B-D-galactopyranoside (IPTG). Expressed cells were further incubated for 4 h at 37 °C. These cells were harvested by centrifugation at 11,355× *g*. Harvested cells were suspended in a buffer A (20 mM Tris-HCl, pH 7.9, and 500 mM NaCl) containing 5% glycerol and lysed by ultrasonication. Lysed cells were centrifugated for 1 h at 28,306× *g*. After centrifugation, the supernatant containing soluble moiety of protein was loaded on a Ni^{2+} affinity chromatography column (Bio-Rad, Hercules, CA, USA), which was pre-equilibrated with buffer A. Loaded protein was washed with buffer A containing 50 mM imidazole to remove impurities. Then, the remaining protein, bound to the Ni^{2+} column, was eluted using an imidazole gradient (100–800 mM) and fractionized. The purity of eluted protein in each fraction was checked using sodium dodecyl sulfate polyacrylamide gel electrophoresis (SDS–PAGE). Finally, selected high-purity protein was loaded on a size-exclusion chromatography column HiLoad 16/600 Superdex 200 prep-grade column (GE Healthcare, Chicago, IL, USA), pre-equilibrated with final buffer (20 mM Tris, pH 8.0, and 500 mM NaCl). Final eluted protein was concentrated to 20 mg/mL using an

Amicon Ultra centrifugal filter unit (Millipore, Burlington, MA, USA). The purity of the final protein solution was verified by SDS-PAGE. Mutants of *B. cereus* MazF were expressed and purified by the same procedure as native ones.

4.3. Crystallization, Data Collection, and Processing

The final protein solution was screened for crystallization using crystallization kit Wizard (Rigaku Reagents, Bainbridge Island, WA, USA) and Index (Hampton Research, Aliso Viejo, CA, USA). Protein solution (1 μL) was mixed with 1 μL of each buffer solution in the crystallization kit. Crystals of *B. cereus* MazF were grown in the transparent 96-well plate using sitting-drop vapor diffusion method at 20 °C. Only one crystallized well contained the hit solution of 100 mM Sodium Citrate, pH 5.6, and 1.0 M ammonium phosphate monobasic. Cryo-protection was achieved by addition of 20% glycerol to the hit solution. After cryo-protection, crystals were immediately cooled in liquid nitrogen prior to data collection. The data were collected using an Rayonix MX-300 HE CCD detector at BL44XU of the Spring-8, Japan. All raw data of crystal were scaled and processed by XDS [44]. A set of data from a hit crystal was used to solve the structure at 2.00 Å resolution and to refine 7BXY (code name of *B. cereus* MazF). *PHENIX* [45] was first used to do molecular replacement and automatically build the model, and *COOT* [46] was utilized to yield the starting model for refinement. The R_{work}/R_{free} values [47] of the final model were obtained using *REFMAC* and *PHENIX* [45,48]. The validation of overall geometry was achieved using *MolProbity* [49]. PyMOL (PyMOL Molecular Graphics System, Version 1.2r3pre, Schrödinger, LLC, NY, USA) was used to generate figures in this study.

4.4. Mutiple Sequence Alignment

To conduct sequence alignment of *B. cereus* MazF and four MazFs whose structures have been identified, the sequence information of five proteins was browsed using Uniprot [50]. Alignments of amino acid residues were carried out using Clustal W [51] and visualized using ESPript 3.0 [52]. The consensus value was set to 0.85 and %Equivalent was used in similarity mode. In visualization, structural information of *B. cereus* MazF was used as top secondary structures and the description and sequence numbering on the topside correspond to *B. cereus* MazF.

4.5. In Vitro Ribonuclease Assay

To confirm the ribonuclease activity of *B. cereus* MazF, an RNase Alert Kit (IDT, Coralville, IA, USA) was purchased and used following the manufacturer's protocol. Equal aliquot RNA substrate (5 μL) was interacted with 2 μM, 4 μM, 8 μM, and 16 μM concentrations of *B. cereus* MazF in a final purification buffer. For ribonuclease activity test, 10 μM aliquots of wild-type *B. cereus* MazF and its mutants were prepared. Then, the resulting fluorescence units (RFU) were detected using a SPECTRAmax GEMINI XS spectro-fluorometer (Marshall Scientific, Hampton, NH, USA) at 37 °C by emission fluorescence at 520 nm upon excitation at 490 nm. Assay setting was performed on 384-well opaque plate, and all of experiments were performed in triplicate.

4.6. Site-Directed Mutagenesis

The mutations in *B. cereus* MazF (R25A, T48A, R25A+T48A) were performed using *B. cereus* MazF in pET28a (+) as template. Single-site mutations were performed in one step, and double-site mutations were performed in a step-wise manner. Reaction components were mixed and subjected to PCR machine guided by the manufacturer's protocol (EZchange Site-Directed Mutagenesis Kit, Enzynomics, Daejeon, Korea). Each plasmid was transformed into *E. coli* XL10-Gold competent cells (Agilent Technologies, Santa Clara, CA, USA), and the resulting inserted genes were verified through DNA sequencing.

Author Contributions: Conceptualization, S.-M.K. and J.S.K.; Methodology, S.-M.K. and J.S.K.; Data curation, S.-M.K. and J.S.K.; Formal analysis, S.-M.K., J.S.K and C.-M.K.; Writing, original draft preparation, S.-M.K., J.S.K.

and D.-H.K.; Writing: review and editing, S.-M.K., J.S.K., D.-H.K and B.-J.L.; All the authors approved the final version of the manuscript. All authors have read and agreed to the published version of the manuscript.

Funding: This research was funded by a National Research Foundation of Korea (NRF) grant funded by the Korean government (grant numbers 2018R1A2A1A19018526, 2018R1A5A2024425, 2019R1C1C1002128, and 2019R1I1A1A01057713) and the 2019 BK21 Plus Project for Medicine, Dentistry and Pharmacy. This work was also funded by International Program (ICR-20-05) of IPR, Osaka University.

Acknowledgments: We thank the beamline (BL) staff members at Pohang Light Source (BL-5C, BL-7A and BL11C), Korea, and SPring-8 (BL44XU) under the Collaborative Research Program of Institute for Protein Research, Osaka University (proposal No. 2019A6972).

Conflicts of Interest: The authors declare they have no conflict of interest.

References

1. Page, R.; Peti, W. Toxin-antitoxin systems in bacterial growth arrest and persistence. *Nat. Chem. Biol.* **2016**, *12*, 208–214. [CrossRef]
2. Yamaguchi, Y.; Park, J.H.; Inouye, M. Toxin-antitoxin systems in bacteria and archaea. *Annu. Rev. Genet.* **2011**, *45*, 61–79. [CrossRef] [PubMed]
3. Lee, K.Y.; Lee, B.J. Structure, Biology, and Therapeutic Application of Toxin-Antitoxin Systems in Pathogenic Bacteria. *Toxins* **2016**, *8*, 305. [CrossRef] [PubMed]
4. Ogura, T.; Hiraga, S. Mini-F plasmid genes that couple host cell division to plasmid proliferation. *Proc. Natl. Acad. Sci. USA* **1983**, *80*, 4784–4788. [CrossRef] [PubMed]
5. Jurenaite, M.; Markuckas, A.; Suziedeliene, E. Identification and Characterization of Type II Toxin-Antitoxin Systems in the Opportunistic Pathogen Acinetobacter baumannii. *J. Bacteriol.* **2013**, *195*, 3165–3172. [CrossRef]
6. Pandey, D.P.; Gerdes, K. Toxin-antitoxin loci are highly abundant in free-living but lost from host-associated prokaryotes. *Nucleic Acids Res.* **2005**, *33*, 966–976. [CrossRef]
7. Sengupta, M.; Austin, S. Prevalence and significance of plasmid maintenance functions in the virulence plasmids of pathogenic bacteria. *Infect. Immun.* **2011**, *79*, 2502–2509. [CrossRef]
8. Kwan, B.W.; Valenta, J.A.; Benedik, M.J.; Wood, T.K. Arrested protein synthesis increases persister-like cell formation. *Antimicrob. Agents Chemother.* **2013**, *57*, 1468–1473. [CrossRef]
9. Goeders, N.; Van Melderen, L. Toxin-Antitoxin Systems as Multilevel Interaction Systems. *Toxins* **2014**, *6*, 304–324. [CrossRef]
10. Mutschler, H.; Gebhardt, M.; Shoeman, R.L.; Meinhart, A. A novel mechanism of programmed cell death in bacteria by toxin-antitoxin systems corrupts peptidoglycan synthesis. *PLoS Biol.* **2011**, *9*, e1001033. [CrossRef]
11. Unoson, C.; Wagner, E.G.H. A small SOS-induced toxin is targeted against the inner membrane in Escherichia coli. *Mol. Microbiol.* **2008**, *70*, 258–270. [CrossRef]
12. Goeders, N.; Chai, R.; Chen, B.H.; Day, A.; Salmond, G.P.C. Structure, Evolution, and Functions of Bacterial Type III Toxin-Antitoxin Systems. *Toxins* **2016**, *8*, 282. [CrossRef]
13. Lobato-Marquez, D.; Diaz-Orejas, R.; Garcia-del Portillo, F. Toxin-antitoxins and bacterial virulence. *FEMS Microbiol. Rev.* **2016**, *40*, 592–609. [CrossRef]
14. Harms, A.; Brodersen, D.E.; Mitarai, N.; Gerdes, K. Toxins, Targets, and Triggers: An Overview of Toxin-Antitoxin Biology. *Mol. Cell* **2018**, *70*, 768–784. [CrossRef]
15. Wang, X.X.; Lord, D.M.; Cheng, H.Y.; Osbourne, D.O.; Hong, S.H.; Sanchez-Torres, V.; Quiroga, C.; Zheng, K.; Herrmann, T.; Peti, W.; et al. A new type V toxin-antitoxin system where mRNA for toxin GhoT is cleaved by antitoxin GhoS. *Nat. Chem. Biol.* **2012**, *8*, 855–861. [CrossRef]
16. Aakre, C.D.; Phung, T.N.; Huang, D.; Laub, M.T. A Bacterial Toxin Inhibits DNA Replication Elongation through a Direct Interaction with the beta Sliding Clamp. *Mol. Cell* **2013**, *52*, 617–628. [CrossRef] [PubMed]
17. Yamaguchi, Y.; Inouye, M. mRNA interferases, sequence-specific endoribonucleases from the toxin-antitoxin systems. *Prog. Mol. Biol. Transl. Sci.* **2009**, *85*, 467–500. [CrossRef] [PubMed]
18. Aizenman, E.; Engelberg-Kulka, H.; Glaser, G. An Escherichia coli chromosomal "addiction module" regulated by guanosine [corrected] 3′,5′-bispyrophosphate: A model for programmed bacterial cell death. *Proc. Natl. Acad. Sci. USA* **1996**, *93*, 6059–6063. [CrossRef] [PubMed]
19. Marianovsky, I.; Aizenman, E.; Engelberg-Kulka, H.; Glaser, G. The regulation of the Escherichia coli mazEF promoter involves an unusual alternating palindrome. *J. Boil. Chem.* **2001**, *276*, 5975–5984. [CrossRef] [PubMed]

20. Kedzierska, B.; Lian, L.Y.; Hayes, F. Toxin-antitoxin regulation: Bimodal interaction of YefM-YoeB with paired DNA palindromes exerts transcriptional autorepression. *Nucleic Acids Res.* **2007**, *35*, 325–339. [CrossRef]
21. Brantl, S.; Muller, P. Toxin(-)Antitoxin Systems in Bacillus subtilis. *Toxins* **2019**, *11*, 262. [CrossRef] [PubMed]
22. Leplae, R.; Geeraerts, D.; Hallez, R.; Guglielmini, J.; Dreze, P.; Van Melderen, L. Diversity of bacterial type II toxin-antitoxin systems: A comprehensive search and functional analysis of novel families. *Nucleic Acids Res.* **2011**, *39*, 5513–5525. [CrossRef] [PubMed]
23. Anantharaman, V.; Aravind, L. New connections in the prokaryotic toxin-antitoxin network: Relationship with the eukaryotic nonsense-mediated RNA decay system. *Genome Biol.* **2003**, *4*, R81. [CrossRef] [PubMed]
24. Schifano, J.M.; Edifor, R.; Sharp, J.D.; Ouyang, M.; Konkimalla, A.; Husson, R.N.; Woychik, N.A. Mycobacterial toxin MazF-mt6 inhibits translation through cleavage of 23S rRNA at the ribosomal A site. *Proc. Natl. Acad. Sci. USA* **2013**, *110*, 8501–8506. [CrossRef] [PubMed]
25. Vesper, O.; Amitai, S.; Belitsky, M.; Byrgazov, K.; Kaberdina, A.C.; Engelberg-Kulka, H.; Moll, I. Selective translation of leaderless mRNAs by specialized ribosomes generated by MazF in Escherichia coli. *Cell* **2011**, *147*, 147–157. [CrossRef] [PubMed]
26. Ehling-Schulz, M.; Lereclus, D.; Koehler, T.M. The Bacillus cereus Group: Bacillus Species with Pathogenic Potential. *Microbiol. Spectr.* **2019**, *7*. [CrossRef] [PubMed]
27. Bottone, E.J. Bacillus cereus, a volatile human pathogen. *Clin. Microbiol. Rev.* **2010**, *23*, 382–398. [CrossRef] [PubMed]
28. Fernandez-Garcia, L.; Blasco, L.; Lopez, M.; Bou, G.; Garcia-Contreras, R.; Wood, T.; Tomas, M. Toxin-Antitoxin Systems in Clinical Pathogens. *Toxins* **2016**, *8*, 227. [CrossRef]
29. Ahn, D.H.; Lee, K.Y.; Lee, S.J.; Park, S.J.; Yoon, H.J.; Kim, S.J.; Lee, B.J. Structural analyses of the MazEF4 toxin-antitoxin pair in Mycobacterium tuberculosis provide evidence for a unique extracellular death factor. *J. Biol. Chem.* **2017**, *292*, 18832–18847. [CrossRef]
30. Krissinel, E.; Henrick, K. Inference of macromolecular assemblies from crystalline state. *J. Mol. Biol.* **2007**, *372*, 774–797. [CrossRef]
31. Zorzini, V.; Buts, L.; Sleutel, M.; Garcia-Pino, A.; Talavera, A.; Haesaerts, S.; De Greve, H.; Cheung, A.; van Nuland, N.A.; Loris, R. Structural and biophysical characterization of Staphylococcus aureus SaMazF shows conservation of functional dynamics. *Nucleic Acids Res.* **2014**, *42*, 6709–6725. [CrossRef]
32. Hoffer, E.D.; Miles, S.J.; Dunham, C.M. The structure and function of Mycobacterium tuberculosis MazF-mt6 toxin provide insights into conserved features of MazF endonucleases. *J. Biol. Chem.* **2017**, *292*, 7718–7726. [CrossRef]
33. Chen, R.; Tu, J.; Liu, Z.; Meng, F.; Ma, P.; Ding, Z.; Yang, C.; Chen, L.; Deng, X.; Xie, W. Structure of the MazF-mt9 toxin, a tRNA-specific endonuclease from Mycobacterium tuberculosis. *Biochem. Biophys. Res. Commun.* **2017**, *486*, 804–810. [CrossRef]
34. Simanshu, D.K.; Yamaguchi, Y.; Park, J.H.; Inouye, M.; Patel, D.J. Structural basis of mRNA recognition and cleavage by toxin MazF and its regulation by antitoxin MazE in Bacillus subtilis. *Mol. Cell* **2013**, *52*, 447–458. [CrossRef]
35. Zorzini, V.; Mernik, A.; Lah, J.; Sterckx, Y.G.; De Jonge, N.; Garcia-Pino, A.; De Greve, H.; Versees, W.; Loris, R. Substrate Recognition and Activity Regulation of the Escherichia coli mRNA Endonuclease MazF. *J. Biol. Chem.* **2016**, *291*, 10950–10960. [CrossRef] [PubMed]
36. Kamada, K.; Hanaoka, F.; Burley, S.K. Crystal structure of the MazE/MazF complex: Molecular bases of antidote-toxin recognition. *Mol. Cell* **2003**, *11*, 875–884. [CrossRef]
37. Guillen Schlippe, Y.V.; Hedstrom, L. A twisted base? The role of arginine in enzyme-catalyzed proton abstractions. *Arch. Biochem. Biophys.* **2005**, *433*, 266–278. [CrossRef] [PubMed]
38. Dunican, B.F.; Hiller, D.A.; Strobel, S.A. Transition State Charge Stabilization and Acid-Base Catalysis of mRNA Cleavage by the Endoribonuclease RelE. *Biochemistry* **2015**, *54*, 7048–7057. [CrossRef] [PubMed]
39. Gogos, A.; Mu, H.; Bahna, F.; Gomez, C.A.; Shapiro, L. Crystal structure of YdcE protein from Bacillus subtilis. *Proteins* **2003**, *53*, 320–322. [CrossRef] [PubMed]
40. Kang, S.M.; Kim, D.H.; Jin, C.; Lee, B.J. A Systematic Overview of Type II and III Toxin-Antitoxin Systems with a Focus on Druggability. *Toxins* **2018**, *10*, 515. [CrossRef]
41. Cook, G.M.; Robson, J.R.; Frampton, R.A.; McKenzie, J.; Przybilski, R.; Fineran, P.C.; Arcus, V.L. Ribonucleases in bacterial toxin-antitoxin systems. *Biochim. Biophys. Acta* **2013**, *1829*, 523–531. [CrossRef] [PubMed]

42. Kang, S.M.; Kim, D.H.; Lee, K.Y.; Park, S.J.; Yoon, H.J.; Lee, S.J.; Im, H.; Lee, B.J. Functional details of the Mycobacterium tuberculosis VapBC26 toxin-antitoxin system based on a structural study: Insights into unique binding and antibiotic peptides. *Nucleic Acids Res.* **2017**, *45*, 8564–8580. [CrossRef] [PubMed]
43. Kim, D.H.; Kang, S.M.; Park, S.J.; Jin, C.; Yoon, H.J.; Lee, B.J. Functional insights into the Streptococcus pneumoniae HicBA toxin-antitoxin system based on a structural study. *Nucleic Acids Res.* **2018**, *46*, 6371–6386. [CrossRef] [PubMed]
44. Kabsch, W. Xds. *Acta Crystallogr. D Biol. Crystallogr.* **2010**, *66*, 125–132. [CrossRef]
45. Adams, P.D.; Afonine, P.V.; Bunkoczi, G.; Chen, V.B.; Davis, I.W.; Echols, N.; Headd, J.J.; Hung, L.W.; Kapral, G.J.; Grosse-Kunstleve, R.W.; et al. PHENIX: A comprehensive Python-based system for macromolecular structure solution. *Acta Crystallogr. D Biol. Crystallogr.* **2010**, *66*, 213–221. [CrossRef]
46. Emsley, P.; Lohkamp, B.; Scott, W.G.; Cowtan, K. Features and development of Coot. *Acta Crystallogr. D Biol. Crystallogr.* **2010**, *66*, 486–501. [CrossRef]
47. Brunger, A.T. Free R value: A novel statistical quantity for assessing the accuracy of crystal structures. *Nature* **1992**, *355*, 472–475. [CrossRef]
48. Murshudov, G.N.; Skubak, P.; Lebedev, A.A.; Pannu, N.S.; Steiner, R.A.; Nicholls, R.A.; Winn, M.D.; Long, F.; Vagin, A.A. REFMAC5 for the refinement of macromolecular crystal structures. *Acta Crystallogr. D Biol. Crystallogr.* **2011**, *67*, 355–367. [CrossRef]
49. Chen, V.B.; Arendall, W.B.; Headd, J.J.; Keedy, D.A.; Immormino, R.M.; Kapral, G.J.; Murray, L.W.; Richardson, J.S.; Richardson, D.C. MolProbity: All-atom structure validation for macromolecular crystallography. *Acta Crystallogr. D Biol. Crystallogr.* **2010**, *66*, 12–21. [CrossRef]
50. UniProt, C. UniProt: A worldwide hub of protein knowledge. *Nucleic Acids Res.* **2019**, *47*, D506–D515. [CrossRef]
51. Thompson, J.D.; Higgins, D.G.; Gibson, T.J. CLUSTAL W: Improving the sensitivity of progressive multiple sequence alignment through sequence weighting, position-specific gap penalties and weight matrix choice. *Nucleic Acids Res.* **1994**, *22*, 4673–4680. [CrossRef] [PubMed]
52. Robert, X.; Gouet, P. Deciphering key features in protein structures with the new ENDscript server. *Nucleic Acids Res.* **2014**, *42*, W320–W324. [CrossRef] [PubMed]

Article

Molecular and Structural Basis of Cross-Reactivity in *M. tuberculosis* Toxin–Antitoxin Systems

Himani Tandon [1], Akhila Melarkode Vattekatte [1,2,3,4,5], Narayanaswamy Srinivasan [1,*] and Sankaran Sandhya [1,*]

1 Molecular Biophysics Unit, Indian Institute of Science, Bangalore 560012, India; himani@iisc.ac.in (H.T.); akhila17589@gmail.com (A.M.V.)
2 Biologie Intégrée du Globule Rouge UMR_S1134, INSERM, Université Paris, Université de la Réunion, Université des Antilles, F-75739 Paris, France
3 Laboratoire d'Excellence GR-Ex, F-75739 Paris, France
4 Faculté des Sciences et Technologies, Saint Denis Messag, F-97715 La Réunion, France
5 Institut National de la Transfusion Sanguine (INTS), F-75739 Paris, France
* Correspondence: ns@iisc.ac.in (N.S.); sandhyas@iisc.ac.in (S.S.); Tel.: +91-80-2293-2837 (N.S. & S.S.)

Received: 14 May 2020; Accepted: 23 June 2020; Published: 29 July 2020

Abstract: *Mycobacterium tuberculosis* genome encodes over 80 toxin–antitoxin (TA) systems. While each toxin interacts with its cognate antitoxin, the abundance of TA systems presents an opportunity for potential non-cognate interactions. TA systems mediate manifold interactions to manage pathogenicity and stress response network of the cell and non-cognate interactions may play vital roles as well. To address if non-cognate and heterologous interactions are feasible and to understand the structural basis of their interactions, we have performed comprehensive computational analyses on the available 3D structures and generated structural models of paralogous *M. tuberculosis* VapBC and MazEF TA systems. For a majority of the TA systems, we show that non-cognate toxin–antitoxin interactions are structurally incompatible except for complexes like VapBC15 and VapBC11, which show similar interfaces and potential for cross-reactivity. For TA systems which have been experimentally shown earlier to disfavor non-cognate interactions, we demonstrate that they are structurally and stereo-chemically incompatible. For selected TA systems, our detailed structural analysis identifies specificity conferring residues. Thus, our work improves the current understanding of TA interfaces and generates a hypothesis based on congenial binding site, geometric complementarity, and chemical nature of interfaces. Overall, our work offers a structure-based explanation for non-cognate toxin-antitoxin interactions in *M. tuberculosis*.

Keywords: toxin–antitoxin; *M. tuberculosis*; bacteria; pathogenesis; protein–protein interactions; cross-talk; protein interface

Key Contribution: Structural analysis of toxin-antitoxin interactions, based on available crystal structures and structural models, enables exploration of non-cognate toxin–antitoxin interactions in *M. tuberculosis*.

1. Introduction

For a long time, the general function of toxin–antitoxin (TA) systems, was believed to be only that of plasmid addiction. However, in the past decade, other roles have been proposed for these systems [1–4]. While many studies have reported their role in persistence and cell death [5–7], it is not clear to what extent these interactions are coupled in the context of their cellular function. Presence of more than 80 TA systems in *Mycobacterium tuberculosis* raises the possibility of a complex network of interactions among TA pairs and their cellular targets [8]. Additionally, an intriguing and exciting possibility is cross-talk (or cross- interactions) among non-cognate toxins and antitoxins. By cross-talk, we mean that if toxins X_1 and X_2 and their respective cognate antitoxins Y_1 and Y_2 are expressed in the cell at the same time and share similar features, then in principle, Y_1 can neutralize X_2 or Y_2 can neutralize X_1. While the cognate interactions between X1 and X2 and Y1 and Y2 are expected, non-cognate or cross interaction between X1–Y2 or X2–Y1 are generally speculative. It is not fully understood whether the phenomenon of cross-talk is beneficial to the bacteria. Physical evidence of such interactions is sparse and include reports of chimeric MazF toxins in *Escherichia coli* that could activate endogenous $MazF_{K\text{-}12.}$ This is mediated through binding of the chimera to $MazE_{K\text{-}12}$ antitoxin, likely due to a competition with the endogenous $MazF_{K\text{-}12}$ toxin [9]. Such non-cognate interactions could disturb the delicate balance between cognate TA pairs and could result in accumulation of free toxin in the cell. It has been speculated that such non-native interactions may even lead to degeneration of these systems [8]. Nevertheless, although the possibility and existence of cross-talk seem feasible, very little experimental evidences are currently available to demonstrate their existence.

Many studies in the past have attempted to capture cross-talk between toxin–antitoxin systems in various organisms and it appears that antitoxins are very specific to their cognate toxins. For example, no cross-talk was observed between toxins and antitoxins of VapBC systems in *M. tuberculosis* [10]. However, when mutations were introduced in the wild-type VapB1 and VapB2 antitoxins from non-typeable *Haemophilus influenzae*, the mutant antitoxins showed relaxed specificity and could neutralize the non-cognate toxins [11]. Direct observation of cross-talk has so far been reported among RelBEs, MazEFs, and VapBC-MazEF systems in *M. tuberculosis*. It was observed that RelBs neutralize non-cognate RelEs through direct interactions [12]. Cross-talk has been reported between MazE9-F6 from *M. tuberculosis* [13]. Interestingly, in their study, Zhu et al. observed cross-interactions between VapB40–MazF6, VapB27–MazF6, VapB40–MazF9, and VapB27–MazF9; thereby, suggesting that toxins and antitoxins from different families can physically interact [13]. However, Ramirez et al. did not observe interactions among any of the MazEFs and suggested that cross-talk via interactions is unlikely to occur for this family of Tas, especially MazF6 [14]. Although more recently, cross-talk was reported between MazF7 and MazE9 from *M. tuberculosis* and for four SprG/SprF type I toxin-antitoxin in *Staphylococcus aureus* [15,16]. Additionally, physical interplay between cognate and non-cognate MazEF and RelBE systems from *Bifidobacterium longum* has also been reported [17]. While the Kis antitoxin from *E. coli* has been shown to inhibit CcdB, albeit with lower efficiency than cognate CcdA, the reverse was not observed [18]. Taken together, despite the presence of multiple TA systems in different organisms, cross-talk has been reported for very few of them, including *M. tuberculosis*. More experiments need to be conducted to convincingly demonstrate that this is a commonly occurring phenomenon. The growing interest in understanding the molecular basis of toxin–antitoxin interactions has seen a steady rise in the number of toxin–antitoxin structures in the protein data bank. As many as 85 non-redundant structures of cognate TA complexes from various organisms are available in the protein data bank (PDB) [19]. The availability of such structural templates provides an opportunity to analyze interactions between homologous toxin and antitoxin pairs of unknown structure. In this study, we have performed computational analyses on available 3D structures and generated homology models of cognate *M. tuberculosis* TA systems to understand the structural basis of their interactions and employed these rules to generate hypothesis on their potential for cross-talk. Through detailed analysis of structural regions of physical binding (interface) between a toxin and an antitoxin we studied the underlying rules that define their interaction. Next, we probed for the potential for cross-talk/cross-reactivity by computing the energetics involved with the structural models of non-cognate TA complexes. In addition, we have performed detailed evaluation of the stereochemical compatibility at the interface of non-cognate TA complexes. We predict that the TA interfaces of VapBC11–VapBC15 and VapBC4–VapBC5 systems are conducive to non-cognate interactions. On the other hand, a majority of the non-cognate TA systems appear unamenable to form a complex due to incompatibility at the TA interfaces. Our detailed studies are useful to capture specificity conferring residues for select TA systems studied here. We anticipate that our findings will be useful in the design of mutations that could relax the specificity of cognate TA interactions in *M. tuberculosis*. We believe that our methodology is simple and can be extended to TA systems from other organisms as and when more structures that are either experimentally solved or confidently modeled become available. We also discuss the challenges involved in modelling TA systems for such studies.

2. Results

2.1. Comparisons of TA Interfaces Predict Cases for Potential Cross-Talk and Reveal Reasons for Insulation in Others

To understand the structural basis of regulation of cross-talk among *M. tuberculosis* TA systems, we first analyzed the experimentally solved structures in detail. Based on the sequence similarity between VapC toxin sequences in *M. tuberculosis*, we had earlier proposed that they may be grouped into sub-clusters suggesting close relationships among few TA systems [20]. Thus, it is reasonable to hypothesize that the toxins and the antitoxins within a sub-group have the potential to interact with non-cognate partners, provided they are expressed in the cell at the same time. To test this hypothesis, we employed the sub-cluster alignments derived from our earlier study. The second largest sub-cluster (i.e., sub-cluster2) (Figure S1) from that study was considered for analyses because experimentally determined structures are available for at least three TA pairs (VapBC2, PDB code: 3h87; VapBC11, PDB code: 6a7v; and VapBC15, PDB code: 4chg) and a toxin (VapC21, PDB code: 5sv2) [21–24]. An alignment of the toxins revealed high structural and sequence similarity (Figure 1A) but the alignment between the antitoxins showed low sequence conservation and poor structural alignment (Figure 1B).

The interfaces of three of these TA systems were compared to probe for similarities in their interacting modes (Table 1).

Table 1. Interface similarity between different cognate VapBC structures.

Complex Structures to be Compared (Vap)	IS-Score	z-Score	% Seq. Identity	No. of Aligned Contacts	RMSD (Å)	*p*-Value
BC2–BC11	0.25	5	10	24	3.11	3.2×10^{-3}
BC2–BC15	0.24	5	5	23	3.27	3×10^{-3}
BC11–BC15	0.52	23	39	48	1.98	4.2×10^{-11}
BC4–BC5	0.81	45	40	70	0.89	2.5×10^{-19}

Similarity between the interfaces of VapBC complexes was calculated using iAlign, an interface alignment method for the structural alignment of protein–protein interfaces. IS-score is the interface similarity score, which not only measures geometric distance, but also the conservation of interfacial contact patterns. The higher the IS-score is, the higher is the interfacial similarity. % sequence identity between the interface residues of the two complexes being compared is also provided. The calculations take into consideration, the residue identities from both toxin and antitoxin chains. RMSD gives the deviation between the interface structure. *p*-value assigns the statistical significance to the score (cut-off: 10^{-5}).

It was observed that out of the three toxin–antitoxin systems from sub-cluster2, only VapBC11 and VapBC15 showed high similarity between their interfaces (IS-score: 0.52). The structures of VapBC11and VapBC15 are shown in Figure 2A with interface residues marked as sticks in different colour. A pairwise sequence comparison between the toxins VapC11 and VapC15 shows conservation of many antitoxin-binding residues (Figure 2B). Interestingly, a pairwise alignment between antitoxins VapB11 and VapB15 also shows conservation of a majority of the toxin binding residues (Figure 2C). Since our results suggest the possibility of cross-talk between VapBC11 and VapBC15, we compared the electrostatic potential surfaces of both toxins and antitoxins (Figure 2D).

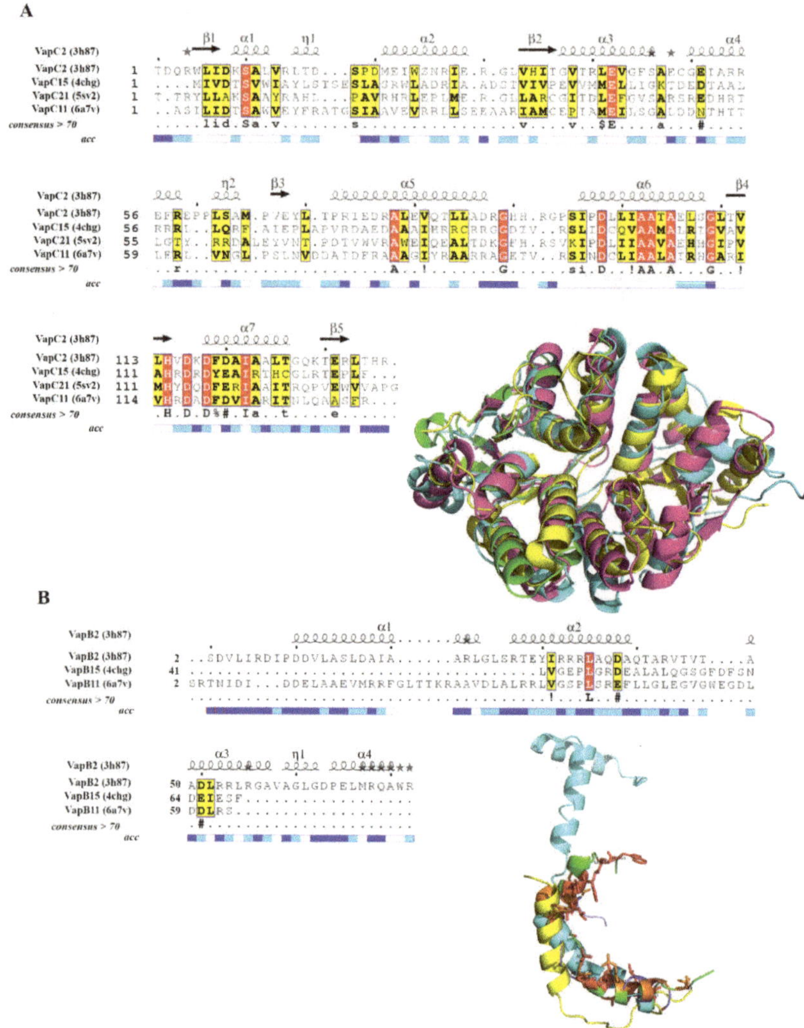

Figure 1. Alignments of VapC toxins and VapB antitoxins. (**A**) Alignment of VapC toxins of known structures (PDB codes mentioned in the figure) from sub-cluster2 is shown. Fully conserved residues are shown in red and conservatively substituted residues are shown in yellow columns. Secondary structure elements of the first structure are depicted on top of the alignment and corresponding solvent accessibility at the bottom with dark blue implying solvent accessible regions, cyan showing partially accessible regions and white showing inaccessible regions. High conservation can be appreciated from the alignment. Aligned VapC structures are shown in the inset. (**B**) Alignment of VapB antitoxins of known structures. Same colour scheme as in (**A**) is followed. Alignments were rendered using ESPript [25]. Antitoxins show poor sequence conservation. Aligned VapB structures are shown in the inset.

VapC11 and VapC15 showed similar electrostatic surfaces. Likewise, their cognate antitoxins VapB11 and VapB15 also showed similarities in their surfaces but with some differences observed at the C-terminus, as shown by change in charge distribution (discussed in Section 2.6). Therefore, non-cognate TA complexes (VapB11–VapC15 and VapB15–VapC11) were in-silico generated based on VapBC15 and VapBC11 as templates (Figure S2). Side-chains were optimized and structures were energy minimized. The calculated interface similarity score showed high similarity in the interfaces between cognate VapBC15 and modelled VapB11–VapC15 (Is-score: 0.57) and cognate VapBC11 and modelled VapB15–VapC11 (IS-score: 0.54) (Table 2). Interestingly, for the non-cognate pair VapB11–VapC15, the interaction energy (−19.1 Kcal/mol) was found to be comparable to the interaction energy of the cognate VapBC15 TA pair (−21.0 Kcal/mol). This suggests that the modelled non-cognate TA pair was reliable with no clashes between the toxins and antitoxins. However, it is important to note that the interaction energy for the non-cognate pair VapB15–VapC11 was found to be −10.7 Kcal/mol, as compared to −35.5 Kcal/mol in VapBC11. The reasons for this disparity are intriguing given the interface similarity score, as we discuss later.

Table 2. Interface similarity between non-cognate VapBC pairs.

Complex Structure	IS-Score	z-Score	% Seq. Identity	No. of Aligned Contacts	RMSD (Å)	*p*-Value	Shape Complementarity for Non-Cognate Pair (S_c)
BC15–B11C15	0.57	24	70	40	1.62	3.8×10^{-11}	0.67
BC11–B15C11	0.54	23	76	38	1.46	8.3×10^{-11}	0.60
BC4–B5C4	0.81	49	75	79	0.25	2.2×10^{-21}	0.67
BC5–B4C5	0.83	49	83	80	0.23	2.0×10^{-21}	0.65

The higher the IS-score, more is the interfacial similarity. % sequence identity between the interface residues is also provided. The calculations take into consideration the residue identities from both toxin and antitoxin chains. RMSD gives the deviation between the interface structure. *p*-value assigns the statistical significance to the score (cut-off: 10^{-5}). The S_c statistics measures the geometric surface complementarity of protein-protein interfaces.

A

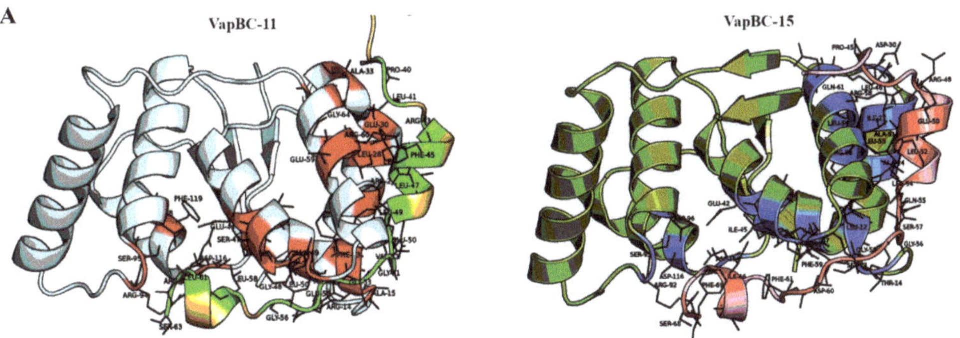

Figure 2. *Cont.*

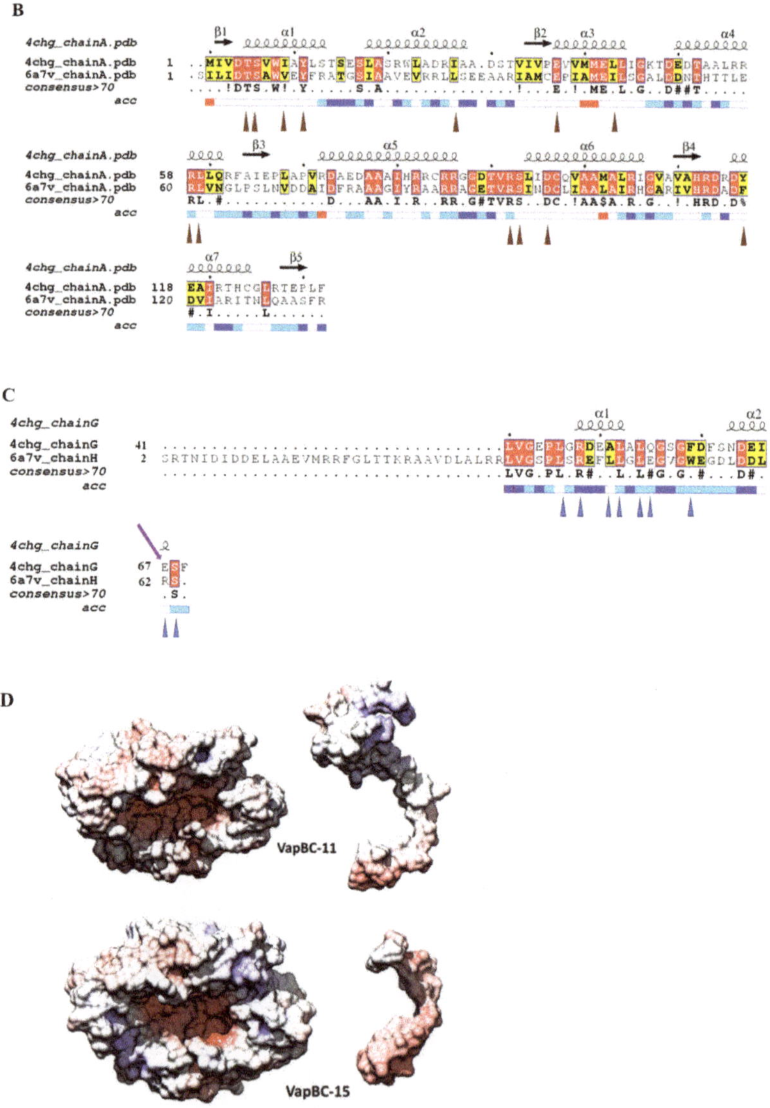

Figure 2. Comparison of VapBC11 and VapBC15 interfaces. (**A**) Cartoon rendering of VapBC11 (6a7v) and VapBC15 (4chg) structures. VapC11 toxin is shown in cyan, VapB11 antitoxin in wheat and residues lining the interfaces in red for VapC11 and light-green for VapB11. VapC15 toxin is shown in dark-green, VapB15 antitoxin in light-pink colour, residues lining VapBC15 interface in blue for VapC15 toxin and in dark-red color for VapB15 antitoxin. Residues at the interface are labelled for clarity. (**B**) Structure-guided pairwise alignment for VapC15 (4chg:A) and VapC11 (6a7v:A). Secondary structural elements of VapC15 are shown on top of the alignment and corresponding solvent accessibility at the bottom. Fully conserved residues are

shown in red and conservatively substituted residues are shown in yellow columns. Residues marked with triangles are the topologically equivalent, antitoxin-binding residues in VapC15 and VapC11. (**C**) Structure-guided pairwise alignment for VapB15 (4chg:G) and VapB11 (6a7v:H). Here also, residues marked with triangles are topologically equivalent, toxin-binding residues in VapB15 and VapB11. Interface residue comparisons in Section 2.6 suggest that the residue position marked with a purple arrow (E67) is the likely specificity conferring site in VapB15. Alignments were rendered using ESPript [25]. (**D**) The electrostatic potential surface for VapBC11 is shown on the top and for VapBC15 is shown at the bottom. Dark blue implying solvent accessible regions, cyan showing partially accessible regions, and white showing inaccessible regions. Both toxins and antitoxins show high similarity in their potential surface, suggesting similar surface interactions. The contouring of the surfaces is between ±10 kBT/e.

2.2. *Templates Could Be Identified for Toxins at a Higher Confidence than Antitoxins*

Since, very few experimentally determined structures are available for TA systems from *M. tuberculosis*, we attempted computational modelling of those toxins and antitoxins for which no experimental structures are available. Multiple sequence alignments of uncharacterized query sequences with homologues of known functions are useful in annotating their potential functional roles. The confidence of predicting functional residues in such proteins increases, if they can be assigned to the same fold associated with the homologues, at high confidence. Since the sequence identity of the toxins and antitoxins with their respective homologues was not high, templates/folds were recognized for the toxins and antitoxins using different approaches (details in Section 5) (Table S1). We observed that while folds could be predicted for toxins, the confidence of association was low for antitoxins.

As shown in the pie chart in Figure S3A, for 16% of the cases, a known fold could be assigned with high confidence and sequence identity >30%. For 68% of the cases, a known fold could be assigned with high confidence but at sequence identities <30%. We considered such assignments, despite low sequence identities, since our earlier analysis of toxin sequences showed that catalytic residues and few other residues with structural roles are conserved despite low sequence identities [20]. Nine percent of the cases were assigned to a fold with low confidence and sequence identity <30%, and for ~7% of the cases no folds could be assigned. Next, the homology models were generated for individual toxins based on the identified templates (Table S1). The aim was to enrich the TA sequence data with structural information so that it would further help in identifying and mapping the interfaces of toxin–antitoxin interactions. Nearly 46 VapC toxins and 9 MazF toxins were modelled, with 38 structures reported with appreciable validation scores from ProQ [26] and HARMONY [27].

The results showed that fold assignments for antitoxins was difficult and challenging on account of low sequence conservation (Table S1 and Figure S3B). For most antitoxins the structural assignment was ambiguous, as folds could be recognized only for 37 out of 81 antitoxins with a low confidence. Moreover, since the C-terminal toxin-binding domain of many antitoxins is known to be intrinsically disordered in the absence of toxin, it was observed that those low confidence assignments were further limited to the N-terminal DNA-binding domains. Therefore, it was infeasible to model antitoxins in their free forms and the idea of modelling antitoxins through independent template recognition was not taken forward.

2.3. *High Confidence Complexes Could Be Generated for Select TA Systems*

Since antitoxins could not be modelled in their free form, they were modelled in complex with their cognate toxin partners. We used template-based docking to model the TA complex for six TA pairs (details in Section 5). For these six TA pairs, the templates are toxin–antitoxin systems either from *M. tuberculosis* or other micro-organisms. Details of the queries and their templates along with sequence identity, query coverage, e-values, and shape complementarity are listed in Table S2. The list includes three VapBC (VapBC3, VapBC4, and VapBC21) TA systems and 3 MazEF (MazEF3, MazEF6, and MazEF9) systems. The alignments of the query TA pairs with their templates is shown in Figure S4. The interfaces were extrapolated from the templates. The side-chain optimized, energy minimized structures of the complexes are shown in Figure S5.

2.4. Assessment of the Modelled Complexes Reveals Strengths and Limitations of Homology Modelling of TA Systems

Predicted toxin–antitoxin interfaces were assessed for the six modelled TA complexes. It can be observed from Table S2 that while the sequence identity between toxin and template for VapC4, VapC21, MazF3, MazF6, and MazF9 was >25%, it was <20% for VapC3. Among the antitoxin targets and templates, the sequence identities were <20% for VapB3 and MazE3. These sequence identities lie within the twilight zone of 20–30% and hence we considered energy values, structural features, and absence of repulsive interactions to ensure the compatibility of interfaces, in addition to the geometric extrapolation from the templates. To achieve this, the S_c score and interaction energy (described in Section 5) were calculated for the modelled structures (Table 3). The S_c statistic suggested high shape complementarity for VapBC4, VapBC21, MazEF6, and MazEF9. For MazEF3, the Sc statistic suggested high shape complementarity with one MazF3 chain and low shape complementarity for the other chain. This could have been due to the slight displacement of the antitoxin with respect to the other chain. VapBC3 showed a S_c score of 0.59, suggesting that the interface complementarity was not high for the model. S_c score for the crystal structures VapBC11, VapBC15, VapBC5, and VapBC2 was calculated as 0.737, 0.739, 0.68, and 0.69, respectively.

Table 3. S_c statistics and interaction energy values for modelled complexes.

TA System	S_c Statistics	Interaction Energy (Kcal/mol)
VapBC3	0.59	-
VapBC4	0.64	−35.04
VapBC21	0.64	−15.83
MazEF3	0.68 & 0.445	−17.2
MazEF6	0.67 & 0.67	−22.2
MazEF9	0.62 & 0.65	−16.2

S_c statistics measures the geometric surface complementarity of protein–protein interfaces. The two values for MazEF complexes show the compatibility of each MazF chain with the MazE antitoxin. The interaction energy between toxin and antitoxin was calculated using AnalyseComplex module from FoldX package [28]. A lower interaction energy score of the complex is suggestive of a more stable complex.

While this study was being conducted, crystal structures for toxins VapC21 (PDB code: 5sv2), MazF3 (PDB code: 5uct), MazF6 (PDB code: 5hk0), and MazF9 (PDB code: 5hjz) [29,30] became available in the PDB [24,31]. Therefore, we compared the modelled toxin structures with their crystal structures to assess if these structures could be used to replace the toxins in the modelled TA complexes (Figure S6). RMSD between the modelled and crystal structures are shown in Table S3. Though the global RMSD is small, local structural differences were observed between them, e.g., the difference in conformation of the loop between the β1 and β2 strands (Figure S6). This loop is responsible for the switch between the antitoxin-bound and substrate-bound conformation of MazF toxins [32,33]. An attempt to generate MazEF6 and MazEF9 complexes with the experimental structures resulted in backbone clashes, suggesting that MazF6 and MazF9 were likely to undergo a conformational change in the β - β2 loop region to accommodate the cognate antitoxin, as reported in earlier studies [32,33]. Although MazEF3 and VapBC21 complexes could be generated using the crystal structure of the toxin, multiple rounds of energy minimization were required to remove side-chain clashes indicating the importance of correct side-chain geometry at the interfaces. This showed that modelling of the correct complex is dependent on the conformational state of the template proteins. Hence, it is reasonable to model toxins and antitoxins as pairs, so that the interface regions are correctly identified, which otherwise may not be accessible in the toxin/antitoxin structures alone. To further assess the compatibility of modelled TA pairs, the electrostatic potential surfaces of modelled toxins and antitoxins were compared (Figure S7). For VapBC21 complex, the surface of VapC21 was found to be moderately positively charged, and that of VapB21, which fits into the cavity of VapC21, was found to be moderately negatively charged, and hence is suggestive of the compatibility between VapB21 and VapC21 (Figure S7A). Similarly, the surface of VapC4 and VapB4 was also found to be compatible with each other (Figure S7B). The electrostatic potential surface for MazF3 showed a negatively charged surface patch near the active site and MazE3 showed a positively charged surface patch in its toxin binding region (Figure S7C). In the structure of the complex, these two regions were found to rightly dock with each other. In contrast, the MazF6 and MazF9 structures showed a positively charged patch on the surface, while their cognate antitoxins MazE6 and MazE9 showed a negatively charged patch suggesting compatibility between their toxins and antitoxins surfaces as well (Figure S7D,E).

2.5. Use of High Confidence Modelled Complexes to Explore Cross-Reactivity

Our comparative studies on the interface of modelled TA complexes and their solved structures showed that deriving models for these complexes is non-trivial. Our studies demonstrated that interpretations from the model are best limited to proposing potential interface residues but cannot be extended to reliably predict the binding mode of interaction between toxin and antitoxin. Therefore, we used only the high confidence modelled structures for further analysis.

Sub-cluster 6 was the other cluster from our grouped toxin alignments that had members with solved crystal structures (Figure S1) [20]. We were unable to derive a reasonably confident model for any of the other cluster members, for the reasons mentioned earlier. Among the members of this sub-cluster, the crystal structure for VapBC5 (PDB code: 3dbo) was available and a high confidence model could be generated for VapBC4 [34]. Before using modelled VapBC4 for the analysis, residues at the interfaces were verified from the existing literature. Mutational analysis has previously shown that residues Asp64, Trp48, Thr66, Leu72, Gln78, Ile55, Leu58, Val59, Leu61, Gly62, Leu 68, Glu71, Glu74, Thr79, Asp81, and Asp82 of VapB4 lie at the VapBC4 interface [35]. In the modelled VapBC4, the majority of these residues lay at the interface improving confidence in further analysis with this model (Table S4). Therefore, we compared the interfaces of VapBC4 and VapBC5 complexes and found high similarity indicated by the IS-score of 0.81 (Table 1 and Figure S8A). A pairwise comparison of VapC4 and VapC5 toxins showed conservation of many antitoxin-binding residues. Similarly, a pairwise comparison of VapB4 and VapB5 antitoxins showed conservation of many toxin-binding residues (Figure S8B,C). The electrostatic potential surfaces of both the Vap toxins and antitoxins were found to be similar (Figure S8D). These results raise the possibility of cross-talk among the two TA systems. Therefore, non-cognate complexes (VapB4–C5 and VapB5–C4) were generated in-silico based on the cognate crystal structure of VapBC4 (Figure S8E). Side-chains for the non-cognate pairs were optimized, and structures were energy minimized. A high similarity was observed between the interface of cognate complex VapBC4 and the modelled VapB5–VapC4, and cognate VapBC5 and the modelled VapB4–VapC5 (Table 2). Intriguingly, for these non-cognate pairs, theoretical binding energy calculations showed comparable energies between VapBC4 and VapB5–C4 (−35 Kcal/mol and −39 Kcal/mol, respectively). In contrast, interaction energy scores for VapB4–C5 was −26 Kcal/mol as compared to −34 Kcal/mol for VapBC5. Theoretically, these results predict a strong binding between the non-cognate pairs. While no experiments have been reported that refute or claim non-cognate interaction between VapC5 and VapB4, Jin et al. demonstrated earlier the inability to rescue growth defect of VapC4 by VapB5 [35]. This motivated us to further probe the discrepancy between computational predictions and experimental results. We also attempted to uncover the reason behind the large difference between interaction energies of the cognate VapBC5 and non-cognate VapC5–VapB4.

To this end, we analyzed the similarities/differences in interactions at the interface of cognate VapBC5 and VapBC4 complexes to determine if any important hot-spot interactions were lacking in the non-cognate TA models. Hot-spots are residues which when mutated to alanine, change the binding energy of the complex by more than 2 Kcal/mol [36]. Interestingly, we observed that almost all hot-spot interactions were preserved in the non-cognate VapB5–VapC4 and VapB4–VapC5 pairs when compared to cognate VapBC4 and VapBC5 pairs, respectively (Table S4). The only exceptions were the interactions between Arg12 (VapB5) and Glu60 (VapC5) in the VapBC5 complex, and Phe21 and Val14 in the VapBC4 complex (Figure S8C and Table S4). In case of non-cognate VapB5–VapC4 complex, Val14 was substituted by Arg14 and hence formed unfavorable interactions with VapC4. Since the spatial proximity of non-bonded atoms does not have a drastic effect on the overall interaction energy, we obtained comparable energies between the VapBC4 and the VapB5–VapC4 complex. Indeed, Val14 has been shown as critical for VapB4 to rescue the growth defect of VapC4 [35]. This goes on to suggest that interaction energies may not be the only criteria to assess the feasibility of a non-cognate complex and a detailed analysis of the residue properties at the interface is also important. For the non-cognate VapB4–VapC5 complex, it was observed that two negatively charged residues viz., Glu12 of VapB4 and Glu60 of VapC5, came in proximity at the interface. This is the reason behind difference in the binding energies of cognate VapBC5 and non-cognate VapB4–VapC5 and is likely to disrupt the interface. It is known that a single residue mutation can render interactions in a TA complex non-specific [11]. Hence, we believe that Arg12 (VapB5) is a likely specificity conferring residue in the VapBC5 complex and Val14 (VapB4) in the VapBC4 complex. It is to be noted that residue numbers for this case are mentioned according to the residue numbers shown for sequence alignments in Figure S8B,C.

2.6. *In-Silico Point Mutation of Antitoxin Residues Is Predicted to Relax the Specificity*

Although our analysis predicted that a non-cognate VapB15–VapC11 interaction would be favourable, with a negative interaction energy, it was much weaker than the cognate pair. Therefore, we explored the reasons for these weak interactions. The pairwise alignment between VapB15 and VapB11 showed a small yet significant difference between the topologically equivalent residues; Glu at position 67 in VapB15, and Arg at position 62 in VapB11 (Figure 2C). Residue Glu67 in VapB15 coordinated with the positively charged Mn^{2+} and Mg^{2+} ions in the catalytic site of VapC15. Arg62 in VapC11 (which is solved without a metal ion) bound with the negatively charged aspartate residues. This difference was also reflected in the electrostatic potential surfaces of C-terminals of VapB15 and VapB11 (Figure 2D). We found that because Arg62 in VapB11 could favorably interact with aspartate/glutamate of the active site in VapC15, the interaction energy between VapB11 and VapC15 was close to the cognate VapBC15 pair. However, the negatively charged Glu67 in VapB15 contributed unfavorably to the negatively charged active site of VapC11 toxin, rendering an energetically less favourable non-cognate TA pair in comparison to the cognate VapBC11 TA pair. This led to the identification of a residue position, which was likely the specificity conferring site for VapB15 and was in-silico mutated to Arg67 in VapB15C11 complex (Figure S9A).

In-silico point mutation in FoldX is known to alter only the side chain with no effect on the backbone conformation. Rotamer conformation was optimized and the mutated complex was energy minimized till no further improvement in energy. The rationale behind this mutation was to check if E67R mutation can lower the energy of VapB15–VapC11 complex. Such residues, which contribute strongly to the interaction energies between two proteins, are termed as super-hotspot residues and confer strength and specificity to the protein complex [37]. Indeed, it was observed that E67R mutation contributes favorably and a marked improvement in the interaction energy (−21.7 Kcal/mol) was observed over wild-type VapB15–VapC11 complex (−10.1 Kcal/mol). This difference is well above the error rate of FoldX and can be considered reliable. Hence, it is hypothesized that residue Glu67 was likely a specificity conferring residue in VapB and a single point mutation could relax its specificity. Similarly, for VapB4–VapC5 complex, Glu12 of VapB5 was mutated to Arginine (E12R) (Figure S9B). The mutated complex was energy minimized and the interaction energies were calculated. It was observed that the mutation E12R contributed favorably and further lowered the energy of the (R12)VapB4–VapC5 complex to −30.3 Kcal/mol over the wild-type VapB4–VapC5 complex (−26 Kcal/mol). This lends further support to Arg12 being the specificity conferring residue in VapB5. Indeed, experiments have been performed previously on *H. influenzae* VapBC1/VapBC2, where a single or double mutation in the antitoxin was found to relax the specificity of antitoxin towards its cognate toxin [11]. So, for both non-cognate complexes, we present testable hypothesis, which could be helpful in designing mutations to relax the specificity in TA systems.

2.7. Modelled Non-Cognate TA Pairs that Fail to Show Cross-Talk in Experiments Differ in Their Interfaces

Ramage et al. demonstrated the inability of select TA systems to cross-react through experiments [10]. However, the growth rescue experiments did not explain the basis for this lack of interaction. We wanted to offer a structural basis of why this interaction was not observed, and therefore performed a computational analysis to be used as control for our study. Out of those few, VapBC2 and VapBC11 belong to the sub-cluster2. Since, the experimental structures are available for VapBC2 and VapBC11, their interfaces were analyzed to explain the basis of the absence of cross-reactivity. As depicted in Figure 1, VapB2 and VapB11 show poor conservation. From Table 1, it can be observed that the interface similarity between VapBC2 and VapBC11 is very low with poor statistical scores. Despite meagre sequence and interface similarity, a brute-force structure for non-cognate VapB11–VapC2 was modelled using VapBC2 structure as template. The electrostatic potential surface and interface comparison of VapC2 and VapB11 showed incompatibility as well (Table 4 and Figure 3).

Table 4. Interface similarity between non-cognate VapBC pairs known not to interact.

Complex Structure	IS-Score	z-Score	% Seq. Identity	RMSD (Å)	p-Value	Shape Complementarity (Sc)
BC2–B11C2	0.3	9.5	No residue from antitoxin	2.9	7×10^{-5}	0.60
BC11–B2C11	0.25	5.7	No residue from antitoxin	3.18	3.3×10^{-3}	0.59

IS-score is the interface similarity score, which not only measures geometric distance, but also the conservation of interfacial contact patterns. The higher the IS-score, more is the interfacial similarity. *p*-value assigns the statistical significance to the score (cut-off: 10^{-5}).

A closer inspection of the interface residues contributed by VapB2 and VapB11 revealed conservation of only one residue. Interestingly, the majority of the antitoxin residues at the interfaces showed incompatible physicochemical properties advocating for dissimilarity and incompatible interactions between this non-cognate pair. Similarly, the interface similarity between the non-cognate TA pair VapB2–VapC11 and VapBC11 was also found to be low (Table 4).

2.8. MazEF Systems from M. tuberculosis Show Weak Signals for Cross-Reactivity

There are 10 MazEF TA systems reported in *M. tuberculosis* so far [38]. BLASTP searches revealed that while the majority of the MazF toxins are homologous to each other, their corresponding antitoxins do not identify each other. Hence, unlike VapB antitoxins that could be clustered into small groups [20], MazE antitoxins could not be grouped together. Out of the 10, experimental structures are available for five MazF toxins (MazF3 (PDB code: 5uct), MazF4 (PDB code: 5xe2), MazF6 (PDB code: 5hk0), MazF7 (PDB code: 5wyg), and MazF9 (PDB code: 5hjz)) and two MazEF pairs (MazEF4, PDB code: 5xe3 and MazEF7, PDB code: 6a6x) [16,29–31,39]. A structural alignment of MazF toxins revealed high overall similarity among them with few conserved antitoxin-binding residues suggesting similarity in their interfaces (Figure 4A). However, it is known that the length of the loop region between β1 and β2 strands defines the conformational state of MazF protein which is specifically modulated by the cognate antitoxins. Moreover, a structure-guided sequence alignment of MazE antitoxins clearly showed poor residue conservation, suggesting specific interactions with cognate toxins (Figure 4B).

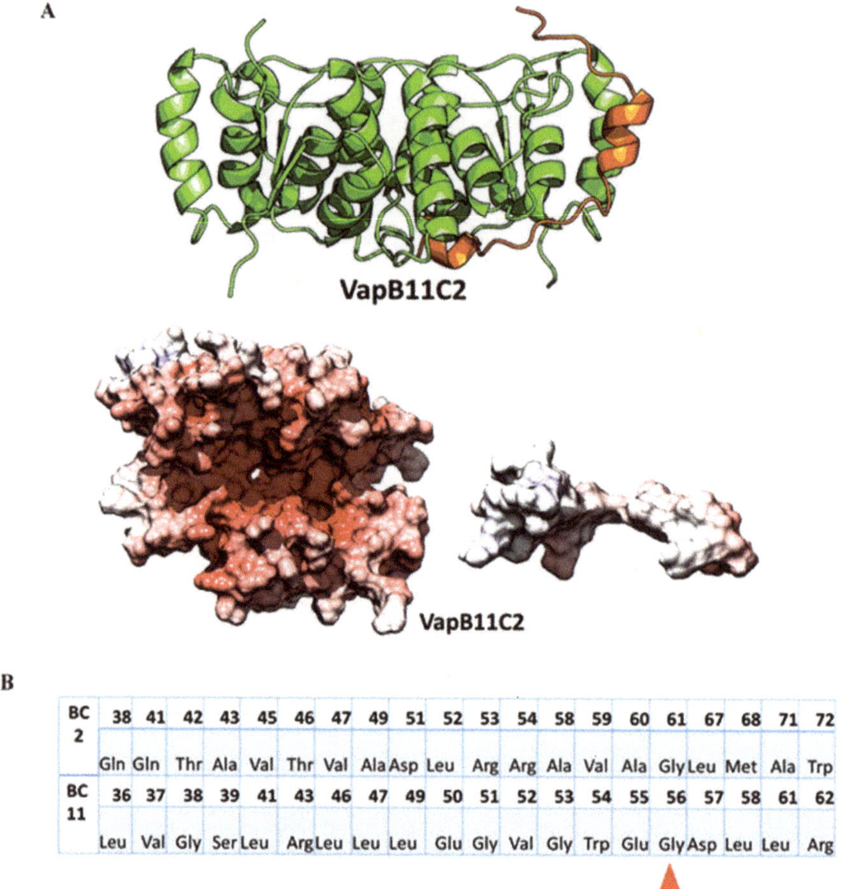

BC 2	38	41	42	43	45	46	47	49	51	52	53	54	58	59	60	61	67	68	71	72
	Gln	Gln	Thr	Ala	Val	Thr	Val	Ala	Asp	Leu	Arg	Arg	Ala	Val	Ala	Gly	Leu	Met	Ala	Trp
BC 11	36	37	38	39	41	43	46	47	49	50	51	52	53	54	55	56	57	58	61	62
	Leu	Val	Gly	Ser	Leu	Arg	Leu	Leu	Leu	Glu	Gly	Val	Gly	Trp	Glu	Gly	Asp	Leu	Leu	Arg

Figure 3. Interface comparison for VapBC complex known to not interact. (**A**) In-silico model for VapB11–VapC2, a non-cognate pair known to not interact is rendered as cartoon. The electrostatic potential surface comparison reveals a strongly negative potential for VapC2 while a less positive or almost neutral surface for VapB11, suggesting incompatibility of the surface. (**B**) shows comparison of the topologically equivalent interface residues in cognate VapB2 and VapB11 antitoxins. Interestingly, the majority of the antitoxin residues at the interfaces show opposite physicochemical properties except for one residue (Gly, red triangle), which is identical among the interface residues contributed by antitoxins.

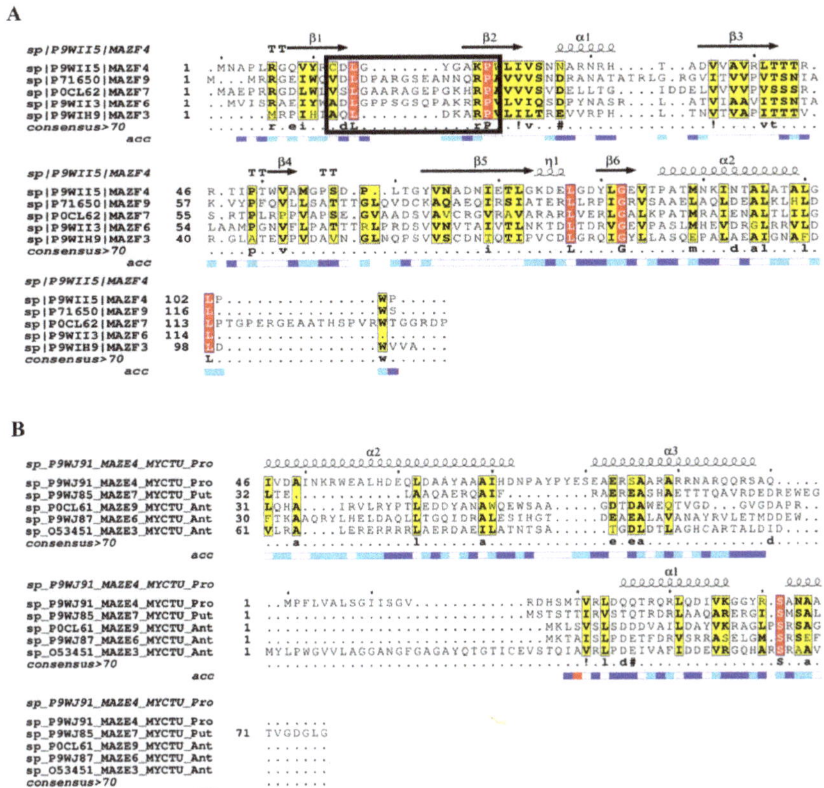

Figure 4. Structure-guided alignments of MazF toxins and MazE antitoxins. (**A**) Structure-guided alignment of MazF toxins show overall high similarity with many residues conserved. Secondary structural elements of MazEF4 are shown on top of the alignment and corresponding solvent accessibility at the bottom with dark blue implying solvent accessible regions, cyan showing partially accessible regions, and white showing inaccessible regions. Fully conserved residues are shown in red and conservatively substituted residues are shown in yellow columns. Sequence of loop region between β1-β2 is marked in black box. (**B**) Structure-guided alignment of MazE antitoxins show less conservation as compared to the toxins. Alignments were rendered using ESPript [25].

Our sequence-based analysis therefore does not predict any signals for potential cross-talk among these systems. However, since high confidence complex structures could be modelled for MazEF3, MazEF6, and MazEF9, they were used along with MazEF4 and MazEF7 crystal structures to compare their interfaces (Table 5). It was observed that apart from MazEF3/MazEF4 and MazEF6/MazEF9, all other pairs showed no similarity between their interfaces. For MazEF3/4 and MazEF6/9 also, the similarities (0.45 and 0.38, respectively) were only modest. But since these scores included the contribution from both toxin and antitoxin, and the p-value statistics associated with these scores was good (7.5×10^{-8} and 2.4×10^{-7}, respectively), the cross-talk can only be predicted (though with low confidence) for these two pairs. For the rest, cross-talk seems dubious.

Table 5. Interface similarity between cognate MazEF pairs.

Complex Structure to be Compared (Maz)	IS-Score	z-Score	% Seq. Identity	No. of Aligned Contacts	RMSD (Å)	p-Value
EF4-EF6	0.16	1.5	35	7	2.38	1.9×10^{-1}
EF4-EF3	0.45	16.3	30	18	2.39	7.5×10^{-8}
EF4-EF7	0.12	−0.4	10	3	1.78	0.8×10^{-1}
EF4-EF9	0.16	0.6	0	8	2.75	3.9×10^{-1}
EF3-EF6	0.15	0.84	25	4	2.32	3.4×10^{-1}
EF3-EF7	0.13	0.34	11	4	3.52	5.0×10^{-1}
EF6-EF7	0.20	3.75	4	12	3.00	2.3×10^{-2}
EF6-EF9	0.38	15.3	7	20	1.56	2.4×10^{-7}
EF3-EF9	0.14	0.12	0	8	3.81	5.8×10^{-1}
EF7-EF9	0.18	2.67	20	13	3.13	6.6×10^{-2}

Interface similarity was calculated using iAlign. IS-score is the interface similarity score. Higher the IS-score, more is the interfacial similarity. % sequence identity between the interface residues is also provided. The calculations take into consideration the residue identities from both toxin and antitoxin chains. RMSD gives the deviation between the interface structure. P-value assigns the statistical significance to the score (cut-off: 10^{-5}).

Existence of cross-talk among MazEF TA systems is long debated. Very recently, cross-talk between MazF9 and MazE7 has been reported [16]. Zhu et al. observed non-cognate interactions between MazF6 and MazE9 [13] but this claim was refuted by Ramirez et al. as they could not capture interactions between any MazEF systems [14]. Nevertheless, attempts were made to generate non-cognate MazEF complexes, viz., MazE3–MazF4/MazE4–MazF3 and MazE6–MazF9/MazE9–MazF6. However, all attempts were futile and ended up with very high energy non-cognate complexes primarily due to van der Waal's clashes. An exception was MazE4–MazF3, which showed interaction energy of −1.95 Kcal/mol between one MazE4–MazF3 chain. However, here too, a high interaction energy was observed for the other chain (2.0 Kcal/mol) likely due to an incompatible toxin/antitoxin conformation with the non-cognate partner.

3. Discussion

Cross-reactivity (or cross-talk) between *M. tuberculosis* toxin–antitoxin systems is anticipated due to presence of paralogous toxin–antitoxin systems and their co-expression in the cell under different stress conditions [40,41]. One would expect these TA systems to be involved in manifold interactions to manage the stress response network of the cell. The delicate balance between toxins and antitoxins regulate the expression of a given TA system [8,42]. It can be expected that if cross-talk exists, the ratio of cognate and non-cognate antitoxins will also be critical for regulating toxin activity. Intriguingly, except for a few dispersed studies, there exists no direct and confident evidence in favor of cross-talk among toxin-antitoxin systems. The present study is an attempt to resolve the existing debate on occurrence of cross-talk among *M. tuberculosis* TA systems from a purely computational point-of-view. The aim of this work is to provide structural insights into the principles behind cognate and non-cognate toxin-antitoxin interactions. A systematic 3D structural analysis has been carried out to explore the feasibility for cross-talk and propose testable hypotheses.

Availability of structures is a prerequisite for such an analysis. However, the experimentally solved structures are available only for a handful of TA complexes or toxins alone in *M. tuberculosis*. Hence, attempts were made to model as many high-confidence structures for TA systems either in complex or free form. Due to low sequence identities (mostly within the twilight zone) between the query toxins or antitoxins with their templates, utmost care was taken to model 46 VapCs and 9 MazF toxins. Antitoxin modelling was especially very difficult due to poor sequence conservation and presence of intrinsic disorder, which limits the applicability of modelling exercise. Further, toxins like MazFs are known to undergo conformational changes upon binding to their cognate antitoxins. These observations prompted us to generate models of toxins and antitoxins in their complexed form rather than as individual toxins or antitoxins. Such attempts may not always yield completely accurate models but provide reasonably accurate representation of the interfaces. Therefore, template-based modelling approach was utilized to model six TA complexes for which the templates and queries showed high similarity for both toxins and antitoxins. We believe that stoichiometry is important for biological function, especially for transcriptional control of TA expression, but the experimental information on this is not available for many TA systems. Since our aim was to analyze the TA interfaces, we kept the stoichiometry for the models to be same as the templates under the assumption that homologous protein–protein complexes are likely to interact in the same manner [43].

Since the TA models were generated using templates with sequence identities between 17% and 34%, they were used to assess the similarity among TA interfaces along with experimental structures with caution. Through a detailed structural analysis, we found that VapBC11–VapBC15 and VapBC4–VapBC5 shared high interface similarity (IS-score: 0.52 and 0.81, respectively). Their electrostatic potential surfaces were found compatible and high similarity was observed between the VapC11/VapC15 (31%), VapB11/VapB15 (52%), VapC4/VapC5 (37%), and VapB4/VapB5 (39%) sequences. In-silico generated non-cognate TA pairs were suggestive of potential cross talk between VapB11–VapC15, VapB4–VapC5, and VapB5–C4 due to comparable interaction energies with VapBC15 (−19.1 Kcal/mol and −21.0 Kcal/mol, respectively), VapBC5 (−26 Kcal/mol and −34 Kcal/mol, respectively), and VapBC4 (−35 Kcal/mol and −39 Kcal/mol, respectively). However, only a low stability structural model (with weak interaction energy) was obtained for VapB15–VapC11. VapB antitoxins are known to block the toxic effect of VapC toxins by binding to their active site and a high compatibility exists between the VapB and VapC electrostatic surfaces in cognate pairs. However, in the case of VapB15/VapC11 non-cognate pairing, this compatibility is compromised by the presence of two negatively charged amino acids at the VapB15/VapC11 interface. So, even though the interaction energy is negative (−10.1 kcal/mol), the two negative charges will repel each other at the active site, and hence we propose that this non-cognate interaction may not be stable enough to neutralize the given toxin.

Interestingly, we observed that the in-silico point mutation of E67R in the VapB15/VapC11 complex yielded a low energy complex and is predicted to relax the specificity of VapB15. This result provides insights into the reasons for insulation of cross-talk among TA systems. We have earlier reported from the sequence alignments of TAs that while toxins show conservation of many antitoxin binding residues, this does not hold true for antitoxins in each cluster [20]. This was reflected in the interface analysis of VapBC11/VapBC2 systems, which are experimentally shown not to interact with non-cognate pairs. However, in the case of VapBC4/VapBC5, although our theoretical analysis revealed clear and strong signals for cross-reactivity, it has been shown that VapC4 could not be neutralized by VapB5 [35]. Further analysis revealed that even though the theoretical energies were comparable to the cognate complexes, a single residue may be important for maintaining the specificity as shown by marked improvement in the interaction energy of VapB4–C5 non-cognate complex upon mutation of E12R residue in VapB5. Though further experiments can test our hypothesis, this certainly suggests that caution should be exercised while extrapolating the results from theoretical analysis. We summarize that the reasonably comparable and good energy values predicted through modelling efforts need not always imply congenial interactions. These studies must be necessarily accompanied with studies of the interface to recognize scenarios such as drastic changes to charges of residues at the interface, potential short contacts through replacements with sterically incompatible residues, absence of critical hydrogen bond forming residues, or change in van der Waal energies accumulated over many residues at the interface. All these factors can contribute to weaker non-cognate interactions despite satisfactory complementarity or energy scores. It is important to note that in addition to these theoretical estimates, several other factors can govern the failure to observe non-cognate TA interactions. These include factors such as expression of the TA systems. Cognate TAs are known to co-express; however, non-cognate TA need to be expressed at the same time for cross reactivity to occur. Another factor is the stoichiometry, i.e., how many individual units are required for formation of the TA and then for cross-reactivity. The number of toxin and antitoxin molecules present in a studied system can influence the outcome of their interaction, provided they have a compatible interaction. If a non-cognate interaction is predicted to be favorable, factors governing the number of toxin and antitoxin molecules in the system can play a decisive role in determining the experimental outcome of such a prediction. Further, mutation of crucial residues and unstructured regions becoming structured or ordered on physical interaction, can impact the binding strength. Modelling of the TA complex can only estimate the potential interaction energy of the complex and scenarios such as changes to the fold of a protein on binding cannot be suitably accounted. Lastly, the accuracy of computational models relies on sequence identity. Expected accuracies are low when identity of the sequence with its template is low [44].

For the four MazEF complexes (three models and one crystal structures), while the majority of putative non-cognate interfaces showed no interface similarities, only MazE4–MazF3 and MazE9–MazF6 showed weak similarities in their interfaces suggesting potential of cross-talk among these systems. So far, there are no confirmed reports available for cross-talk among MazEF systems in *M. tuberculosis* apart from MazE9–MazF6, which is also debatable. Indeed, absence of cross-talk seems to be favored by nature because even a slight imbalance of toxin–antitoxin systems can be catastrophic for the bacterial cells. While the results here suggest that majority of the non-cognate TA complexes are incompatible, cross-reactivity could be predicted for few. Taken together, this study provides general set of rules and testable hypothesis for cross-talk and further experiments can elucidate if this non-cognate binding among TAs is functional and can neutralize the non-cognate toxin in vitro.

4. Conclusions

In the present study, we have employed a computational approach to determine the structural basis for cognate TA interactions and devised rules that are useful to predict cases where non-cognate interactions are feasible for *M. tuberculosis*. We demonstrate, through the case studies of VapBC11–VapBC15 and VapBC4–VapBC5, that in addition to interaction energy values of the predicted non-cognate TA complexes, compatibility of residues at the interaction interface is a key factor in governing the formation of the TA complex (whether cognate/non-cognate as the case may be). We show that in some cases, although interaction interfaces may seem compatible overall and interaction energy scores are favourable, the local environment of residues mediating interaction are critical in predicting whether a non-cognate TA interaction is possible. We elaborate on the structural basis of why non-cognate interactions have failed in earlier experiments that have also attempted to examine this.

5. Materials and Methods

5.1. Comparative Modelling of Toxins, Antitoxins and TA Complexes

It is observed that toxins in general share between 16 to 80% sequence identity with their homologues, while antitoxins show poor sequence conservation (20 to 50%) [20]. Consequently, a search for homologues for toxin or antitoxin queries in PDB could not always identify hits that could serve as templates in homology modelling for several of the toxins and antitoxins. Therefore, to identify a suitable template for the toxins or antitoxins, a multi-step procedure was used (Figure 5).

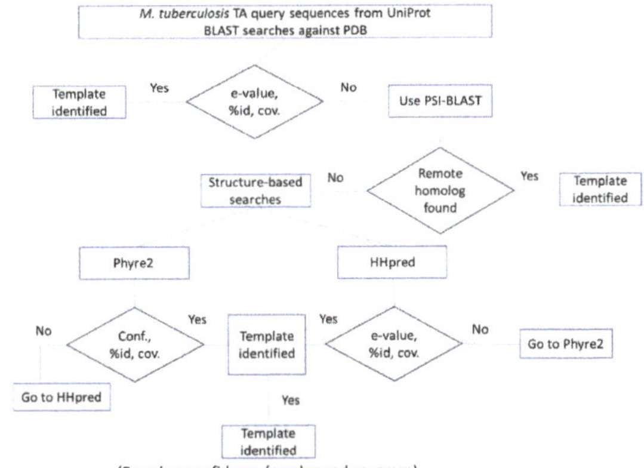

Figure 5. Step-by-step procedure to identify templates for toxins/antitoxins. *M. tuberculosis* TA sequences from UniProt are first searched against PDB using BLAST. If a given hit qualifies the e-value cut-off of 10^{-4} and query coverage >70%, then the template is identified. If no hits are found, PSI-BLAST is used for remote homologue search against PDB. If no template is found, then more comprehensive fold assignment methods such as Phyre2 and HHpred are used, both of which are HMM alignment-based methods.

Query toxins and antitoxins from *M. tuberculosis*, for which crystal structure information was not available in the PDB, were identified. Their sequences were obtained from the UniProt database and queried against the protein data bank (PDB) using BLASTP or PSI-BLAST [45]. If a hit was found at an e-value $< 10^{-4}$ and query coverage $> 70\%$, it was selected as a suitable template. If no hits were identified in BLAST searches, rigorous profile HMM-based search strategies, Phyre2 [46] and HHpred [47], were employed to identify a suitable template. A confidence score > 90% from Phyre2 and e-value $< 10^{-4}$ from HHpred were employed as the cut-off criteria. Query coverage >70% was used for both Phyre2 and HHpred. Once the template was identified, Modeller v9.14 was used to model the query protein [48]. Models were generated based on the available co-ordinates deposited in the PDB and only the aligned regions were considered for model building. Since the stoichiometry for many TA systems was not known, it was kept same as the templates while modelling. 100 models were generated per query and the model with lowest DOPE score was selected as final model. Side chains for the final model were optimized using SCWRL4.0 rotamer library [49]. This side-chain optimized model was further subjected to quality check using ProQ [26] and HARMONY [27]. Homology searches for antitoxin sequences did not result in the identification of suitable templates in many instances. Since antitoxins could not be confidently modelled in the unbound form, template-based modelling approach was used to generate the TA complex. In this approach, two proteins, A and B, for which the complex structure was to be modelled, were used as queries in searches for a template. If their respective homologues, A′ and B′, were also known to interact, and their complex structure was readily available, then AB complex was modelled based on A′B′. If this criterion was not met, we did not attempt to build the model for the TA pair AB.

5.2. Assessment of Toxin–Antitoxin Interfaces

Depending on the sequence identity, a modelled query complex is anticipated to possess interfacial features like the template. This is the underlying assumption of the approach called template-based modelling (docking) of the protein–protein complexes [50–52]. However, this assumption may not always predict correct interfaces. Hence, to obtain high confidence models, the generated complexes were further assessed for interface compatibility using three criteria:

- Surface complementarity was checked using S_c statistics, which measures the geometric surface complementarity of protein–protein interfaces [53]. S_c depends on the relative shape of the surfaces with respect to each other and on the extent to which the interaction brings individual elements of opposing surfaces into proximity. The score ranges between 0 and 1 and the threshold is generally decided based on the shape complementarity between antigen–antibody interfaces, where the weakest shape complementarity interface is reported with S_c values between 0.64 and 0.69.
- Electrostatic surface complementarity was assessed by calculating the electrostatic potential of the proteins. Hydrogens were added to the individual proteins. Charges were assigned to the residues using PDB2PQR [54] plugin and electrostatics calculation was performed using APBS [55] plugin in Chimera 1.13.1. The molecular surface was then color based on electrostatic potential and scaled between ± 10 k_BT and manually inspected.
- Interaction energy between toxin and antitoxin was calculated using AnalyseComplex module from FoldX package [28]. FoldX force-field is empirical in nature with terms for de-solvation energies, coulombic interactions, van der Waal's forces, hydrogen bonding, entropic changes, and others. All the structures were energy minimized with GROMACS v5.1 using CHARMM27 force-field and steepest descent method for either 50,000 steps or till convergence. A dodecahedron box with TIP3P water molecules was defined around the protein and the system was neutralized by adding counter ions prior to minimization. The structures were further repaired for any distorted geometry using RepairPDB module from FoldX prior to energy calculations. The complex structures were minimized iteratively till no further improvement in the energy values.

5.3. Using Structures to Explore Cross-Reactivity between Non-Cognate Toxins and Antitoxins

Toxins cluster into distinct sub-clusters based on their sequence identity [20]. To predict if a non-cognate toxin–antitoxin pair can interact and form a stable complex, we selected the clusters of paralogous toxins with the highest number of known crystal structures for further analysis was chosen. From each of the sub-clusters of 2 and 6, structures of the cognate TA pairs (Figure S1) in each cluster were analyzed and their interfacial regions were compared using iAlign [56]. TA pairs that showed high interface similarity were further probed for cross-reactivity. A pairwise alignment between the toxin and antitoxin sequences of such complexes within each sub-cluster was individually analyzed to identify conserved interactions between cognate TA pairs. In-silico non-cognate complexes were generated with cognate structures as template using Modeller v9.14 [48]. 100 models were generated and side-chains were optimized using SCWRL4.0 [49] rotamer library for the model with least DOPE score. All the in-silico complexes were energy minimized with GROMACS v5.14 as mentioned in the earlier section [57]. Further, interaction energy was calculated, and the shape complementarity of the non-cognate pair was analyzed. Electrostatic potential at the interfaces of non-cognate toxins and antitoxins were compared with cognate TA pairs. As a control, available structures for TA pairs that are experimentally known not to interact were analyzed in a similar fashion to understand the reasons behind insulation of cross-talk between those pairs.

5.4. Prediction of Hotspot Residues at the Interface and In-Silico Mutations Using FoldX

After analyzing non-cognate pairs of VapC11–B15 and VapBC4–BC5, residues were identified which could behave as the specificity conferring residue in VapB15 and VapB5. These results were further supported by hotspot predictions using the AlaScan module from the FoldX package [28]. $\Delta\Delta G$ cut-off of 2 Kcal/mol was used. These residues were in-silico mutated using the PositionScan module from the

FoldX package. The mutated complex structure was energy minimized and the interaction energies were calculated using AnalyseComplex module of FoldX.

Supplementary Materials: The following are available online at http://www.mdpi.com/2072-6651/12/8/481/s1, Figure S1: sub-cluster alignments, Figure S2: in-silico modelled non-cognate VapBC complexes, Figure S3: fold assignment/template prediction for toxins and antitoxins, Figure S4: alignment of query toxins and antitoxins with their templates, Figure S5: homology models for toxin–antitoxin complexes, Figure S6: superposition of modelled and crystal structures of different MazF toxins, Figure S7: electrostatic potential surfaces for toxin–antitoxin models, Figure S8: comparison of VapBC5 and VapBC4 interfaces, Figure S9: in-silico point mutations in VapBs enhance binding, Table S1: template information for TA systems in *M. tuberculosis*, Table S2: details of templates and query TA complexes, Table S3: RMSD between modelled and crystal structures, and Table S4: hotspot interactions between cognate and non-cognate VapBC4–BC5 pairs.

Author Contributions: Conceptualization, H.T., S.S., and N.S.; methodology, H.T., S.S., and N.S.; validation, H.T.; formal analysis, H.T.; investigation, H.T. and A.M.V.; resources, H.T.; data curation H.T.; writing—original draft preparation, H.T.; writing—review and editing, H.T., A.M.V., N.S., and S.S.; supervision, S.S. and N.S.; project administration, N.S. and S.S.; and funding acquisition, N.S. All authors have read and agreed to the published version of the manuscript.

Funding: N.S. acknowledges funding for infrastructural support from the following agencies or programs of the Government of India—MHRD (Ministry of Human Resource Development), DST-FIST (Department of Science and Technology-Fund for improvement of S and T infrastructure), UGC (University Grants Commission) Center for Advanced Study, Mathematical Biology program of the Department of Science and Technology, and the DBT-IISc (Department of Biotechnology-Indian Institute of Science) Partnership Program. N.S. acknowledges funding received from DBT-COE (Department of Biotechnology-Centre of Excellence) and CEFIPRA. N.S. is a J.C. Bose National Fellows. S.S. acknowledges DBT-IISc partnership program for post-doctoral funding. H.T. is a DST-INSPIRE fellow.

Acknowledgments: Authors acknowledge Raghavan Varadarajan, MBU, IISc., and Ramandeep Singh, THSTI, Faridabad, for useful discussions.

Conflicts of Interest: The authors declare no conflict of interest. The funders had no role in the design of the study; in the collection, analyses, or interpretation of data; in the writing of the manuscript, or in the decision to publish the results.

References

1. Hayes, F.; Van Melderen, L. Toxins-Antitoxins: Diversity, Evolution and Function. *Crit. Rev. Biochem. Mol. Biol.* **2011**, *46*, 386–408. [CrossRef]
2. Kedzierska, B.; Hayes, F. Emerging Roles of Toxin-Antitoxin Modules in Bacterial Pathogenesis. *Molecules* **2016**, *21*, 790. [CrossRef]
3. Wen, Y.; Behiels, E.; Devreese, B. Toxin-Antitoxin Systems: Their Role in Persistence, Biofilm Formation, and Pathogenicity. *Pathog. Dis.* **2014**, *70*, 240–249. [CrossRef]
4. Lobato-Márquez, D.; Díaz-Orejas, R.; García-del Portillo, F. Toxin-Antitoxins and Bacterial Virulence. *FEMS Microbiol. Rev.* **2016**, *40*, 592–609. [CrossRef]
5. Coray, D.S.; Wheeler, N.E.; Heinemann, J.A.; Gardner, P.P. Why so Narrow: Distribution of Anti-Sense Regulated, Type I Toxin-Antitoxin Systems Compared with Type II and Type III Systems. *RNA Biol.* **2017**, *14*, 275–280. [CrossRef] [PubMed]
6. Leplae, R.; Geeraerts, D.; Hallez, R.; Guglielmini, J.; Drze, P.; Van Melderen, L. Diversity of Bacterial Type II Toxin-Antitoxin Systems: A Comprehensive Search and Functional Analysis of Novel Families. *Nucleic Acids Res.* **2011**, *39*, 5513–5525. [CrossRef] [PubMed]
7. Van Melderen, L.; De Bast, M.S. Bacterial Toxin-Antitoxin Systems: More than Selfish Entities? *PLoS Genet.* **2009**, *5*, e1000437. [CrossRef] [PubMed]
8. Goeders, N.; Van Melderen, L. Toxin-Antitoxin Systems as Multilevel Interaction Systems. *Toxins* **2013**, *6*, 304–324. [CrossRef]
9. Park, J.H.; Yoshizumi, S.; Yamaguchi, Y.; Wu, K.P.; Inouye, M. ACA-Specific RNA Sequence Recognition Is Acquired via the Loop 2 Region of MazF MRNA Interferase. *Proteins Struct. Funct. Bioinform.* **2013**, *81*, 874–883. [CrossRef]

10. Ramage, H.R.; Connolly, L.E.; Cox, J.S. Comprehensive Functional Analysis of *Mycobacterium tuberculosis* Toxin-Antitoxin Systems: Implications for Pathogenesis, Stress Responses, and Evolution. *PLoS Genet.* **2009**, *5*, e1000767. [CrossRef]
11. Walling, L.R.; Butler, J.S. Structural Determinants for Antitoxin Identity and Insulation of Cross Talk between Homologous Toxin-Antitoxin Systems. *J. Bacteriol.* **2016**, *198*, 3287–3295. [CrossRef] [PubMed]
12. Yang, M.; Gao, C.; Wang, Y.; Zhang, H.; He, Z.G. Characterization of the Interaction and Cross-Regulation of Three *Mycobacterium tuberculosis* RelBE Modules. *PLoS ONE* **2010**, *5*, e10672. [CrossRef] [PubMed]
13. Zhu, L.; Sharp, J.D.; Kobayashi, H.; Woychik, N.A.; Inouye, M. Noncognate *Mycobacterium tuberculosis* Toxin-Antitoxins Can Physically and Functionally Interact. *J. Biol. Chem.* **2010**, *285*, 39732–39738. [CrossRef] [PubMed]
14. Ramirez, M.V.; Dawson, C.C.; Crew, R.; England, K.; Slayden, R.A. MazF6 Toxin of *Mycobacterium tuberculosis* Demonstrates Antitoxin Specificity and Is Coupled to Regulation of Cell Growth by a Soj-like Protein. *BMC Microbiol.* **2013**, *13*, 240. [CrossRef] [PubMed]
15. Riffaud, C.; Pinel-Marie, M.L.; Pascreau, G.; Felden, B. Functionality and Cross-Regulation of the Four SprG/SprF Type I Toxin-Antitoxin Systems in *Staphylococcus aureus*. *Nucleic Acids Res.* **2019**, *47*, 1740–1758. [CrossRef]
16. Chen, R.; Tu, J.; Tan, Y.; Cai, X.; Yang, C.; Deng, X.; Su, B.; Ma, S.; Liu, X.; Ma, P.; et al. Structural and Biochemical Characterization of the Cognate and Heterologous Interactions of the MazEF-Mt9 TA System. *ACS Infect. Dis.* **2019**, *5*, 1306–1316. [CrossRef]
17. Wei, Y.; Li, Y.; Yang, F.; Wu, Q.; Liu, D.; Li, X.; Hua, H.; Liu, X.; Wang, Y.; Zheng, K.; et al. Physical and Functional Interplay between MazF 1Bif and Its Noncognate Antitoxins from *Bifidobacterium longum*. *Appl. Environ. Microbiol.* **2017**, *83*. [CrossRef]
18. Smith, A.B.; López-Villarejo, J.; Diago-Navarro, E.; Mitchenall, L.A.; Barendregt, A.; Heck, A.J.; Lemonnier, M.; Maxwell, A.; Díaz-Orejas, R. A Common Origin for the Bacterial Toxin-Antitoxin Systems ParD and Ccd, Suggested by Analyses of Toxin/Target and Toxin/Antitoxin Interactions. *PLoS ONE* **2012**, *7*, e46499. [CrossRef]
19. Berman, H.M.; Westbrook, J.; Feng, Z.; Gilliland, G.; Bhat, T.N.; Weissig, H.; Shindyalov, I.N.; Bourne, P.E. The Protein Data Bank. *Struct. Bioinform.* **2005**, 181–198. [CrossRef]
20. Tandon, H.; Sharma, A.; Wadhwa, S.; Varadarajan, R.; Singh, R.; Srinivasan, N.; Sandhya, S. Bioinformatic and Mutational Studies of Related Toxin–Antitoxin Pairs in *Mycobacterium tuberculosis* Predict and Identify Key Functional Residues. *J. Biol. Chem.* **2019**, *294*, 9048–9063. [CrossRef]
21. Min, A.B.; Miallau, L.; Sawaya, M.R.; Habel, J.; Cascio, D.; Eisenberg, D. The Crystal Structure of the Rv0301-Rv0300 VapBC-3 Toxin-Antitoxin Complex from *M. tuberculosis* Reveals a Mg2+ Ion in the Active Site and a Putative RNA-Binding Site. *Protein Sci.* **2012**, *21*, 1754–1767. [CrossRef] [PubMed]
22. Deep, A.; Tiwari, P.; Agarwal, S.; Kaundal, S.; Kidwai, S.; Singh, R.; Thakur, K.G. Structural, Functional and Biological Insights into the Role of *Mycobacterium tuberculosis* VapBC11 Toxin–Antitoxin System: Targeting a TRNase to Tackle Mycobacterial Adaptation. *Nucleic Acids Res.* **2018**, *46*, 11639–11655. [CrossRef] [PubMed]
23. Das, U.; Pogenberg, V.; Subhramanyam, U.K.T.; Wilmanns, M.; Gourinath, S.; Srinivasan, A. Crystal Structure of the VapBc-15 Complex from *Mycobacterium tuberculosis* Reveals a Two-Metal Ion Dependent Pin-Domain Ribonuclease and a Variable Mode of Toxin-Antitoxin Assembly. *J. Struct. Biol.* **2014**, *188*, 249–258. [CrossRef] [PubMed]
24. Jardim, P.; da Silva Santos, I.C.; Barbosa, J.A.R.G.; de Freitas, S.M.; Valadares, N.F. Crystal Structure of VapC21 from *Mycobacterium tuberculosis* at 1.31 Å Resolution. *Biochem. Biophys. Res. Commun.* **2016**, *478*, 1370–1375. [CrossRef]
25. Robert, X.; Gouet, P. Deciphering Key Features in Protein Structures with the New ENDscript Server. *Nucleic Acids Res.* **2014**, *42*, W320–W324. [CrossRef]
26. Wallner, B.; Elofsson, A. Can Correct Protein Models Be Identified? *Protein Sci.* **2003**, *12*, 1073–1086. [CrossRef] [PubMed]
27. Pugalenthi, G.; Shameer, K.; Srinivasan, N.; Sowdhamini, R. HARMONY: A Server for the Assessment of Protein Structures. *Nucleic Acids Res.* **2006**, *34*, W231–W234. [CrossRef] [PubMed]

28. Schymkowitz, J.; Borg, J.; Stricher, F.; Nys, R.; Rousseau, F.; Serrano, L. The FoldX Web Server: An Online Force Field. *Nucleic Acids Res.* **2005**, *33* (Suppl. 2), W382–W388. [CrossRef]
29. Yen, T.J.; Brennan, R.G. Crystal Structure of *M. tuberculosis* MazF-Mt3 (Rv1991c) in Complex with RNA. Ph.D. Thesis, Rutgers University-School of Graduate Studies, Rutgers, NJ, USA, 2017. [CrossRef]
30. Yen, T.J.; Brennan, R.G. Structure of *M. tuberculosis* MazF-Mt1 (Rv2801c) in Complex with RNA. *ACS Infect. Dis.* **2017**. [CrossRef]
31. Hoffer, E.D.; Miles, S.J.; Dunham, C.M. The Structure and Function of *Mycobacterium tuberculosis* MazF-Mt6 Toxin Provide Insights into Conserved Features of MazF Endonucleases. *J. Biol. Chem.* **2017**, *292*, 7718–7726. [CrossRef]
32. Zorzini, V.; Mernik, A.; Lah, J.; Sterckx, Y.G.J.; De Jonge, N.; Garcia-Pino, A.; De Greve, H.; Verse, W.; Loris, R. Substrate Recognition and Activity Regulation of the *Escherichia Coli* MRNA Endonuclease MazF. *J. Biol. Chem.* **2016**, *291*, 10950–10960. [CrossRef] [PubMed]
33. Kamada, K.; Hanaoka, F.; Burley, S.K. Crystal Structure of the MazE/MazF Complex: Molecular Bases of Antidote-Toxin Recognition. *Mol. Cell* **2003**, *11*, 875–884. [CrossRef]
34. Miallau, L.; Faller, M.; Chiang, J.; Arbing, M.; Guo, F.; Cascio, D.; Eisenberg, D. Structure and Proposed Activity of a Member of the VapBC Family of Toxin-Antitoxin Systems. *J. Biol. Chem.* **2009**, *284*, 276–283. [CrossRef]
35. Jin, G.; Pavelka, M.S.; Butler, J.S. Structure-Function Analysis of VapB4 Antitoxin Identifies Critical Features of a Minimal VapC4 Toxin-Binding Module. *J. Bacteriol.* **2015**, *197*, 1197–1207. [CrossRef] [PubMed]
36. Thorn, K.S.; Bogan, A.A. ASEdb: A Database of Alanine Mutations and Their Effects on the Free Energy of Binding in Protein Interactions. *Bioinformatics* **2001**, *17*, 284–285. [CrossRef] [PubMed]
37. Vishwanath, S.; Sukhwal, A.; Sowdhamini, R.; Srinivasan, N. Specificity and Stability of Transient Protein–Protein Interactions. *Curr. Opin. Struct. Biol.* **2017**, *44*, 77–86. [CrossRef]
38. Sala, A.; Bordes, P.; Genevaux, P. Multiple Toxin-Antitoxin Systems in *Mycobacterium tuberculosis*. *Toxins* **2014**, *6*, 1002–1020. [CrossRef]
39. Chen, R.; Tu, J.; Liu, Z.; Meng, F.; Ma, P.; Ding, Z.; Yang, C.; Chen, L.; Deng, X.; Xie, W. Structure of the MazF-Mt9 Toxin, a TRNA-Specific Endonuclease from *Mycobacterium tuberculosis*. *Biochem. Biophys. Res. Commun.* **2017**, *486*, 804–810. [CrossRef]
40. Agarwal, S.; Tiwari, P.; Deep, A.; Kidwai, S.; Gupta, S.; Thakur, K.G.; Singh, R. System-Wide Analysis Unravels the Differential Regulation and in Vivo Essentiality of Virulence-Associated Proteins B and C Toxin-Antitoxin Systems of *Mycobacterium tuberculosis*. *J. Infect. Dis.* **2018**, *217*, 1809–1820. [CrossRef]
41. Gupta, A.; Venkataraman, B.; Vasudevan, M.; Gopinath Bankar, K. Co-Expression Network Analysis of Toxin-Antitoxin Loci in *Mycobacterium tuberculosis* Reveals Key Modulators of Cellular Stress. *Sci. Rep.* **2017**, *7*, 1–14. [CrossRef]
42. Frampton, R.; Aggio, R.B.M.; Villas-Bôas, S.G.; Arcus, V.L.; Cook, G.M. Toxin-Antitoxin Systems of *Mycobacterium smegmatis* Are Essential for Cell Survival. *J. Biol. Chem.* **2012**, *287*, 5340–5356. [CrossRef]
43. Aloy, P.; Ceulemans, H.; Stark, A.; Russell, R.B. The Relationship between Sequence and Interaction Divergence in Proteins. *J. Mol. Biol.* **2003**, *332*, 989–998. [CrossRef] [PubMed]
44. Yazhini, A.; Srinivasan, N. How Good Are Comparative Models in the Understanding of Protein Dynamics? *Proteins Struct. Funct. Bioinform.* **2020**, *88*, 874–888. [CrossRef]
45. Altschul, S.F.; Madden, T.L.; Schäffer, A.A.; Zhang, J.; Zhang, Z.; Miller, W.; Lipman, D.J. Gapped BLAST and PSI-BLAST: A New Generation of Protein Database Search Programs. *Nucleic Acids Res.* **1997**, *25*, 3389–3402. [CrossRef] [PubMed]
46. Kelley, L.A.; Mezulis, S.; Yates, C.M.; Wass, M.N.; Sternberg, M.J.E. The Phyre2 Web Portal for Protein Modeling, Prediction and Analysis. *Nat. Protoc.* **2015**, *10*, 845–858. [CrossRef]
47. Hildebrand, A.; Remmert, M.; Biegert, A.; Söding, J. Fast and Accurate Automatic Structure Prediction with HHpred. *Proteins Struct. Funct. Bioinform.* **2009**, *77* (Suppl. 9), 128–132. [CrossRef]
48. Webb, B.; Sali, A. Comparative Protein Structure Modeling Using MODELLER. *Curr. Protoc. Bioinforma.* **2016**, *2016*, 5.6.1–5.6.37.
49. Krivov, G.G.; Shapovalov, M.V.; Dunbrack, R.L. Improved Prediction of Protein Side-Chain Conformations with SCWRL4. *Proteins Struct. Funct. Bioinform.* **2009**, *77*, 778–795. [CrossRef] [PubMed]

50. Baspinar, A.; Cukuroglu, E.; Nussinov, R.; Keskin, O.; Gursoy, A. PRISM: A Web Server and Repository for Prediction of Protein-Protein Interactions and Modeling Their 3D Complexes. *Nucleic Acids Res.* **2014**, *42*(W1), W285–W289. [CrossRef]
51. Kundrotas, P.J.; Zhu, Z.; Janin, J.; Vakser, I.A. Templates Are Available to Model Nearly All Complexes of Structurally Characterized Proteins. *Proc. Natl. Acad. Sci. USA* **2012**, *109*, 9438–9441. [CrossRef] [PubMed]
52. Rajagopala, S.V.; Sikorski, P.; Kumar, A.; Mosca, R.; Vlasblom, J.; Arnold, R.; Franca-Koh, J.; Pakala, S.B.; Phanse, S.; Ceol, A.; et al. The Binary Protein-Protein Interaction Landscape of *Escherichia coli*. *Nat. Biotechnol.* **2014**, *32*, 285–290. [CrossRef] [PubMed]
53. Lawrence, M.; Colman, P. Shape Complementarity at Protein–Protein Interfaces. *J. Mol. Biol.* **1993**, *234*, 946–950. [CrossRef] [PubMed]
54. Dolinsky, T.J.; Nielsen, J.E.; McCammon, J.A.; Baker, N.A. PDB2PQR: An Automated Pipeline for the Setup of Poisson-Boltzmann Electrostatics Calculations. *Nucleic Acids Res.* **2004**, *32*, W665–W667. [CrossRef]
55. Baker, N.A.; Sept, D.; Joseph, S.; Holst, M.J.; McCammon, J.A. Electrostatics of Nanosystems: Application to Microtubules and the Ribosome. *Proc. Natl. Acad. Sci. USA* **2001**, *98*, 10037–10041. [CrossRef]
56. Gao, M.; Skolnick, J.; Rost, B. IAlign: A Method for the Structural Comparison of Protein-Protein Interfaces. *Bioinformatics* **2011**, *27*, 2259–2265. [CrossRef]
57. Christen, M.; Hünenberger, P.H.; Bakowies, D.; Baron, R.; Bürgi, R.; Geerke, D.P.; Heinz, T.N.; Kastenholz, M.A.; Kräutler, V.; Oostenbrink, C.; et al. The GROMOS Software for Biomolecular Simulation: GROMOS05. *J. Comput. Chem.* **2005**, *26*, 1719–1751. [CrossRef] [PubMed]

Review

Evaluating the Potential for Cross-Interactions of Antitoxins in Type II TA Systems

Chih-Han Tu †, Michelle Holt †, Shengfeng Ruan † and Christina Bourne *

Department of Chemistry and Biochemistry, University of Oklahoma, Norman, OK 73019, USA; chihhantu@ou.edu (C.-H.T.); mnholt1217@ou.edu (M.H.); shengfengruan@ou.edu (S.R.)

* Correspondence: cbourne@ou.edu; Tel.: +1-405-325-5348

† All authors contributed equally to this article.

Received: 31 May 2020; Accepted: 19 June 2020; Published: 26 June 2020

Abstract: The diversity of Type-II toxin–antitoxin (TA) systems in bacterial genomes requires tightly controlled interaction specificity to ensure protection of the cell, and potentially to limit cross-talk between toxin–antitoxin pairs of the same family of TA systems. Further, there is a redundant use of toxin folds for different cellular targets and complexation with different classes of antitoxins, increasing the apparent requirement for the insulation of interactions. The presence of Type II TA systems has remained enigmatic with respect to potential benefits imparted to the host cells. In some cases, they play clear roles in survival associated with unfavorable growth conditions. More generally, they can also serve as a "cure" against acquisition of highly similar TA systems such as those found on plasmids or invading genetic elements that frequently carry virulence and resistance genes. The latter model is predicated on the ability of these highly specific cognate antitoxin–toxin interactions to form cross-reactions between chromosomal antitoxins and invading toxins. This review summarizes advances in the Type II TA system models with an emphasis on antitoxin cross-reactivity, including with invading genetic elements and cases where toxin proteins share a common fold yet interact with different families of antitoxins.

Keywords: cognate interactions; cross-interactions; molecular insulation; toxin; antitoxin; TA systems; addiction; anti-addiction

Key Contribution: Current models posit TA systems can be used for both cellular addiction to mobile extra-chromosomal DNA, as well as for anti-addiction functions. This review examines known cases of antitoxin cross-interactions with non-cognate toxins, and explores the potential for additional cross-interactions based on conserved structures of toxin families.

1. Introduction

Toxin–antitoxin (TA) systems are abundant in bacterial and archaeal chromosomes as well as extra-chromosomal genetic elements including plasmids, phages, and transposons [1–7]. TA systems have been the subject of numerous reviews that describe their typically bicistronic operon encoding a toxin that targets essential cellular process and the cognate neutralizing antitoxin [8–16]. Toxin proteins are able to manipulate their bacterial host cells in powerful ways. This drives great interest in understanding their functions and the potential to utilize them for biotechnology and bacterial control strategies [17–26].

Depending on the molecular identity of the antitoxin, as well as the mechanism by which it neutralizes the toxin, the known TA systems have been classified into six different types. While the toxin is typically a protein, the antitoxin is either a noncoding RNA (in Type I and III) or a protein (in Type II, IV, V and VI) [15,18,25,27–30]. Limited numbers of Type III TA pairs [31–34] and IV TA pairs have been identified [35–37]. Type V systems identified to date are the GhoST system in *Escherichia coli* K strains and the orphan OrtT toxin from *Salmonella* [38–40]. Similarly, the Type VI system is currently comprised of the Soc system in *Caulobacter crescentus* [41]. However, continuing surprises are challenging the canonical TA system paradigms. These include the integration of some TA systems into alarmone signaling pathways [42,43] and recognition of alternatives to protease-dependent regulation such as acetylation of antitoxins and/or chaperone protection [44–46], in addition to widespread variations on the bicistronic antitoxin–toxin arrangement [44,47]. In comparison, Type I and II TA systems are well-studied, and Type II systems appear in thousands of bacterial loci that are accessible and searchable through tailored web portals; to-date, the TADB webserver lists more than 6000 Type II systems with approximately 10% of these located on plasmids [1,2,7,48–51]. In Type II systems the antitoxin partner plays a dual role by neutralizing the cognate toxin and by mediating transcriptional regulation through binding to its promoter, making it the lynchpin of Type II TA system functions [52] (Figure 1). As such, many of the proposed functions of these systems are predicated on the ability of antitoxins to potentially neutralize multiple homologous toxins, as explored in the current review.

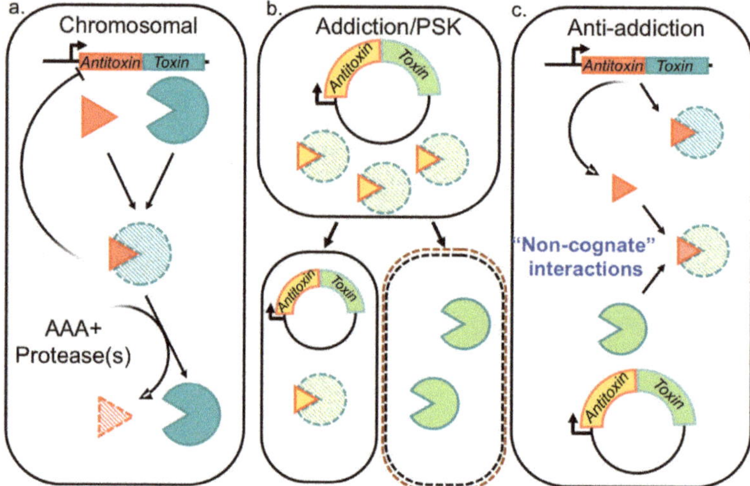

Figure 1. Paradigms for Type II toxin–antitoxin (TA) system functions span roles in (**a**) physiology mediated by chromosomally-encoded systems, in addition to potential roles (**b**) as addiction modules on mobile genetic elements. An alternative may arise (**c**) when multiple TA systems are present within the same cell, giving rise to an anti-addiction role. Experimental evidence has demonstrated that chromosomally-encoded systems can protect individual cells from external stressors (**a**). When TA systems are encoded on mobile genetic elements (pictured on a plasmid) they can function as "Addiction" elements that enforce retention (also referred to as post-segregation killing, or PSK) (**b**). A hybrid paradigm combines these two functionalities, wherein a chromosomally-encoded antitoxin (or toxin) can neutralize an invading toxin, thus providing an anti-addiction function (**c**).

2. Paradigms for Type II TA Systems

Type II TA systems were originally recognized as mediators of plasmid addiction, also termed "post-segregational killing" (PSK) (Figure 1b) [53–57]. Subsequent genome sequencing efforts identified them throughout bacteria and archaea [1,2,7], with those presenting some levels of similarity to the plasmidic versions but not other related genomic versions referred to as "xenologs" to highlight their likelihood of arising through horizontal gene transfer [58]. Analysis of xenolog distribution led to insight that these TA systems are a form of bacterial immunity with implications for protection from invading genetic material [28,59–61]. Therefore, the roles of given Type II TA systems described in Figure 1 remain unsettled and are likely overlapping.

2.1. Functions Attributed to Chromosomal Type II TA Systems

The prevailing evidence favors the stochastic accumulation of TA systems through horizontal gene transfer, with a predominance of phage-derived systems at "hotspots" of genetic diversity within a given bacterial species [2,7,58,62–65]. As such, their potential role has been a subject of intense debate. For many, it has been demonstrated that they are located within integrated (at least previously) mobile genetic elements, and further, some of these function as addiction modules for those integrated elements (see Section 2.2, below). Some have proposed that integration of a TA system into a bacterial chromosome has allowed it to be co-opted to maintain normal physiological homeostasis in response to environmental or other changes in growth conditions, sometimes termed "domestication" or linked to "accessory" genomic content [8,66]. Further, for some TA systems there likely exists a functional overlap of these ideas, encompassing both genetic addiction and the potential for usefulness to the cells (Figure 1) [27,67].

Some investigations have documented an attenuation of chromosomally-encoded TA system toxicity. This has arisen either through sequence changes resulting in lower affinity for interaction with the cellular target [68–72] or by attenuation of expression via changes in the promoter regions [73–75]. A loss of function, termed "degeneration", has also been observed for some systems, such as a chromosomal CcdAB system in *E. coli* [62]. Attenuation of toxicity provides an opportunity for toxins to regulate their cellular target without killing the host cell. While this led to the long-touted idea of TA system involvement in persister cell generation, it can also be more generalized simply as the ability of cells to withstand external stressors by reducing metabolism or protecting critical cellular targets from damage [16,76,77].

Gyrase-targeting systems CcdAB and ParDE, each originally identified as mediators of PSK, have been documented to provide useful advantages to their host cells. The CcdB toxin, carried on the F plasmid and the chromosomal xenolog in *E. coli*, both were observed to increase cellular survival to heat, anti-gyrase compounds, and other antibiotics [68,78]. An analogous protective effect from anti-gyrase antibiotics was observed for the chromosomal ParE from *Pseudomonas aeruginosa* [71]. The protection of host cells from thermal stress has also been observed for the ParE toxin carried on the RK2 plasmid [79]. *Caulobacter crescentus* carries three functional chromosomal ParDE systems, and these were observed to offer protection to stressors, as did one of a ParDE system from *Mycobacteria tuberculosis* [70,80]. We note that many studies rely on "over-expression" of toxin proteins to analyze the cellular impacts, resulting in non-native concentrations. However, effects can be observed at very low levels of toxin protein expression barely detectable by Western blotting, leading to the suggestion these could be reasonable estimates for the free toxin of an "activated" system [69]. Overall, it seems that gyrase-targeting TA systems may occupy a specialized niche that bridges protection at low-levels of expression and/or with attenuated toxicity, to a higher-level toxicity at prolonged exposure or concentration levels. Given that FicT toxins have also been identified as modulating both DNA gyrase and topoisomerase IV via adenylation [81], it remains to be determined if they will have similar impacts to heat tolerance or target protection.

Protection from thermal stress is noted for multiple members of RNA-degrading ("RNase") type toxins, particularly those of the RelE-type fold. The YoeB toxin from *E. coli* protects thermal stress, and further, this is dependent on the proteases needed for antitoxin degradation, implying this protection arises from some direct action of the toxin [82]. Studies in the gram-positive bacteria *Streptococcus pneumonia* demonstrated genetic deletions of two *yoeB* loci and related RNase toxins were less hardy when exposed to oxidative conditions, while complemented mutants recapitulated wild type survival levels [83]. The structurally related GraT toxin from *Pseudomonads* similarly protects from temperature stresses but appears to function in response to lower rather than elevated temperatures [72]. Subsequent work with GraT highlighted the overall depression in metabolism in response to this toxin [84]. YafQ, again within the same structural class as the RelE/ParE family, was also found to be protective for bacterial growth at sub-optimal temperatures [85].

Other classes of toxin activities, particularly those that impact RNA lifetimes, have also proven to impart beneficial functions to host physiology. Three RelE toxins in the genome of *Mycobacterium tuberculosis* were identified as upregulated at the transcript level in response to altered oxygen levels or limiting nitrogen [86]. This study linked RNase toxin activation to proteome alteration by two-dimensional gel electrophoresis and mass spectrometry analysis of differential products. The RelBE TA system has been documented to be activated in response to nutritional stress [87], likely triggered in natural settings by high cell density [88] and similar to findings with *Acinetobacter baumannii* and *P. aeruginosa* HigBA systems [89,90]. These systems in *Mycobacterium tuberculosis* also provide protection to antibiotics but, importantly, did not induce persister cell formation [76]. It remains unclear if this is the "native" use for these; or, in other words, if the bacterial cells experience these conditions as part of their normal growth cycle and in turn utilize these toxins to slow growth and thus hedge survival. Further, considerable controversy still surrounds many earlier findings for specific details of the RNase toxin–starvation models [12,13,91].

Many of these studies rely on a reductive approach in defined growth conditions, whereas the use of TA systems within a natural ecological setting has been harder to access. The TA systems YefM-YoeB, Hha-YbaJ (Hha-TomB), and PasTI (also named RatA [92] to reflect its Ribosome Association toxin, or inhibition) harbored on the chromosome of extraintestinal pathogenic *E. coli* (ExPEC) strains were noted to promote increases in recoverable viable bacteria from the bladder and kidneys of a mouse model of infection [93]. The deletion of MazEF systems in *M. tuberculosis* reduced virulence in an animal model, as did Vap systems in *Haemophilus influenza* [94,95]. The Hha-TomB Type II system, which impacts translation through ribosomal interactions, has also been associated to increased *Salmonella typhimurium* survival in infection models [96]. Other studies have highlighted a role for TA systems in the survival of *Salmonella* within macrophages [97]. Subsequent studies have complicated this model and suggested that the effect resulted from slowed bacterial metabolism, which is a known impact from TA system activation [16,77]. Similar complexities that produce different outcomes have been highlighted for the MqsRA and MazEF systems in *E. coli* [98–103]. The impacts of chromosomal TA systems on their host cells are clearly affected by very specific interactions and by interwoven pathways. These have been well reviewed elsewhere, and are brought up here to illustrate the unsettled nature of these on-going studies and evolving conclusions [104–107].

The continued discovery of new TA systems highlights their broad capacity for integration into bacterial cells. The *Streptococcus pneumoniae* chromosomal PezAT system, a xenolog of the Epsilon-Zeta system [108], is located within an integrated pathogenicity island, and when deleted induced phenotypic changes that produced both beneficial (harder to lyse) and negative (more sensitive to cell wall antibiotics) effects for the cell [109]. The ParST, an mART-type that transfers an ADP-ribosyl group onto an enzyme involved in phosphoribosyl pyrophosphate synthetase, induces bacteriostasis when transplanted from its native *Sphingobium* host to *E. coli* [110]. This study notes the widespread distribution of the ParST system with enrichment in Proteobacterial classes. The diverse and highly integrated nature of TA systems and bacterial physiology is exemplified by the recently published work demonstrating some alarmone synthases are housed as TA loci [43]. These systems, named toxSAS, deviate from the traditional canon in encoding multiple antitoxins per synthase, or toxin, with both cognate and universal neutralization interactions [43]. It seems, then, that the diversity and functions of TA systems are expansive and may yet reveal new secrets of bacterial growth in the coming years.

2.2. Plasmid Selection and/or Addiction via Type II TA Systems

As mentioned above, chromosomal Type II TA systems were almost certainly acquired by invading genetic material. For example, a ParDE system in *P. aeruginosa* is located within the Pf1 prophage [66], and the widely-studied RelBE system in *E. coli* is located within a Qin prophage [63]. The *Vibrio cholera* superintegron on chromosome II, including an integrated and conjugative element (ICE) called SXT, is enriched with both antibiotic resistance cassettes as well as Type II TA systems [111,112]. *Neisseriaceae* and *Klebsiella* species have similar integrated genetic elements, including Type-IV Secretion System components connected to TA systems [47,113], as well as other polymorphic toxin systems resembling Type II TA systems [104,114–116]. The chromosomal localization of these TA systems coupled with demonstrated toxicity in the absence of the antitoxin strongly implies these likely functioned as an addiction system for loss of the genetic material, and as are mentioned here in the "addiction" category rather than above with other chromosomal TA systems (see Table 1).

The idea of plasmid addiction, also referred to as PSK, is generally found with low copy number plasmids and is predicated on the shorter half-life of the neutralizing antitoxin, allowing daughter cells to inherit portions of the parental cytoplasmic material including more stable toxins (Figure 1b) [57,117–120]. While this is also feasible for phage to utilize and thus mediate infections, as recently described for the Pf1 prophage in *P. aeruginosa* [121,122], this is more commonly used by Type III TA systems and therefore Table 1 focuses on well characterized examples of plasmid-based Type II systems. When the host bacterial cell lacks a corresponding mechanism to neutralize the inherited toxin proteins, a negative impact on growth is realized. However, some bacteria encode chromosomal antitoxins that can neutralize the plasmidic counterparts, referred to as "anti-addiction" (see Section 2.3, below).

Type II TA systems are widespread on plasmids where they mediate either addiction or plasmidic competition [8,12,123–125]. Further, because TA systems select for plasmid maintenance, they also then contribute to spread of AMR [8,11]. For example, the pUM505 plasmid (an IncI-type) from *P. aeruginosa* contains a pathogenicity island with an encoded *pumAB* TA system (a RelBE homolog) and numerous resistance genes, including a ciprofloxacin-modifying enzyme CrpC [126]. Strains containing this plasmid were more virulent in both *Caenorhabditis elegans* and mouse model infections. A recent analysis of plasmids carried in *Klebsiella* strains found a strong association of Type II TA systems and resistance genes, with enrichment for ParE, ParE-like, CcdB, and Vap-type PIN domain toxins within IncA/C- and IncH-type plasmids [47].

A recent preprint report nicely summarizes TA systems found on different Inc plasmids in *Enterobacteriaceae* [5]. For *Klebsiella pneumonia* strains, they can contain up to 11 different Inc types of plasmids, with approximately half of these being IncF-types [5]. These included 27 different Type-II families with *ccdAB* and *pemIK* the most common, consistent with previous reports on the pOXA-48 IncL/M-type plasmid [5,127] as well as previous work on *E. coli*-derived plasmids [128]. Interestingly, when compared to their counterparts in *E. coli*, the *ccdAB* loci show greater sequence divergence, while the *pemIK* systems are relatively well conserved [5]. Previous reports on IncX-type plasmids noted enrichment for RelE/ParE-type TA systems [123].

The literature presents at least one well-documented case of a CcdAB xenolog in *E. coli* O157:H7 that is neutralized by its plasmidic counterpart from the F plasmid, whereas the plasmid-borne toxin is only neutralized by the same (cognate) plasmidic antitoxin [73]. These homologous systems co-exist stably in the population, perhaps mediating a reverse addiction wherein the plasmid is retained specifically to neutralize the chromosomally-integrated copy of the toxin. What is clear is that this type of addiction requires an antitoxin protein to interact with multiple different toxins, and that this standard PSK "addiction" model would not typically work if chromosomal antitoxin xenologs could cross-interact with plasmid-derived toxins, thus requiring a type of directional insulation for cross-interactions.

Table 1. Plasmidic TA systems characterized in the literature.

TA System	Plasmid	Host Bacteria	Chromosomal Homolog/Xenolog	Citations
ParDE	RK2 */RP4 *	*(Enterobacteriaceae) Escherichia coli*	ParDE	[55,79,129]
RelBE	P307, IncB/C	*E. coli*	RelBE	[130,131]
VapBC	IncF	*E. coli*		[131]
StbDE	R485	*Morganella morganii*		[132]
CcdAB, Kid/Kis (PemKI)	F, R1/R100, pCHP91	*E. coli, Erwinia chrysanthemi, Staphylococcus*	MazEF	[118,133–136]
PhD-Doc	Bacteriophage P1	*E. coli*		[137]
HigBA	pRTS1	*Proteus vulgaris*		[138]
Tad-ata	pAMI2	*Paracoccus aminophilus*		[139]
Axe-txe	pRUM	*Enterococcus faecalis*	YoeB-YefM	[140,141]
TasA-TasB	pGI1	*Bacillus thuringiensis*		[125]
VapBC	pMYSH6000	*Shigella flexneri*		[142]
AtaRT	pB171-like plasmid	*E.coli*	AtaRT	[143]
Omega-epsilon-zeta	pSM19035 *, pVER1/2	*Streptococcus pyogenes. Enterococcus faecium*	PezAT	[141,144]
RelBE, HigBA, HTH/GNAT, SplTA	p3ABAYE	*Acinetobacter baumannii*		[145]
YacAB	PWR100	*E. coli*		[146]

* Indicates a broad-host range plasmid.

2.3. *Type II TA Systems Mediating Anti-Addition through Antitoxin Cross-Interactions*

An alternative outcome to the Addiction model discussed in Section 2.2 is Anti-addiction, wherein chromosomally encoded TA systems protect the host bacteria against PSK mediated by their plasmid-encoded counterparts through cross-interactions (Figure 1c) [7,59,63]. Recent work highlights an analogous model for Tn3 transposons [4] and prophage sequences [64]. Other experiments have highlighted that PSK is not necessarily the driving force for retention of TA systems, and instead it is due to plasmid competition, such that a TA system with homologs carried on two plasmids would "compete" for cross-toxin neutralization, leading to "survival" or retention of the "winning" plasmid [147]. Additional support comes from observations that PSK does not actually result in complete sterilization of the culture, but rather a decreased viability that recovers over time [148].

Anti-addiction was clearly demonstrated for the *Erwinia chrysanthemi* chromosomal antitoxin CcdA, which interacts with and neutralizes the incoming F1 plasmidic CcdB toxin [59]. A similar modality is mediated by the phage-derived protein Dmd, which serves as an antitoxin for the RnlA and LsoA toxins in *E. coli* [149]. Anti-addiction is closely linked to "abortive infection" of phages, which center more on the Type IV type of TA system including ToxIN and AbiEI [31,35,150–152]. The presence of orphan antitoxins that appear to encode a protein, such as used in Type II TA systems, readily supports the anti-addiction model, but this remains to be demonstrated as a common usage of these orphans [13].

This model of TA system functions is dependent on the ability of antitoxin proteins to cross-react, such that they can neutralize the toxin on invading genetic material. Given the dogma of cognate toxin–antitoxin interactions, the current review will revisit examples of known non-cognate interactions (with some previously reviewed in [28]) as well as the feasibility of this occurring in selected systems.

3. Conservation of Type II TA System Folds and Cognate Antitoxin Interactions

Numerous experimentally determined structures are available for Type II TA systems, and these confirm that although sequence conservation is low, their structures are highly conserved and can be used to group them into superfamilies [28,51,153]. These classifications highlight the modular nature of toxin family interactions with different antitoxin families, consistent with propagation by horizontal gene transfer [28,154]. Antitoxins are the critical regulatory component for Type II systems, wherein they (typically) contain an N-terminal DNA binding motif used for autoregulation of the TA operon, and a less structured C-terminal domain to neutralize the toxin's activity. Selected systems that have the antitoxin domains reversed, such that the N-terminal region mediates toxin neutralization [89,155–159], while some antitoxins are limited to only the toxin-binding domain [160]. Antitoxins neutralize toxins by either by blocking the active site or by causing conformation changes that prevent toxin interaction with the cellular target [51,154,161].

3.1. Toxin Families Share Conserved Folds but Interact with Different Families of Antitoxins

The Pfam database utilizes Hidden Markov-models to associate similar protein families exhibiting a conserved fold or annotated function into "clans" [162]. When viewed for Type II TA systems, there are three large mostly toxin-containing clans, in addition to three common antitoxin-containing clans (Figure 2). A key feature of these is the mix-and-match nature of different antitoxin families with toxins from the same family, and vice versa. The toxin proteins are categorized into a PIN-type clan, encompassing Vap and Fit family toxins, a RelE/ParE-type clan, and a CcdB/PemK-type clan. Among these, the PIN-type VapBC systems and the RelBE-like systems appear to be the two largest families [49]. Some toxins are self-contained families that do not correlate into a larger clan (at least to date), including those involved in post-translation modifications such as the HEPN, FIC, and GNAT types (for these, see other recent reviews [7,12,27]). Previous work demonstrated the strong conservation of DNA binding motifs, generally a helix-turn-helix (HTH, see Figure 2) or ribbon-helix-helix (RHH, see "CL0057", Figure 2), in addition to the Abr-like DNA binding domain for different antitoxin families (Figure 2) [28,154,163].

The CL0280 group of toxins contains families that interact with either CL0132 or CL0057, depending on the toxin family. The toxins in this family all share a PIN domain that forms a compact RNA-binding with three highly conserved acidic residues required for metal-dependent endonuclease activity [153,164–166]. Within this fold, the VapC-type toxins found in *M. tuberculosis*, *Shigella flexneri* and *Rickettsia felis* have minor variations in numbers of specific secondary structure elements [167,168]. This fold also encompasses the FitB toxin family [169]. While the VapC and FitB toxins share structural homology, their cognate antitoxins, VapB and FitA, are located in different Pfam clans. These mediate similar yet distinct interactions their cognate toxins, with distinct structures at the more N-terminal part of FitA as compared to the C-terminus of VapB (red versus tan ribbons, Figure 3a). The VapB antitoxin is part of the RHH antitoxin family (CL0057), while the FitA antitoxins more closely correlate with the Mnt-like repressors (CL0132). Within this family, the conserved toxin fold thus interacts with antitoxins from two different structural Pfam "clans" [168–170].

Overall, toxins segregated in the CL0136 group are paired with distinct antitoxin families in the CL0057 group with a few notable exceptions. This toxin group is comprised of the RelE/ParE family of toxins containing a shared microbial RNase fold but variability in the specific active site amino acids as well as extensions at the C-terminus [13,171,172]. While most of the toxin members mediate RNA cleavage, as recently reviewed [13], the ParE toxins are unique in inhibiting DNA gyrase through an as yet unknown mechanism [71,80,173]. Despite the shared fold and mechanism, subfamilies appear to exist with both ParE-types and RelE-types, including a wide range of RelE-like RNases including HigB, YoeB, YafQ, BrnT, and MqsR (Figure 2) [27,171,172,174]. This family also includes an integrated phage-derived tripartite ParE system that appears to not mediate RNA cleavage or DNA gyrase inhibition [160,175]. Interactions with antitoxins span five different specific families in CL0057, as well as antitoxin members of CL0136 and CL0123 (Figure 2) [1,51,174]. These interactions are mediated by analogous surfaces among these distinct families, yet with distinct sequence differences and, in particular, the ParD antitoxins have longer C-terminal regions versus the RelB-type antitoxins (Figure 3b).

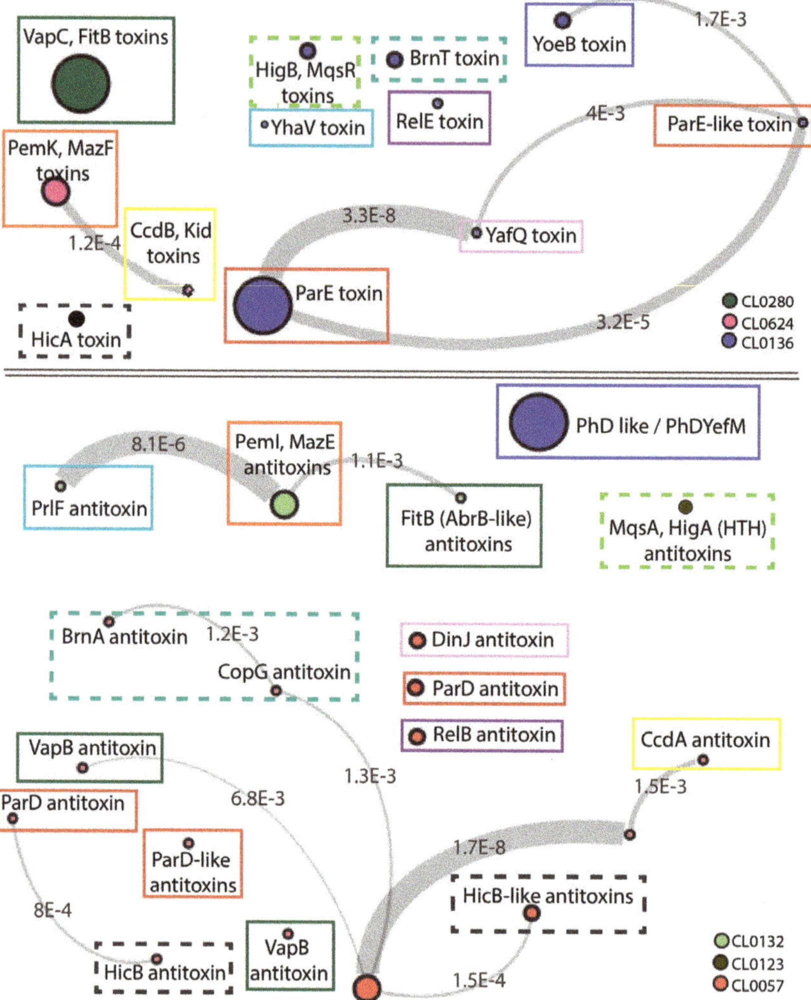

Figure 2. TA families can be grouped into Pfam "clans" (here connected by grey lines) based on Hidden Markov models, although toxin families can interact with antitoxin families from disparate clans. The force-directed diagram indicates the families that toxins and antitoxins belong to providing a predictive effect for noncognate toxin-antitoxin interaction. Patterns are adapted from the Pfam database, where the size of the circle corresponds to the number of sequences within a given family while the color corresponds to the clan. Colored boxes (dashed or solid lines) indicate cognate toxin (top) and antitoxin (bottom) pairs. The linkage between each circle indicates the E value between related families, with a lower limit of 10^{-3} to be considered significant.

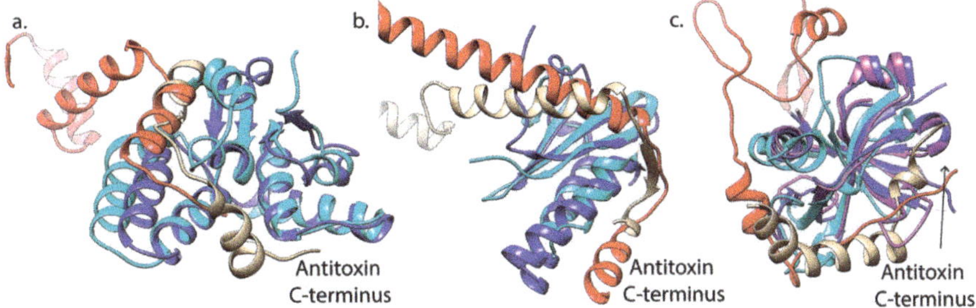

Figure 3. Shared toxin structures are neutralized by distinct antitoxin family structures, with variation evident in the C-terminal ends of the antitoxins. (**a**) Superposition of VapC (cyan) and FitB (blue) toxins reveals an overlap in interaction surfaces with antitoxin VapB (tan) and FitA (red). (PDB IDs 3DBO [167] and 2H1O [169]) (**b**) Superposition of RelE (cyan) and ParE (blue) toxins reveals an overlap in interaction surfaces with antitoxin RelB (tan) and ParD (red). (PDB IDs 4FXE [176] and 3KXE [177]) (**c**) Superposition of Kis toxin (light blue) with MazF (cyan) and CcdB (blue) toxins reveals an overlap in interaction surfaces with antitoxin MazE (tan) and CcdA (red). (PDB IDs 1M1F [178], 1UB4 [179], and 3HPW [180]).

The CL0624 toxin group similarly interacts with antitoxins in either CL0132 or CL0057. This Pfam "clan" of toxins includes the CcdAB, Kis-Kid, PemIK, and MazEF Type II TA systems. CcdB toxins act by inhibiting DNA gyrase, although using distinct mechanisms as the ParE-type toxins [181–183]. CcdB toxins have a striking structural similarity with the toxins Kid and PemK, which are endoribonuclease encoded by the Kis-Kid (*parD*) TA system found on the R1 plasmid and the PemIK system found on the R100 plasmid [133–135,184,185]. This structural family is further expanded by PemIK xenologs in *Bacillus anthracis* [186] and *E. coli* (named ChpAB) [187]. The MazF toxin is a structurally homologous RNase, although its interaction with the antitoxin is distinct among this clan (Figure 2) [179]. The MazE antitoxin consisting of a looped–hinge–helix (LHH) fold, the N-terminal region of Kis antitoxin has a unique LHH fold [188], while the CcdA antitoxin has an RHH fold [189]. It is clear that cross-interactions with non-cognate toxins in theory could occur, as the antitoxins are classified based on the DNA binding domain rather than the toxin-binding domain, and the toxin surfaces complexed with antitoxin are largely overlapping (Figure 3c).

3.2. Interactions with Cognate Antitoxins

For type II TA systems, the neutralization between the toxin and the antitoxin mediate direct protein-protein interaction. Under normal condition, the antitoxin neutralizes its cognate toxin as well as its expression to prevent its toxicity, whereas under environmental stressors, the antitoxins are believed to follow a common proteolytic degradation to release toxins, allowing it to kill cells or return cells to a dormant state [12,154,163,190]. The insulation within homologous systems has been well documented [190–192], such that the direct protein-protein interactions of toxins and antitoxins, the central mechanism of toxin control for Type II TA systems, appear to be highly specific [175,193,194].

This is exemplified within *Caulobacter crescentus,* which contains four chromosomal RelBE systems in addition to three functional ParDE systems; while these toxin families have structural similarity, they are neutralized by different classes of antitoxins [70,154]. Using a deletion approach, each cognate pair was demonstrated to have no cross-reactivity by virtue of a lack of survival when cognate antitoxins were deleted [70]. Similarly, seventeen TA systems in *V. cholera* were demonstrated to have no cross-reactivity of antitoxins [195]. Further, the relatively unique tripartite systems paaA-ParE found in integrated prophage regions in *E. coli* O157:H7 also do not cross-react [175].

M. tuberculosis encodes up to 55 different cognate VapBC systems, which were the subject of a recent study that, guided by available crystal structures, made predictions on the amino acids in the interface of each cognate pair [196]. They were able to identify sub-clusters within both toxins and antitoxins that were more likely to contain cross-interacting pairs. These predictions corroborate previous work that experimentally demonstrated high insulation thus limited crosstalk between these systems in different Vap sub-clusters [197]. Within a sub-cluster, however, possibility for cross-talk increases [196], and has previously been predicated on the identity of the C-terminal (30 amino acids) of at least a few of these VapB antitoxins [198]

4. Feasibility for Cross-Interactions of Type II TA Systems

A recent review provided a concise view of the models for TA system functions [12], and we will attempt to not repeat those here. The Anti-Addiction/Plasmid Competition model presents an attractive explanation for the long-sought functional significance of chromosomal TA systems. Based on this, we can draw the following suppositions that should be fulfilled for this model: That the chromosomal TA system will have a "match" on some invading genetic material (transposons, phage, integrons, and plasmids), and, that the chromosomal antitoxin will match the invading toxin well enough to neutralize it.

Instead, our intent is to more closely examine some of these ideas from the structural and molecular standpoint, and in particular, is predicated on the feasibility of antitoxin cross-reactions required to fulfill prevailing models of TA system addiction and anti-addiction functions.

4.1. Examples of Antitoxin Cross-Reactivity

Many TA systems, despite having low sequence similarity, exhibit a similar folding structure [25,153,154,199]. Typically these structures enforce interactions limited to cognate partners, providing insulation from cross-reactivity. However, some examples of cross-interactions have been noted to occur both between chromosomal and plasmid-borne TA systems.

Such complex cross-regulations were observed between three different *M. tuberculosis* RelBE-like modules, RelBE (Rv1246c-Rv1247c), RelFG (Rv2865-Rv2866) and RelJK (Rv3357-Rv3358). Using in vitro and cell survival assays, it was demonstrated that the RelB antitoxin can cross-neutralize the non-cognate RelG toxin, but RelB can also enhance the toxicity of the RelK toxin in cell survival assays, although the molecular basis for this remains unclear. On the other hand, RelF, the antitoxin of RelG, is able to enhance the toxicity of RelE which causes severe inhibition on bacterial growth compared to the set only expressing RelE [191]. A similar form of cross-interaction was noted for two RelB proteins encoded in the *Y. pestis* CO92 genome, although these only differ by three amino acids [200]. Cross-reactivity has been documented for the CcdAB system carried on the F plasmid [59,73]. The *Erwinia chrysanthemi* chromosomal CcdA antitoxin interacts and neutralizes the incoming F1 CcdB toxin that would otherwise kill the cell [59]. Subsequent studies noted that amino acid changes Asn 69 to Tyr in the chromosomal CcdA antitoxin and the plasmid-derived Tyr 8 Arg in the CcdB toxin affects cross-interactions, resulting in weak plasmid-derived toxin binding to chromosomal antitoxin [68,184].

Interestingly, the Kid and MazE antitoxins are able to mutually interact and partially neutralize the toxicity mediated by the non-cognate family [188]. Similarly, the CcdA and Kis antitoxins are able to cross-interact with non-cognate toxins from non-cognate family members (Kid and CcdB, respectively) [176]. CcdA binding enhances the endoribonuclease activity of Kid by triggering a conformational change that promotes interaction with its target RNA, while the Kis antitoxin effectively neutralizes the toxicity of the CcdB toxin [184]. The different effect of antitoxins on the non-cognate toxins results from their overlapping yet distinct binding sites along the toxin (Figure 3c), and as well as from potential differences in their DNA binding regions [184,188,189].

The VapBC systems in *M. tuberculosis* are highly specific for their cognate pairs; a given VapB antitoxin is not able to neutralize a non-cognate VapC toxin [190]. This was also demonstrated for different ParDE families that interact in a highly specific manner [161]. However, both studies identify the determinants for antitoxin recognition as well as the specificity and insulation of crosstalk between different TA systems. Mutating a single tryptophan amino acid in the VapB1 antitoxin from non-typeable *Haemophilus influenza* renders it to antagonize both its cognate VapC1 toxin and its non-cognate VapC2 toxin [190]. In the ParDE family, switching antitoxin residues 60, 61 and 64 in ParD3 is sufficient to alter the specificity from the cognate ParE3 to non-cognate ParE2 [161]. Those results suggested the possibility of breakage to the insulation of crosstalk and specificity between different TA systems.

Other experiments were able to generate lab-derived cross-reaction of antitoxins. Chromosomal antitoxins MazF and ChpB were mutated and constructs were selected for their ability to then neutralize plasmid-derived PemK toxin located on R100/R1 plasmids [201]. Similarly, using chemical mutagenesis of the ChpBI system that then selected for cross-neutralization against the Kis toxin and noted this mutated version could also still neutralize the ChpK toxin [201]. Chromosomal MazE is a homolog of Kis on the R1 plasmid, and can neutralize the plasmid-derived Kid although the interaction is weaker than with the cognate pair. [188,202] *M. tuberculosis* encodes seven annotated MazEF systems and two additional MazE antitoxin homologs of the *E. coli* systems, although these are not paired with the normal cognate MazF toxin [203]. They identified a "network" of non-cognate interactions, such that one of the tested MazE antitoxins interacts with two different MazF toxins. Another set of non-cognate interactions is completely reciprocated, such that a MazE antitoxin can interact with non-cognate VapC type toxins, and their cognate VapB antitoxins interact with the cognate MazF toxin as well as the non-cognate VapC toxin [203].

4.2. Orphan Antitoxins

Given the widespread nature of TA systems, it is not surprising that many partial systems are annotated. However, antitoxins that lack a cognate toxin may be particularly important for Anti-Addiction functions, as they can provide a source of toxin neutralization independent of any inherent addiction properties themselves. Orphan antitoxins have been noted in bioinformatics studies of bacterial genomes, including pathogens associated with high incidences of antibacterial resistance. The *Bartonella schoenbuchensis* type-IV secretion system is encoded on a conjugative plasmid, pVbh, where it encodes fourteen canonical TA systems as well as four orphan antitoxins [104]. While the TA systems likely mediate PSK, which is an addiction function, the orphan antitoxins are inferred to likely function as anti-addiction modules.

A recent study analyzed the genome sequences of 259 species of *Klebsiella pneumonia* complex strains and was able to predict up to 2253 orphan antitoxins [47]. These were proposed to encode remnants of degraded TA systems, or to be regulators of other (unidentified) TA pairs, or as anti-addiction modules to prevent foreign genetic material from being retained in the cell [47]. The sequences surrounding these orphan antitoxins were screened for similarities to known TA system arrangements, which revealed a high percentage were likely to be degraded from previously intact canonical systems. However, around 20% of those identified appeared to be genuine orphan genes, and further, around half of these open reading frames encoded a protein with canonical features of antitoxins [47].

A similar study mined *Mycobacterium tuberculosis* (Mt) genome sequences and identified both VapB orphan antitoxins and VapC orphan toxins [196]. The orphan toxins were more closely related to other paralogous toxins, whereas the orphan antitoxins were not similar to other known antitoxins. *Acinetobacter baumannii* also encodes numerous orphan antitoxins, as well as pairings of canonical antitoxin or toxins with non-TA system proteins [145]. A subset of the Mt orphan toxins is closely related to TA systems in *Mycobacterium marinum,* implying horizontal gene transfer between the two organisms [196]. However, variations in the active site residue raised questions about the potential for these orphan toxins to remain active [196]; however, it is not clear if these amino acid changes would impact interactions with antitoxins. Similarly, a *parDE* loci in *Caulobacter crescentus* encodes a *parE* pseudogene, resulting in a ParD antitoxin with no apparent mate, technically an "orphan" [70]. A TA system found in *Shigella flexneri* encodes a non-functional toxin, YacB, due to a frameshift mutation causing a premature stop codon [146]. The cognate antitoxin, Orf176, has previously been identified as the YacA antitoxin paired with a YacB toxin on *E. coli* plasmid pWR100 [146]. While it can be considered an "orphan" antitoxin, interestingly it is one of three such orphans that were validated substrates for type-III secretion in *Shigella* [204].

4.3. Predictions of Antitoxin Cross-Reactivity

The idea of cross-interactions in anti-addiction (Figure 1c) is predicated on the ability of one partner to interact with multiple others; the likelihood of this increases as structures or sequences are more conserved. We undertook an examination of 14 available structures, identified using tools at the Protein Databank interface [205,206], for chromosomal RelE-type toxin interactions with their corresponding antitoxins to identify sequence conservation that may indicate the ability for cross-reactions [154,159,174,176,207–216]. However, we note that within this set, the RelBE complex from *E. coli* is represented by two structures (PDB IDs 2KC8 and 4FXE, [176,207], respectively), as is the Doc toxin from the P1 bacteriophage (PDB IDs 3DD7 and 3KH2, [154,210], respectively), limiting the structure set to twelve unique complexes. The overall secondary structure and interaction of cognate RelB-like antitoxins are conserved, with two α-helices separated by a β strand and wherein this β strand typically pairs with a strand from the toxin to form an extended cross-molecule β-sheet.

We utilized the PISA webserver [217] to list the interacting amino acids, visualized these with UCSF Chimera [218], and then inspected the different interactions by eye. Within these structures we identified two sets of antitoxins that have very similar sequences at the toxin binding sites (Figure 4a). In particular, the RelJ antitoxin from *M. tuberculosis* and the *E. coli* YefM antitoxin have high sequence similarity at the region of interaction with toxins (residues 39–77 of RelJ and residues 51–89 of *E. coli* YefM) at 56.4%(or 22 out of 39 amino acids). Further, the distribution of polar versus hydrophobic amino acids at the antitoxin–toxin interfaces are also highly correlated (Figure 4b,c). We also identified that the sequences of *E. coli* HigA and *S. flexneri* HigA are identical throughout the toxin interactions regions. Overall, the complexes adopt the same structure with some minor differences in toxin loops visualized in the crystal structures (PDB IDs 4FXE and 2KC8, [159,216]). Not surprisingly, the HigB toxins from these two organisms are identical [159,216]. These examples lend some support to the idea that a structure-based approach should be able to predict cross-interacting pairs.

With a similar objective in mind, we undertook a search of phage-derived sequences to assess any potential cross-reactive TA systems as compared to *M. tuberculosis,* such as would be expected for an anti-addiction function of chromosomal systems (Figure 1). The toxin and antitoxin sequences for *M. tuberculosis* (Mtb) strain H37Rv were obtained from the toxin-antitoxin database and included the RelBE, HigBA, and VapBC families [49]. The sequences were used as a query to search the Actinobacteriophage (previously "Mycobacteriophage") Database that contains approximately 3400 sequenced phage, of which 1900 are known to originate from Mycobacterial hosts (https://phagesdb.org; [219]). The resulting BlastP resulting sequences were surprisingly poorly matched; as such, any with an E-value of 0.10 or less were aligned with the H37Rv toxin or antitoxin to which a similarity was indicated [176,213,220,221].

Among the antitoxin sequences, Mtb VapB8, VapB13, and VapB40 all had a potential match with the phage database; however, each had only one conserved amino acid likely to be in the interface. Two HigA antitoxins (Rv2021c and Rv3183) presented sequence matches in the E 10^{-3} to E 10^{-4} range, and similarities were limited to approximately 30 amino acids at the toxin-binding interface. For Rv3183, also known as HigA3, 18 of 31 amino acids present an exact match while an addition four are conserved; further, this antitoxin sequence is derived from a *Mycobacterium*-derived phage. However, this appears to be an orphan antitoxin, as the annotated open reading frames on either side of this gene do not contain similarity to the HigB (or any other) toxin. *M. tuberculosis* HigA2, annotated as Rc2021c, matches a sequence derived from a phage originating in Propionibacterium. Similar to the HigA3 antitoxin, this HigA2 antitoxin contains 22 identical and 12 similar amino acids out of 53 total, although these matches are more central to the protein so would be expected to span the toxin binding domain and the DNA binding domain. While it contains a reasonably-sized open reading frame in the toxin position (150 amino acids), the encoded protein failed to match any known sequences in the TADB or any named protein in a Blast search at the NCBI (all were "hypothetical"), indicating perhaps this is a novel toxin or an orphan HigA antitoxin embedded within a different genetic context.

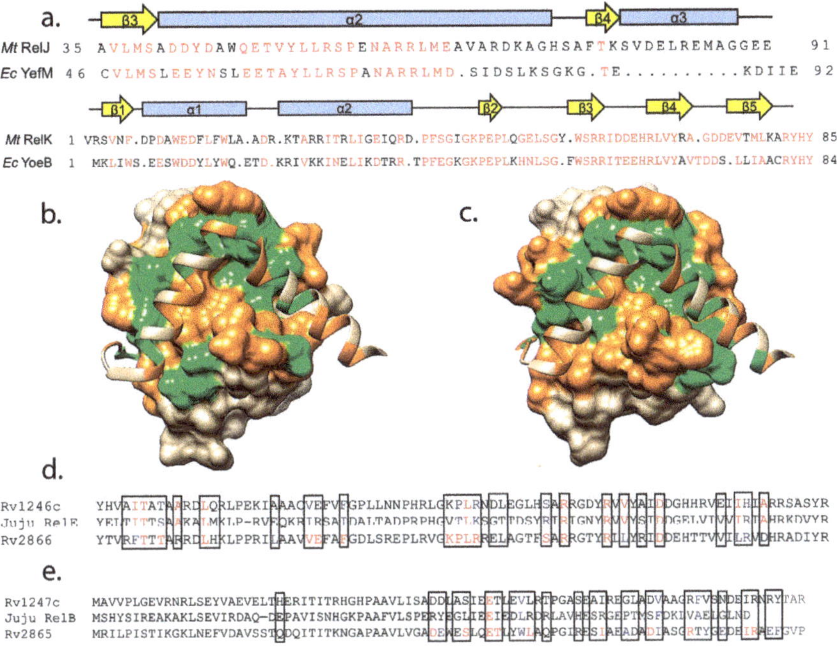

Figure 4. Potential cross-interactions of RelB/YefM antitoxins identified from analysis of sequence and structural conservation among available experimentally determined structures. (**a**) Sequence alignments (top, antitoxins; bottom, toxins) depict conserved amino acids (red), with secondary structure indicated above. (**b**) Surface of *Mycobacteria tuberculosis* toxin RelK with ribbon backbone of *M. tuberculosis* antitoxin RelJ (C-terminus is to the right); coloring is orange for hydrophobic interaction points, and green for polar interaction points. (PDB ID 3OEI, [212]). (**c**) Surface of *E. coli* toxin YoeB with ribbon backbone of *E. coli* antitoxin YefM; coloring as in b. (PDB ID 2A6Q, [211]). (**d**) Sequence alignment for Rv1246c, a RelE toxin found in the chromosome of *M. tuberculosis* and exhibiting similarity to a phage-derived sequences. This is compared to Rv2866, another *M. tuberculosis* RelE toxin with an available crystal structure (PDB ID 3G50, [212]), which was used to delineate likely protein–protein contacts for the RelE toxin and RelB antitoxin (black boxes). Red text indicates conservation, while blue indicates similarity. Note, Rv1246c is colored relative to the phage derived sequence. (**e**) The same type of sequence analysis was carried out for the respective RelB antitoxin sequences (colored and labeled as in d).

Three VapC toxins (VapC5, VapC28, and VapC42) had an E-value of 0.10 or less with the matching phage sequences. The most successful homologs were to an experimentally verified RelE toxin (Rv1246c) with an E-value to the best-matched sequence of 2 E 10^{-5}. This sequence originated from the "Juju" phage, with 33% identity and 50% conservation, as well as strong matches with 33 other phage entries [191,222]. Among these phage, the top twelve matches originated from a Gordonia host, while the remainder are from *Mycobacterium smegmatis*. Each of these phage (where genomes were annotated) contains an open reading frame for a RelB antitoxin just preceding the RelE-type toxin.

These sequences were analyzed for conservation of interaction sites based on the closest matching RelBE complex, the RelBE2 system (Rv2865-2866, PDB ID 3G5O, [212]). Assuming a conservation of interaction sites between RelE toxin and RelB antitoxin, at these sites of interaction the Rv1246c RelE toxin and the Juju phage RelE toxin have the same amino acids at 11 positions, conserved amino acids at 7 positions, and different amino acids at the remaining 9 positions (Figure 4d). When comparing the same chromosomal RelE toxin to its closest match in the PDB (Rv2866) there are 15 identical amino acids, 3 conserved, and 9 different amino acids at the interface (PDB ID 3G5O, [212]).

Given that the match was based on toxin sequences, we speculated that the corresponding RelB antitoxin sequences would be similar. The same type of analysis revealed limited conservation of antitoxins, with identical amino acids at one position, conserved amino acids at only 3 positions, and different amino acids at the remaining 22 positions (Figure 4e). As a marker of comparison, when the chromosomal Rv1246c antitoxin was compared to the crystal structure of Rv2865 there are 10 identical amino acids, 8 conserved, and 8 different amino acids at the toxin-contacting interface (PDB ID 3G5O, [212]).

This leaves an open question of—do toxins with similar antitoxin-interaction amino acids cross-react with disparate antitoxin sequences? Or, another way to phrase this might be, could the *M. tuberculosis* chromosomal RelB antitoxin (Rv1247c) neutralize an invading phage RelE toxin, and in so doing mediate anti-addiction? Further, the recovery of the phage from related Gordonia bacterial species could indicate an anti-addiction function when this chromosomal matching RelE toxin is present, or, simply could represent a sampling bias for environmental phage collection comprising the sequence database [219].

5. Discussion

TA systems are abundant on bacterial chromosomes with a seemingly high insulation from non-homologous interactions. To date, such cross-interactions have been detected when changes to either partner accumulate. Given that the premise of anti-addition offered by chromosomal systems towards mobile genetic elements requires cross-reactivity, it seems likely that it would be present in some integrated chromosomal systems as well. However, only limited cross-reactions have been observed between chromosomal and mobile genetic elements systems, perhaps due to bias in the type of experiments and functions of TA systems pursued.

It is not possible to reliably predict cross-interactions based on the currently available molecular characterizations, particularly if only a few amino acid changes are required in either partner. It may be that the natural variation that leads to such cross-reactivity does not follow such a minimalistic approach, with instead many more mutations providing a cumulative basis for an anti-addiction paradigm. It is clear that a more comprehensive understanding and screening for cross-interactions of non-cognate TA system partners would provide welcome insights into the functional possibilities for these intriguing systems.

Funding: This research was funded by the Oklahoma Center for the Advancement of Science (OCAST), grant number HR17-099.

Acknowledgments: We thank Kevin Snead for helpful input during the early stages of this manuscript.

Conflicts of Interest: The authors declare no conflict of interest. The funders had no role in the design of the study; in the collection, analyses, or interpretation of data; in the writing of the manuscript, or in the decision to publish the results.

References

1. Anantharaman, V.; Aravind, L. New connections in the prokaryotic toxin-antitoxin network: Relationship with the eukaryotic nonsense-mediated RNA decay system. *Genome Biol.* **2003**, *4*, R81. [CrossRef] [PubMed]
2. Pandey, P.D.; Gerde, K.S. Toxin-antitoxin loci are highly abundant in free-living but lost from host-associated prokaryotes. *Nucleic Acids Res.* **2005**, *33*, 966–976. [CrossRef]
3. Melderen, V.L.; Jurenas, D.; Garcia-Pino, A. Messing up translation from the start: How AtaT inhibits translation initiation in *E. coli*. *RNA Biol.* **2018**, *15*, 303–307. [CrossRef]

4. Lima-Mendez, G.; Alvarenga, D.O.; Ross, K.; Hallet, B.; Van Melderen, L.; Varani, A.M.; Chandler, M. Toxin-Antitoxin Gene Pairs Found in Tn3 Family Transposons Appear To Be an Integral Part of the Transposition Module. *mBio* **2020**, *11*, e00452-20. [CrossRef] [PubMed]
5. Wu, Y.A.; Kamruzzaman, M.; Iredell, J.R. Specialised functions of two common plasmid mediated toxin-antitoxin systems, ccdAB and pemIK, in Enterobacteriaceae. *bioRxiv* **2020**. [CrossRef]
6. Huguet, K.T.; Gonnet, M.; Doublet, B.; Cloeckaert, A. A toxin antitoxin system promotes the maintenance of the IncA/C-mobilizable Salmonella Genomic Island 1. *Sci. Rep.* **2016**, *6*, 32285. [CrossRef] [PubMed]
7. Makarova, S.K.; Wolf, Y.I.; Koonin, E.V. Comprehensive comparative-genomic analysis of type 2 toxin-antitoxin systems and related mobile stress response systems in prokaryotes. *Biol. Direct* **2009**, *4*, 19–56. [CrossRef] [PubMed]
8. Fernández-García, L.; Blasco, L.; Lopez, M.; Bou, G.; García-Contreras, R.; Wood, T.; Tomas, M. Toxin-antitoxin systems in clinical pathogens. *Toxins (Basel)* **2016**, *8*, 227–249. [CrossRef]
9. Kedzierska, B.; Hayes, F. Emerging Roles of Toxin-Antitoxin Modules in Bacterial Pathogenesis. *Molecules* **2016**, *21*, 790. [CrossRef]
10. Ramisetty, M.B.C.; Santhosh, R.S. Endoribonuclease type II toxin-antitoxin systems: Functional or selfish? *Microbiology* **2017**, *163*, 931–939. [CrossRef]
11. Yang, E.Q.; Walsh, T.R. Toxin-antitoxin systems and their role in disseminating and maintaining antimicrobial resistance. *FEMS Microbiol. Rev.* **2017**, *41*, 343–353. [CrossRef] [PubMed]
12. Fraikin, N.; Goormaghtigh, F.; van Melderen, L. Type II Toxin-Antitoxin Systems: Evolution and Revolutions. *J. Bacteriol.* **2020**, *202*, e00763-19. [CrossRef] [PubMed]
13. Jurėnas, D.; van Melderen, L. The Variety in the Common Theme of Translation Inhibition by Type II Toxin–Antitoxin Systems. *Front. Genet.* **2020**, *11*, 262. [CrossRef]
14. Hall, M.A.; Gollan, B.; Helaine, S. Toxin-antitoxin systems: Reversible toxicity. *Curr. Opin. Microbiol.* **2017**, *36*, 102–110. [CrossRef]
15. Page, R.; Peti, W. Toxin-antitoxin systems in bacterial growth arrest and persistence. *Nat. Chem. Biol.* **2016**, *12*, 208–214. [CrossRef]
16. Ronneau, S.; Helaine, S. Clarifying the Link between Toxin-Antitoxin Modules and Bacterial Persistence. *J. Mol. Biol* **2019**, *431*, 3462–3471. [CrossRef]
17. Unterholzner, J.S.; Poppenberger, B.; Rozhon, W. Toxin-antitoxin systems: Biology, identification, and application. *Mob. Genet. Elem.* **2013**, *3*, e26219. [CrossRef]
18. Schuster, F.C.; Bertram, R. Toxin-antitoxin systems are ubiquitous and versatile modulators of prokaryotic cell fate. *FEMS Microbiol. Lett.* **2013**, *340*, 73–85. [CrossRef]
19. Beyer, M.H.; Iwai, H. Off-Pathway-Sensitive Protein-Splicing Screening Based on a Toxin/Antitoxin System. *ChemBioChem* **2019**, *20*, 1933–1938. [CrossRef] [PubMed]
20. Klimina, K.M.; Kasianov, A.S.; Poluektova, E.U.; Emelyanov, K.V.; Voroshilova, V.N.; Zakharevich, N.V.; Kudryavtseva, A.V.; Makeev, V.J.; Danilenko, V.N. Employing toxin-antitoxin genome markers for identification of Bifidobacterium and Lactobacillus strains in human metagenomes. *PeerJ* **2019**, *7*, e6554. [CrossRef] [PubMed]
21. Kang, S.M.; Kim, D.H.; Jin, C.; Lee, B.J. A Systematic Overview of Type II and III Toxin-Antitoxin Systems with a Focus on Druggability. *Toxins (Basel)* **2018**, *10*, 515. [CrossRef] [PubMed]
22. Równicki, M.; Pieńko, T.; Czarnecki, J.; Kolanowska, M.; Bartosik, D.; Trylska, J. Artificial Activation of Escherichia coli mazEF and hipBA Toxin-Antitoxin Systems by Antisense Peptide Nucleic Acids as an Antibacterial Strategy. *Front. Microbiol.* **2018**, *9*, 2870. [CrossRef]
23. Lioy, V.S.; Rey, O.; Balsa, D.; Pellicer, T.; Alonso, J.C. A toxin-antitoxin module as a target for antimicrobial development. *Plasmid* **2010**, *63*, 31–39. [CrossRef]
24. Fedorec, A.J.; Ozdemir, T.; Doshi, A.; Ho, Y.K.; Rosa, L.; Rutter, J.; Velazquez, O.; Pinheiro, V.B.; Danino, T.; Barnes, C.P. Two New Plasmid Post-segregational Killing Mechanisms for the Implementation of Synthetic Gene Networks in Escherichia coli. *iScience* **2019**, *14*, 323–334. [CrossRef] [PubMed]
25. Lee, Y.K.; Lee, B.J. Structure, Biology, and Therapeutic Application of Toxin-Antitoxin Systems in Pathogenic Bacteria. *Toxins (Basel)* **2016**, *8*, 305. [CrossRef] [PubMed]
26. Magnuson, R.D. Hypothetical functions of toxin-antitoxin systems. *J. Bacteriol.* **2007**, *189*, 6089–6092. [CrossRef]

27. Harms, A.; Brodersen, D.E.; Mitarai, N.; Gerdes, K. Toxins, Targets, and Triggers: An Overview of Toxin-Antitoxin Biology. *Mol. Cell* **2018**, *70*, 768–784. [CrossRef]
28. Goeders, N.; van Melderen, L. Toxin-antitoxin systems as multilevel interaction systems. *Toxins (Basel)* **2014**, *6*, 304–324. [CrossRef]
29. Fozo, M.E.; Hemm, M.R.; Storz, G. Small Toxic Proteins and the Antisense RNAs That Repress Them. *Microbiol. Mol. Biol. Rev.* **2008**, *72*, 579–589. [CrossRef]
30. Blower, R.T.; Salmond, G.P.; Luisi, B.F. Balancing at survival's edge: The structure and adaptive benefits of prokaryotic toxin-antitoxin partners. *Curr. Opin. Struct. Biol.* **2011**, *21*, 109–118. [CrossRef]
31. Fineran, P.C.; Blower, T.R.; Foulds, I.J.; Humphreys, D.P.; Lilley, K.S.; Salmond, G.P. The phage abortive infection system, ToxIN, functions as a protein–RNA toxin–antitoxin pair. *Proc. Natl. Acad. Sci. USA* **2009**, *106*, 894. [CrossRef] [PubMed]
32. Samson, J.E.; Spinelli, S.; Cambillau, C.; Moineau, S. Structure and activity of AbiQ, a lactococcal endoribonuclease belonging to the type III toxin-antitoxin system. *Mol. Microbiol.* **2013**, *87*, 756–768. [CrossRef] [PubMed]
33. Blower, T.R.; Short, F.L.; Rao, F.; Mizuguchi, K.; Pei, X.Y.; Fineran, P.C.; Luisi, B.F.; Salmond, G.P. Identification and classification of bacterial Type III toxin-antitoxin systems encoded in chromosomal and plasmid genomes. *Nucleic Acids Res.* **2012**, *40*, 6158–6173. [CrossRef] [PubMed]
34. Goeders, N.; Chai, R.; Chen, B.; Day, A.; Salmond, G.P. Structure, Evolution, and Functions of Bacterial Type III Toxin-Antitoxin Systems. *Toxins (Basel)* **2016**, *8*, 282. [CrossRef] [PubMed]
35. Dy, R.L.; Przybilski, R.; Semeijn, K.; Salmond, G.P.; Fineran, P.C. A widespread bacteriophage abortive infection system functions through a Type IV toxin-antitoxin mechanism. *Nucleic Acids Res.* **2014**, *42*, 4590–4605. [CrossRef]
36. Masuda, H.; Tan, Q.; Awano, N.; Wu, K.P.; Inouye, M. YeeU enhances the bundling of cytoskeletal polymers of MreB and FtsZ, antagonizing the CbtA (YeeV) toxicity in Escherichia coli. *Mol. Microbiol.* **2012**, *84*, 979–989. [CrossRef]
37. Masuda, H.; Tan, Q.; Awano, N.; Yamaguchi, Y.; Inouye, M. A novel membrane-bound toxin for cell division, CptA (YgfX), inhibits polymerization of cytoskeleton proteins, FtsZ and MreB, in Escherichia coli. *FEMS Microbiol. Lett.* **2012**, *328*, 174–181. [CrossRef]
38. Wang, X.; Lord, D.M.; Cheng, H.Y.; Osbourne, D.O.; Hong, S.H.; Sanchez-Torres, V.; Quiroga, C.; Zheng, K.; Herrmann, T.; Peti, W.; et al. A new type V toxin-antitoxin system where mRNA for toxin GhoT is cleaved by antitoxin GhoS. *Nat. Chem. Biol.* **2012**, *8*, 855–861. [CrossRef]
39. Barbosa, L.C.B.; dos Santos Carrijo, R.; da Conceição, M.B.; Campanella, J.E.M.; Júnior, E.C.; Secches, T.O.; Bertolini, M.C.; Marchetto, R. Characterization of an OrtT-like toxin of Salmonella enterica serovar Houten. *Braz. J. Microbiol.* **2019**, *50*, 839–848. [CrossRef]
40. Islam, S.; Benedik, M.J.; Wood, T.K. Orphan toxin OrtT (YdcX) of Escherichia coli reduces growth during the stringent response. *Toxins (Basel)* **2015**, *7*, 299–321. [CrossRef]
41. Aakre, C.D.; Phung, T.N.; Huang, D.; Laub, M.T. A bacterial toxin inhibits DNA replication elongation through a direct interaction with the beta sliding clamp. *Mol. Cell* **2013**, *52*, 617–628. [CrossRef]
42. Gabrisko, M.; Barak, I. Evolution of the SpoIISABC Toxin-Antitoxin-Antitoxin System in Bacilli. *Toxins (Basel)* **2016**, *8*, 180. [CrossRef]
43. Jimmy, S.; Saha, C.K.; Kurata, T.; Stavropoulos, C.; Oliveira, S.R.A.; Koh, A.; Cepauskas, A.; Takada, H.; Rejman, D.; Tenson, T.; et al. A widespread toxin-antitoxin system exploiting growth control via alarmone signaling. *Proc. Natl. Acad. Sci. USA* **2020**, *117*, 10500–10510. [CrossRef] [PubMed]
44. Sala, A.; Calderon, V.; Bordes, P.; Genevaux, P. TAC from Mycobacterium tuberculosis: A paradigm for stress-responsive toxin-antitoxin systems controlled by SecB-like chaperones. *Cell Stress Chaperones* **2013**, *18*, 129–135. [CrossRef] [PubMed]
45. Wilcox, B.; Osterman, I.; Serebryakova, M.; Lukyanov, D.; Komarova, E.; Gollan, B.; Morozova, N.; Wolf, Y.I.; Makarova, K.S.; Helaine, S.; et al. Escherichia coli ItaT is a type II toxin that inhibits translation by acetylating isoleucyl-tRNAIle. *Nucleic Acids Res.* **2018**, *46*, 7873–7885. [CrossRef] [PubMed]
46. VanDrisse, M.C.; Parks, A.R.; Escalante-Semerena, J.C. A Toxin Involved in Salmonella Persistence Regulates Its Activity by Acetylating Its Cognate Antitoxin, a Modification Reversed by CobB Sirtuin Deacetylase. *mBio* **2017**, *8*, e00708-17. [CrossRef] [PubMed]

47. Horesh, G.; Fino, C.; Harms, A.; Dorman, M.J.; Parts, L.; Gerdes, K.; Heinz, E.; Thomson, N.R. Type II and type IV toxin-antitoxin systems show different evolutionary patterns in the global Klebsiella pneumoniae population. *Nucleic Acids Res.* **2020**, *48*, 4357–4370. [CrossRef]
48. Yamaguchi, Y.; Park, J.H.; Inouye, M. Toxin-antitoxin systems in bacteria and archaea. *Ann. Rev. Genet.* **2011**, *45*, 61–79. [CrossRef]
49. Xie, Y.; Wei, Y.; Shen, Y.; Li, X.; Zhou, H.; Tai, C.; Deng, Z.; Ou, H.Y. TADB 2.0: An updated database of bacterial type II toxin-antitoxin loci. *Nucleic Acids Res.* **2018**, *46*, D749–D753. [CrossRef]
50. Akarsu, H.; Bordes, P.; Mansour, M.; Bigot, D.J.; Genevaux, P.; Falquet, L. TASmania: A bacterial Toxin-Antitoxin Systems database. *PLoS Comput. Biol.* **2019**, *15*, e1006946. [CrossRef]
51. Leplae, R.; Geeraerts, D.; Hallez, R.; Guglielmini, J.; Dreze, P.; Van Melderen, L. Diversity of bacterial type II toxin-antitoxin systems: A comprehensive search and functional analysis of novel families. *Nucleic Acids Res.* **2011**, *39*, 5513–5525. [CrossRef] [PubMed]
52. Chan, T.W.; Balsa, D.; Espinosa, M. One cannot rule them all: Are bacterial toxins-antitoxins druggable? *FEMS Microbiol. Rev.* **2015**, *39*, 522–540. [CrossRef] [PubMed]
53. Bernard, P.; Couturier, M. Cell killing by the F plasmid CcdB protein involves poisoning of DNA-topoisomerase II complexes. *J. Mol. Biol.* **1992**, *226*, 735–745. [CrossRef]
54. Melderen, V.L.; Bernard, P.; Couturier, M. Lon-dependent proteolysis of CcdA is the key control for activation of CcdB in plasmid-free segregant bacteria. *Mol. Microbiol.* **1994**, *11*, 1151–1157. [CrossRef] [PubMed]
55. Roberts, C.R.; Strom, A.R.; Helinski, D.R. The parDE operon of the broad-host-range plasmid RK2 specifies growth inhibition associated with plasmid loss. *J. Mol. Biol.* **1994**, *237*, 35–51. [CrossRef]
56. Sobecky, P.A.; Easter, C.L.; Bear, P.D.; Helinski, D.R. Characterization of the stable maintenance properties of the par region of broad-host-range plasmid RK2. *J. Bacteriol.* **1996**, *178*, 2086–2093. [CrossRef]
57. Yarmolinsky, M.B. Programmed cell death in bacterial populations. *Science* **1995**, *267*, 836. [CrossRef]
58. Koonin, E.V.; Makarova, K.S.; Aravind, L. Horizontal Gene Transfer in Prokaryotes: Quantification and Classification. *Annu. Rev. Microbiol.* **2001**, *55*, 709–742. [CrossRef]
59. de Bast, S.M.; Mine, N.; van Melderen, L. Chromosomal toxin-antitoxin systems may act as antiaddiction modules. *J. Bacteriol.* **2008**, *190*, 4603–4609. [CrossRef]
60. Makarova, S.K.; Wolf, Y.I.; Koonin, E.V. Comparative genomics of defense systems in archaea and bacteria. *Nucleic Acids Res.* **2013**, *41*, 4360–4377. [CrossRef]
61. Hernández-Arriaga, A.M.; Chan, W.T.; Espinosa, M.; Díaz-Orejas, R. Conditional Activation of Toxin-Antitoxin Systems: Postsegregational Killing and Beyond. *Microbiol. Spectr.* **2014**, *2*, 175–192. [CrossRef] [PubMed]
62. Mine, N.; Guglielmini, J.; Wilbaux, M.; Van Melderen, L. The decay of the chromosomally encoded ccdO157 toxin-antitoxin system in the Escherichia coli species. *Genetics* **2009**, *181*, 1557–1566. [CrossRef]
63. Ramisetty, M.B.C.; Santhosh, R.S. Horizontal gene transfer of chromosomal Type II toxin–antitoxin systems of Escherichia coli. *FEMS Microbiol. Lett.* **2016**, *363*. [CrossRef] [PubMed]
64. Ramisetty, B.C.M.; Sudhakari, P.A. Bacterial 'Grounded' Prophages: Hotspots for Genetic Renovation and Innovation. *Front. Genet.* **2019**, *10*, 65. [CrossRef] [PubMed]
65. Koonin, V.E.; Wolf, Y.I. Genomics of bacteria and archaea: The emerging dynamic view of the prokaryotic world. *Nucleic Acids Res.* **2008**, *36*, 6688–6719. [CrossRef]
66. Andersen, S.B.; Ghoul, M.; Griffin, A.S.; Petersen, B.; Johansen, H.K.; Molin, S. Diversity, Prevalence, and Longitudinal Occurrence of Type II Toxin-Antitoxin Systems of Pseudomonas aeruginosa Infecting Cystic Fibrosis Lungs. *Front. Microbiol.* **2017**, *8*, 1180. [CrossRef] [PubMed]
67. Diaz-Orejas, R.; Espinosa, M.; Yeo, C.C. The Importance of the Expendable: Toxin-Antitoxin Genes in Plasmids and Chromosomes. *Front. Microbiol.* **2017**, *8*, 1479. [CrossRef]
68. Gupta, K.; Tripathi, A.; Sahu, A.; Varadarajan, R. Contribution of the chromosomal ccdAB operon to bacterial drug tolerance. *J. Bacteriol.* **2017**, *199*, e00397-17. [CrossRef]
69. Ames, J.R.; Muthuramalingam, M.; Murphy, T.; Najar, F.Z.; Bourne, C.R. Expression of different ParE toxins results in conserved phenotypes with distinguishable classes of toxicity. *MicrobiologyOpen* **2019**, *8*, e902. [CrossRef]
70. Fiebig, A.; Castro Rojas, C.M.; Siegal-Gaskins, D.; Crosson, S. Interaction specificity, toxicity and regulation of a paralogous set of ParE/RelE-family toxin-antitoxin systems. *Mol. Microbiol.* **2010**, *77*, 236–251. [CrossRef]

71. Muthuramalingam, M.; White, J.C.; Murphy, T.; Ames, J.R.; Bourne, C.R. The toxin from a ParDE toxin-antitoxin system found in Pseudomonas aeruginosa offers protection to cells challenged with anti-gyrase antibiotics. *Mol. Microbiol.* **2018**, *111*, 441–454. [CrossRef] [PubMed]
72. Tamman, H.; Ainelo, A.; Ainsaar, K.; Hõrak, R. A moderate toxin, GraT, modulates growth rate and stress tolerance of Pseudomonas putida. *J. Bacteriol.* **2014**, *196*, 157–169. [CrossRef] [PubMed]
73. Wilbaux, M.; Mine, N.; Guérout, A.M.; Mazel, D.; Van Melderen, L. Functional interactions between coexisting toxin-antitoxin systems of the ccd family in Escherichia coli O157:H7. *J. Bacteriol.* **2007**, *189*, 2712–2719. [CrossRef] [PubMed]
74. Fernandez-Garcia, L.; Kim, J.S.; Tomas, M.; Wood, T.K. Toxins of toxin/antitoxin systems are inactivated primarily through promoter mutations. *J. Appl. Microbiol.* **2019**, *127*, 1859–1868. [CrossRef] [PubMed]
75. Deter, H.S.; Jensen, R.V.; Mather, W.H.; Butzin, N.C. Mechanisms for Differential Protein Production in Toxin-Antitoxin Systems. *Toxins (Basel)* **2017**, *9*, 211. [CrossRef]
76. Singh, R.; Barry, C.E., 3rd; Boshoff, H.I. The three RelE homologs of Mycobacterium tuberculosis have individual, drug-specific effects on bacterial antibiotic tolerance. *J. Bacteriol.* **2010**, *192*, 1279–1291. [CrossRef]
77. Pontes, H.M.; Groisman, E.A. Slow growth determines nonheritable antibiotic resistance in Salmonella enterica. *Sci. Signal.* **2019**, *12*, eaax3938. [CrossRef] [PubMed]
78. Tripathi, A.; Dewan, P.C.; Barua, B.; Varadarajan, R. Additional role for the ccd operon of F-plasmid as a transmissible persistence factor. *Proc. Natl Acad. Sci. USA* **2012**, *109*, 12497–12502. [CrossRef] [PubMed]
79. Kamruzzaman, M.; Iredell, J. A ParDE-family toxin antitoxin system in major resistance plasmids of Enterobacteriaceae confers antibiotic and heat tolerance. *Sci. Rep.* **2019**, *9*, 9872. [CrossRef]
80. Gupta, M.; Nayyar, N.; Chawla, M.; Sitaraman, R.; Bhatnagar, R.; Banerjee, N. The chromosomal parDE2 toxin-antitoxin system of Mycobacterium tuberculosis H37Rv: Genetic and functional characterization. *Front. Microbiol.* **2016**, *7*, 886. [CrossRef]
81. Harms, A.; Stanger, F.V.; Scheu, P.D.; de Jong, I.G.; Goepfert, A.; Glatter, T.; Gerdes, K.; Schirmer, T.; Dehio, C. Adenylylation of Gyrase and Topo IV by FicT Toxins Disrupts Bacterial DNA Topology. *Cell Rep.* **2015**, *12*, 1497–1507. [CrossRef] [PubMed]
82. Janssen, D.B.; Garza-Sanchez, F.; Hayes, C.S. YoeB toxin is activated during thermal stress. *MicrobiologyOpen* **2015**, *4*, 682–697. [CrossRef]
83. Chan, W.T.; Domenech, M.; Moreno-Córdoba, I.; Navarro-Martínez, V.; Nieto, C.; Moscoso, M.; García, E.; Espinosa, M. The Streptococcus pneumoniae yefM-yoeB and relBE Toxin-Antitoxin Operons Participate in Oxidative Stress and Biofilm Formation. *Toxins (Basel)* **2018**, *10*, 378. [CrossRef] [PubMed]
84. Ainelo, A.; Porosk, R.; Kilk, K.; Rosendahl, S.; Remme, J.; Hõrak, R. Pseudomonas putida Responds to the Toxin GraT by Inducing Ribosome Biogenesis Factors and Repressing TCA Cycle Enzymes. *Toxins (Basel)* **2019**, *11*, 103. [CrossRef] [PubMed]
85. Zhao, Y.; McAnulty, M.J.; Wood, T.K. Toxin YafQ Reduces Escherichia coli Growth at Low Temperatures. *PLoS ONE* **2016**, *11*, e0161577. [CrossRef] [PubMed]
86. Korch, S.B.; Malhotra, V.; Contreras, H.; Clark-Curtiss, J.E. The Mycobacterium tuberculosis relBE toxin:antitoxin genes are stress-responsive modules that regulate growth through translation inhibition. *J. Microbiol.* **2015**, *53*, 783–795. [CrossRef]
87. Christensen, K.S.; Gerdes, K. Delayed-relaxed response explained by hyperactivation of RelE. *Mol. Microbiol.* **2004**, *53*, 587–597. [CrossRef]
88. Tashiro, Y.; Kawata, K.; Taniuchi, A.; Kakinuma, K.; May, T.; Okabe, S. RelE-mediated dormancy is enhanced at high cell density in Escherichia coli. *J. Bacteriol.* **2012**, *194*, 1169–1176. [CrossRef]
89. Armalytė, J.; Jurėnas, D.; Krasauskas, R.; Čepauskas, A.; Sužiedėlienė, E. The higBA Toxin-Antitoxin Module From the Opportunistic Pathogen Acinetobacter baumannii—Regulation, Activity, and Evolution. *Front. Microbiol.* **2018**, *9*, 732. [CrossRef]
90. Wood, L.T.; Wood, T.K. The HigB/HigA toxin/antitoxin system of Pseudomonas aeruginosa influences the virulence factors pyochelin, pyocyanin, and biofilm formation. *MicrobiolOpen* **2016**, *5*, 499–511. [CrossRef]
91. Ramisetty, B.; Ghosh, D.; Roy Chowdhury, M.; Santhosh, R.S. What Is the Link between Stringent Response, Endoribonuclease Encoding Type II Toxin-Antitoxin Systems and Persistence? *Front. Microbiol.* **2016**, *7*, 1882. [CrossRef] [PubMed]
92. Zhang, Y.; Inouye, M. RatA (YfjG), an Escherichia coli toxin, inhibits 70S ribosome association to block translation initiation. *Mol. Microbiol.* **2011**, *79*, 1418–1429. [CrossRef] [PubMed]

93. Norton, P.J.; Mulvey, M.A. Toxin-antitoxin systems are important for niche-specific colonization and stress resistance of uropathogenic Escherichia coli. *PLoS Pathog.* **2012**, *8*, e1002954. [CrossRef] [PubMed]
94. Tiwari, P.; Arora, G.; Singh, M.; Kidwai, S.; Narayan, O.P.; Singh, R. MazF ribonucleases promote Mycobacterium tuberculosis drug tolerance and virulence in guinea pigs. *Nat. Commun.* **2015**, *6*, 6059. [CrossRef]
95. Ren, D.; Walker, A.N.; Daines, D.A. Toxin-antitoxin loci vapBC-1 and vapXD contribute to survival and virulence in nontypeable Haemophilus influenzae. *BMC Microbiol.* **2012**, *12*, 263. [CrossRef]
96. Jaiswal, S.; Paul, P.; Padhi, C.; Ray, S.; Ryan, D.; Dash, S.; Suar, M. The Hha-TomB Toxin-Antitoxin System Shows Conditional Toxicity and Promotes Persister Cell Formation by Inhibiting Apoptosis-Like Death in S. *Typhimurium*. *Sci. Rep.* **2016**, *6*, 38204. [CrossRef]
97. Helaine, S.; Cheverton, A.M.; Watson, K.G.; Faure, L.M.; Matthews, S.A.; Holden, D.W. Internalization of Salmonella by macrophages induces formation of nonreplicating persisters. *Science* **2014**, *343*, 204–208. [CrossRef]
98. Fraikin, N.; Rousseau, C.J.; Goeders, N.; Van Melderen, L. Reassessing the Role of the Type II MqsRA Toxin-Antitoxin System in Stress Response and Biofilm Formation: mqsA Is Transcriptionally Uncoupled from mqsR. *mBio* **2019**, *10*, e02678-19. [CrossRef]
99. Gerdes, K.; Christensen, S.K.; Lobner-Olesen, A. Prokaryotic toxin-antitoxin stress response loci. *Nat. Rev. Microbiol.* **2005**, *3*, 371–382. [CrossRef]
100. Kim, Y.; Wang, X.; Zhang, X.S.; Grigoriu, S.; Page, R.; Peti, W.; Wood, T.K. Escherichia coli toxin/antitoxin pair MqsR/MqsA regulate toxin CspD. *Environ. Microbiol.* **2010**, *12*, 1105–1121. [CrossRef]
101. Wade, T.J.; Laub, M.T. Concerns about "Stress-Induced MazF-Mediated Proteins in *Escherichia coli*". *mBio* **2019**, *10*, e00825-19. [CrossRef] [PubMed]
102. Culviner, H.P.; Laub, M.T. Global Analysis of the *E. coli* Toxin MazF Reveals Widespread Cleavage of mRNA and the Inhibition of rRNA Maturation and Ribosome Biogenesis. *Mol. Cell* **2018**, *70*, 868–880. [CrossRef] [PubMed]
103. Nigam, A.; Ziv, T.; Oron-Gottesman, A. Engelberg-Kulka Stress-Induced MazF-Mediated Proteins in *Escherichia coli*. *mBio* **2019**, *10*, e00340-19. [PubMed]
104. Harms, A.; Liesch, M.; Körner, J.; Québatte, M.; Engel, P.; Dehio, C. A bacterial toxin-antitoxin module is the origin of inter-bacterial and inter-kingdom effectors of Bartonella. *PLoS Genet.* **2017**, *13*, e1007077. [CrossRef] [PubMed]
105. Harms, A.; Maisonneuve, E.; Gerdes, K. Mechanisms of bacterial persistence during stress and antibiotic exposure. *Science* **2016**, *354*, aaf4268. [CrossRef] [PubMed]
106. Goormaghtigh, F.; Fraikin, N.; Putrinš, M.; Hallaert, T.; Hauryliuk, V.; Garcia-Pino, A.; Sjödin, A.; Kasvandik, S.; Udekwu, K.; Tenson, T.; et al. Reassessing the Role of Type II Toxin-Antitoxin Systems in Formation of *Escherichia coli* Type II Persister Cells. *mBio* **2018**, *9*, e00640-18. [CrossRef] [PubMed]
107. Goormaghtigh, F.; Fraikin, N.; Putrinš, M.; Hauryliuk, V.; Garcia-Pino, A.; Udekwu, K.; Tenson, T.; Kaldalu, N.; Van Melderen, L. Reply to Holden and Errington, "Type II Toxin-Antitoxin Systems and Persister Cells". *mBio* **2018**, *9*, e01838-18. [CrossRef] [PubMed]
108. Zielenkiewicz, U.; Ceglowski, P. The toxin-antitoxin system of the streptococcal plasmid pSM19035. *J. Bacteriol.* **2005**, *187*, 6094–6105. [CrossRef]
109. Chan, T.W.; Espinosa, M. The Streptococcus pneumoniae pezAT Toxin-Antitoxin System Reduces beta-Lactam Resistance and Genetic Competence. *Front. Microbiol.* **2016**, *7*, 1322. [CrossRef]
110. Piscotta, J.F.; Jeffrey, P.D.; Link, A.J. ParST is a widespread toxin-antitoxin module that targets nucleotide metabolism. *Proc. Natl. Acad. Sci. USA* **2019**, *116*, 826–834. [CrossRef]
111. Yuan, J.; Yamaichi, Y.; Waldor, M.K. The three Vibrio cholerae chromosome II-encoded ParE toxins degrade chromosome I following loss of chromosome II. *J. Bacteriol.* **2011**, *193*, 611–619. [CrossRef] [PubMed]
112. Wozniak, A.R.; Waldor, M.K. A toxin-antitoxin system promotes the maintenance of an integrative conjugative element. *PLoS Genet.* **2009**, *5*, e1000439. [CrossRef] [PubMed]
113. Hughes-Games, A.; Roberts, A.P.; Davis, S.A.; Hill, D.J. Identification of integrative and conjugative elements in pathogenic and commensal Neisseriaceae species via genomic distributions of DNA uptake sequence dialects. *Microb. Genom.* **2020**, *6*, e000372.

114. Berni, B.; Soscia, C.; Djermoun, S.; Ize, B.; Bleves, S. A Type VI Secretion System Trans-Kingdom Effector Is Required for the Delivery of a Novel Antibacterial Toxin in Pseudomonas aeruginosa. *Front. Microbiol.* **2019**, *10*, 1218. [CrossRef]
115. Cao, Z.; Casabona, M.G.; Kneuper, H.; Chalmers, J.D.; Palmer, T. The type VII secretion system of Staphylococcus aureus secretes a nuclease toxin that targets competitor bacteria. *Nat. Microbiol.* **2016**, *2*, 16183. [CrossRef]
116. Whitney, J.C.; Peterson, S.B.; Kim, J.; Pazos, M.; Verster, A.J.; Radey, M.C.; Kulasekara, H.D.; Ching, M.Q.; Bullen, N.P.; Bryant, D.; et al. A broadly distributed toxin family mediates contact-dependent antagonism between gram-positive bacteria. *Elife* **2017**, *6*, e26938. [CrossRef]
117. Gerdes, K.; Bech, F.W.; Jørgensen, S.T.; Løbner-Olesen, A.; Rasmussen, P.B.; Atlung, T.; Boe, L.; Karlstrom, O.; Molin, S.; Von Meyenburg, K. Mechanism of postsegregational killing by the hok gene product of the parB system of plasmid R1 and its homology with the relF gene product of the *E. coli* relB operon. *EMBO J.* **1986**, *5*, 2023–2029. [CrossRef]
118. Ogura, T.; Hiraga, S. Mini-F plasmid genes that couple host cell division to plasmid proliferation. *Proc. Natl. Acad. Sci. USA* **1983**, *80*, 4784–4788. [CrossRef]
119. Aizenman, E.; Engelberg-Kulka, H.; Glaser, G. An *Escherichia coli* chromosomal "addiction module" regulated by guanosine 3′,5′-bispyrophosphate: A model for programmed bacterial cell death. *Proc. Natl. Acad. Sci. USA* **1996**, *93*, 6059–6063. [CrossRef]
120. Jensen, B.R.; Gerdes, K. Programmed cell death in bacteria: Proteic plasmid stabilization systems. *Mol. Microbiol.* **1995**, *17*, 205–210. [CrossRef]
121. Zander, I.; Shmidov, E.; Roth, S.; Ben-David, Y.; Shoval, I.; Shoshani, S.; Danielli, A.; Banin, E. Characterization of PfiT/PfiA toxin-antitoxin system of Pseudomonas aeruginosa that affects cell elongation and prophage induction. *Environ. Microbiol.* **2020**. [CrossRef] [PubMed]
122. Li, Y.; Liu, X.; Tang, K.; Wang, W.; Guo, Y.; Wang, X. Prophage encoding toxin/antitoxin system PfiT/PfiA inhibits Pf4 production in Pseudomonas aeruginosa. *Microb. Biotechnol.* **2020**, *13*, 1132–1144. [CrossRef] [PubMed]
123. Bustamante, P.; Iredell, J.R. Carriage of type II toxin-antitoxin systems by the growing group of IncX plasmids. *Plasmid* **2017**, *91*, 19–27. [CrossRef] [PubMed]
124. Cooper, F.T.; Paixao, T.; Heinemann, J.A. Within-host competition selects for plasmid-encoded toxin-antitoxin systems. *Proc. Biol. Sci.* **2010**, *277*, 3149–3155. [CrossRef] [PubMed]
125. Fico, S.; Mahillon, J. TasA-tasB, a new putative toxin-antitoxin (TA) system from Bacillus thuringiensis pGI1 plasmid is a widely distributed composite mazE-doc TA system. *BMC Genom.* **2006**, *7*, 259. [CrossRef]
126. Hernández-Ramírez, K.C.; Chávez-Jacobo, V.M.; Valle-Maldonado, M.I.; Patiño-Medina, J.A.; Díaz-Pérez, S.P.; Jácome-Galarza, I.E.; Ortiz-Alvarado, R.; Meza-Carmen, V.; Ramírez-Díaz, M.I. Plasmid pUM505 encodes a Toxin-Antitoxin system conferring plasmid stability and increased Pseudomonas aeruginosa virulence. *Microb. Pathog.* **2017**, *112*, 259–268. [CrossRef] [PubMed]
127. Poirel, L.; Bonnin, R.A.; Nordmann, P. Genetic features of the widespread plasmid coding for the carbapenemase OXA-48. *Antimicrob. Agents Chemother.* **2012**, *56*, 559–562. [CrossRef]
128. Mnif, B.; Vimont, S.; Boyd, A.; Bourit, E.; Picard, B.; Branger, C.; Denamur, E.; Arlet, G. Molecular characterization of addiction systems of plasmids encoding extended-spectrum beta-lactamases in *Escherichia coli*. *J. Antimicrob. Chemother.* **2010**, *65*, 1599–1603. [CrossRef]
129. Davis, L.T.; Helinski, D.R.; Roberts, R.C. Transcription and autoregulation of the stabilizing functions of broad-host-range plasmid RK2 in *Escherichia coli*, Agrobacterium tumefaciens and Pseudomonas aeruginosa. *Mol. Microbiol.* **1992**, *6*, 1981–1994. [CrossRef]
130. Gronlund, H.; Gerdes, K. Toxin-antitoxin systems homologous with relBE of *Escherichia coli* plasmid P307 are ubiquitous in prokaryotes. *J. Mol. Biol.* **1999**, *285*, 1401–1415. [CrossRef]
131. Wang, J.; Stephan, R.; Zurfluh, K.; Hächler, H.; Fanning, S. Characterization of the genetic environment of bla ESBL genes, integrons and toxin-antitoxin systems identified on large transferrable plasmids in multi-drug resistant *Escherichia coli*. *Front. Microbiol.* **2014**, *5*, 716. [PubMed]
132. Hayes, F. A family of stability determinants in pathogenic bacteria. *J. Bacteriol.* **1998**, *180*, 6415–6418. [CrossRef]

133. Bravo, A.; de Torrontegui, G.; Diaz, R. Identification of components of a new stability system of plasmid R1, ParD, that is close to the origin of replication of this plasmid. *Mol. Gen. Genet.* **1987**, *210*, 101–110. [CrossRef] [PubMed]
134. Diago-Navarro, E.; Hernandez-Arriaga, A.M.; López-Villarejo, J.; Muñoz-Gómez, A.J.; Kamphuis, M.B.; Boelens, R.; Lemonnier, M.; Díaz-Orejas, R. parD toxin-antitoxin system of plasmid R1–basic contributions, biotechnological applications and relationships with closely-related toxin-antitoxin systems. *FEBS J.* **2010**, *277*, 3097–3117. [CrossRef] [PubMed]
135. Tsuchimoto, S.; Ohtsubo, H.; Ohtsubo, E. Two genes, pemK and pemI, responsible for stable maintenance of resistance plasmid R100. *J. Bacteriol.* **1988**, *170*, 1461–1466. [CrossRef]
136. Bukowski, M.; Hyz, K.; Janczak, M.; Hydzik, M.; Dubin, G.; Wladyka, B. Identification of novel mazEF/pemIK family toxin-antitoxin loci and their distribution in the Staphylococcus genus. *Sci. Rep.* **2017**, *7*, 13462. [CrossRef]
137. Lehnherr, H.; Maguin, E.; Jafri, S.; Yarmolinsky, M.B. Plasmid addiction genes of bacteriophage P1: Doc, which causes cell death on curing of prophage, and phd, which prevents host death when prophage is retained. *J. Mol. Biol.* **1993**, *233*, 414–428. [CrossRef]
138. Tian, Q.B.; Hayashi, T.; Murata, T.; Terawaki, Y. Gene product identification and promoter analysis of hig locus of plasmid Rts1. *Biochem. Biophys Res. Commun.* **1996**, *225*, 679–684. [CrossRef]
139. Dziewit, L.; Jazurek, M.; Drewniak, L.; Baj, J.; Bartosik, D. The SXT conjugative element and linear prophage N15 encode toxin-antitoxin-stabilizing systems homologous to the tad-ata module of the Paracoccus aminophilus plasmid pAMI2. *J. Bacteriol.* **2007**, *189*, 1983–1997. [CrossRef]
140. Grady, R.; Hayes, F. Axe-Txe, a broad-spectrum proteic toxin-antitoxin system specified by a multidrug-resistant, clinical isolate of Enterococcus faecium. *Mol. Microbiol.* **2003**, *47*, 1419–1432. [CrossRef]
141. Rosvoll, T.C.; Pedersen, T.; Sletvold, H.; Johnsen, P.J.; Sollid, J.E.; Simonsen, G.S.; Jensen, L.B.; Nielsen, K.M.; Sundsfjord, A. PCR-based plasmid typing in Enterococcus faecium strains reveals widely distributed pRE25-, pRUM-, pIP501- and pHTbeta-related replicons associated with glycopeptide resistance and stabilizing toxin-antitoxin systems. *FEMS Immunol. Med. Microbiol.* **2010**, *58*, 254–268. [CrossRef] [PubMed]
142. Zhang, Y.X.; Guo, X.K.; Chuan, W.U.; Bo, B.I.; Ren, S.X.; Wu, C.F.; Zhao, G.P. Characterization of a novel toxin-antitoxin module, VapBC, encoded by Leptospira interrogans chromosome. *Cell Res.* **2004**, *14*, 208–216. [CrossRef] [PubMed]
143. Jurėnas, D.; Chatterjee, S.; Konijnenberg, A.; Sobott, F.; Droogmans, L.; Garcia-Pino, A.; Van Melderen, L. AtaT blocks translation initiation by N-acetylation of the initiator tRNAfMet. *Nat. Chem. Biol.* **2017**, *13*, 640. [CrossRef] [PubMed]
144. Wen, Y.; Behiels, E.; Devreese, B. Toxin-Antitoxin systems: Their role in persistence, biofilm formation, and pathogenicity. *Pathog. Dis.* **2014**, *70*, 240–249. [CrossRef] [PubMed]
145. Jurenaite, M.; Markuckas, A.; Suziedeliene, E. Identification and characterization of type II toxin-antitoxin systems in the opportunistic pathogen Acinetobacter baumannii. *J. Bacteriol.* **2013**, *195*, 3165–3172. [CrossRef] [PubMed]
146. McVicker, G.; Tang, C.M. Deletion of toxin-antitoxin systems in the evolution of Shigella sonnei as a host-adapted pathogen. *Nat. Microbiol.* **2016**, *2*, 16204. [CrossRef]
147. Cooper, F.T.; Heinemann, J.A. Postsegregational killing does not increase plasmid stability but acts to mediate the exclusion of competing plasmids. *Proc. Natl. Acad. Sci. USA* **2000**, *97*, 12643–12648. [CrossRef]
148. Song, S.; Wood, T.K. Post-segregational Killing and Phage Inhibition Are Not Mediated by Cell Death Through Toxin/Antitoxin Systems. *Front. Microbiol.* **2018**, *9*, 814. [CrossRef]
149. Otsuka, Y.; Yonesaki, T. Dmd of bacteriophage T4 functions as an antitoxin against *Escherichia coli* LsoA and RnlA toxins. *Mol. Microbiol.* **2012**, *83*, 669–681. [CrossRef]
150. Blower, T.R.; Fineran, P.C.; Johnson, M.J.; Toth, I.K.; Humphreys, D.P.; Salmond, G.P.C. Mutagenesis and functional characterization of the RNA and protein components of the toxIN abortive infection and toxin-antitoxin locus of Erwinia. *J. Bacteriol.* **2009**, *191*, 6029–6039. [CrossRef]
151. Dy, R.L.; Richter, C.; Salmond, G.P.; Fineran, P.C. Remarkable Mechanisms in Microbes to Resist Phage Infections. *Annu. Rev. Virol.* **2014**, *1*, 307–331. [CrossRef] [PubMed]
152. Chen, B.; Akusobi, C.; Fang, X.; Salmond, G.P. Environmental T4-Family Bacteriophages Evolve to Escape Abortive Infection via Multiple Routes in a Bacterial Host Employing "Altruistic Suicide" through Type III Toxin-Antitoxin Systems. *Front. Microbiol.* **2017**, *8*, 1006. [CrossRef] [PubMed]

153. Park, J.S.; Son, W.S.; Lee, B.J. Structural overview of toxin-antitoxin systems in infectious bacteria: A target for developing antimicrobial agents. *Biochim. Biophys. Acta* **2013**, *1843*, 1155–1167. [CrossRef]
154. Arbing, M.A.; Handelman, S.K.; Kuzin, A.P.; Verdon, G.; Wang, C.; Su, M.; Rothenbacher, F.P.; Abashidze, M.; Liu, M.; Hurley, J.M.; et al. Crystal structures of Phd-Doc, HigA, and YeeU establish multiple evolutionary links between microbial growth-regulating toxin-antitoxin systems. *Structure (London)* **2010**, *18*, 996–1010. [CrossRef] [PubMed]
155. Meinhart, A.; Alonso, J.C.; Sträter, N.; Saenger, W. Crystal structure of the plasmid maintenance system epsilon/zeta: Functional mechanism of toxin zeta and inactivation by epsilon 2 zeta 2 complex formation. *Proc. Natl. Acad. Sci. USA* **2003**, *100*, 1661–1666. [CrossRef]
156. Jaen-Luchoro, D.; Aliaga-Lozano, F.; Gomila, R.M.; Gomila, M.; Salva-Serra, F.; Lalucat, J.; Bennasar-Figueras, A. First insights into a type II toxin-antitoxin system from the clinical isolate Mycobacterium sp. MHSD3, similar to epsilon/zeta systems. *PLoS ONE* **2017**, *12*, e0189459. [CrossRef]
157. Fernández-Bachiller, M.I.; Brzozowska, I.; Odolczyk, N.; Zielenkiewicz, U.; Zielenkiewicz, P.; Rademann, J. Mapping Protein-Protein Interactions of the Resistance-Related Bacterial Zeta Toxin-Epsilon Antitoxin Complex (epsilon(2)zeta(2)) with High Affinity Peptide Ligands Using Fluorescence Polarization. *Toxins (Basel)* **2016**, *8*, 222. [CrossRef]
158. Budde, P.P.; Davis, B.M.; Yuan, J.; Waldor, M.K. Characterization of a higBA toxin-antitoxin locus in Vibrio cholerae. *J. Bacteriol.* **2007**, *189*, 491–500. [CrossRef]
159. Yang, J.; Zhou, K.; Liu, P.; Dong, Y.; Gao, Z.; Zhang, J.; Liu, Q. Structural insight into the *E. coli* HigBA complex. *Biochem. Biophys. Res. Commun.* **2016**, *478*, 1521–1527. [CrossRef]
160. Sterckx, Y.G.J.; Jové, T.; Shkumatov, A.V.; Garcia-Pino, A.; Geerts, L.; De Kerpel, M.; Lah, J.; De Greve, H.; Van Melderen, L.; Loris, R. A unique hetero-hexadecameric architecture displayed by the *Escherichia coli* O157 PaaA2-ParE2 antitoxin-toxin complex. *J. Mol. Biol.* **2016**, *428*, 1589–1603. [CrossRef]
161. Aakre, C.D.; Herrou, J.; Phung, T.N.; Perchuk, B.S.; Crosson, S.; Laub, M.T. Evolving new protein-protein interaction specificity through promiscuous intermediates. *Cell* **2015**, *163*, 594–606. [CrossRef] [PubMed]
162. El-Gebali, S.; Mistry, J.; Bateman, A.; Eddy, S.R.; Luciani, A.; Potter, S.C.; Qureshi, M.; Richardson, L.J.; Salazar, G.A.; Smart, A.; et al. The Pfam protein families database in 2019. *Nucleic Acids Res.* **2019**, *47*, D427–D432. [CrossRef]
163. Chan, T.W.; Espinosa, M.; Yeo, C.C. Keeping the Wolves at Bay: Antitoxins of Prokaryotic Type II Toxin-Antitoxin Systems. *Front. Mol. Biosci.* **2016**, *3*, 9. [CrossRef] [PubMed]
164. Senissar, M.; Manav, M.C.; Brodersen, D.E. Structural conservation of the PIN domain active site across all domains of life. *Protein Sci.* **2017**, *26*, 1474–1492. [CrossRef] [PubMed]
165. Matelska, D.; Steczkiewicz, K.; Ginalski, K. Comprehensive classification of the PIN domain-like superfamily. *Nucleic Acids Res.* **2017**, *45*, 6995–7020. [CrossRef]
166. Arcus, V.L.; McKenzie, J.L.; Robson, J.; Cook, G.M. The PIN-domain ribonucleases and the prokaryotic VapBC toxin-antitoxin array. *Protein Eng. Des. Sel.* **2011**, *24*, 33–40. [CrossRef]
167. Miallau, L.; Faller, M.; Chiang, J.; Arbing, M.; Guo, F.; Cascio, D.; Eisenberg, D. Structure and proposed activity of a member of the VapBC family of toxin-antitoxin systems. VapBC-5 from Mycobacterium tuberculosis. *J. Biol. Chem.* **2009**, *284*, 276–283. [CrossRef]
168. Maté, M.J.; Vincentelli, R.; Foos, N.; Raoult, D.; Cambillau, C.; Ortiz-Lombardía, M. Crystal structure of the DNA-bound VapBC2 antitoxin/toxin pair from Rickettsia felis. *Nucleic Acids Res.* **2012**, *40*, 3245–3258.
169. Mattison, K.; Wilbur, J.S.; So, M.; Brennan, R.G. Structure of FitAB from Neisseria gonorrhoeae bound to DNA reveals a tetramer of toxin-antitoxin heterodimers containing pin domains and ribbon-helix-helix motifs. *J. Biol. Chem.* **2006**, *281*, 37942–37951. [CrossRef]
170. Min, A.B.; Miallau, L.; Sawaya, M.R.; Habel, J.; Cascio, D.; Eisenberg, D. The crystal structure of the Rv0301-Rv0300 VapBC-3 toxin-antitoxin complex from M. *tuberculosis reveals a Mg(2)(+) ion in the active site and a putative RNA-binding site. Protein Sci.* **2012**, *21*, 1754–1767.
171. Schureck, M.A.; Dunkle, J.A.; Maehigashi, T.; Miles, S.J.; Dunham, C.M. Defining the mRNA recognition signature of a bacterial toxin protein. *Proc. Natl. Acad. Sci. USA* **2015**, *112*, 13862–13867. [CrossRef] [PubMed]
172. Schureck, M.A.; Repack, A.; Miles, S.J.; Marquez, J.; Dunham, C.M. Mechanism of endonuclease cleavage by the HigB toxin. *Nucleic Acids Res.* **2016**, *44*, 7944–7953. [CrossRef] [PubMed]

173. Yuan, J.; Sterckx, Y.; Mitchenall, L.A.; Maxwell, A.; Loris, R.; Waldor, M.K. Vibrio cholerae ParE2 poisons DNA gyrase via a mechanism distinct from other gyrase inhibitors. *J. Biol. Chem.* **2010**, *285*, 40397–40408. [CrossRef] [PubMed]
174. Brown, B.L.; Grigoriu, S.; Kim, Y.; Arruda, J.M.; Davenport, A.; Wood, T.K.; Peti, W.; Page, R. Three dimensional structure of the MqsR:MqsA complex: A novel TA pair comprised of a toxin homologous to RelE and an antitoxin with unique properties. *PLoS Pathog.* **2009**, *5*, e1000706. [CrossRef] [PubMed]
175. Hallez, R.; Geeraerts, D.; Sterckx, Y.; Mine, N.; Loris, R.; Van Melderen, L. New toxins homologous to ParE belonging to three-component toxin-antitoxin systems in *Escherichia coli* O157:H7. *Mol. Microbiol.* **2010**, *76*, 719–732. [CrossRef]
176. Bøggild, A.; Sofos, N.; Andersen, K.R.; Feddersen, A.; Easter, A.D.; Passmore, L.A.; Brodersen, D.E. The crystal structure of the intact *E. coli* RelBE toxin-antitoxin complex provides the structural basis for conditional cooperativity. *Structure (London)* **2012**, *20*, 1641–1658.
177. Dalton, M.K.; Crosson, S. A conserved mode of protein recognition and binding in a ParD-ParE toxin-antitoxin complex. *Biochemistry* **2010**, *49*, 2205–2215. [CrossRef]
178. Hargreaves, D.; Santos-Sierra, S.; Giraldo, R.; Sabariegos-Jareño, R.; de la Cueva-Méndez, G.; Boelens, R.; Díaz-Orejas, R.; Rafferty, J.B. Structural and functional analysis of the kid toxin protein from *E. coli* plasmid R1. *Structure* **2002**, *10*, 1425–1433.
179. Kamada, K.; Hanaoka, F.; Burley, S.K. Crystal structure of the MazE/MazF complex: Molecular bases of antidote-toxin recognition. *Mol. Cell* **2003**, *11*, 875–884. [CrossRef]
180. De Jonge, N.; Garcia-Pino, A.; Buts, L.; Haesaerts, S.; Charlier, D.; Zangger, K.; Wyns, L.; De Greve, H.; Loris, R. Rejuvenation of CcdB-poisoned gyrase by an intrinsically disordered protein domain. *Mol. Cell* **2009**, *35*, 154–163. [CrossRef]
181. Bahassi, E.M.; O'Dea, M.H.; Allali, N.; Messens, J.; Gellert, M.; Couturier, M. Interactions of CcdB with DNA gyrase. Inactivation of Gyra, poisoning of the gyrase-DNA complex, and the antidote action of CcdA. *J. Biol. Chem.* **1999**, *274*, 10936–10944. [CrossRef] [PubMed]
182. Kampranis, C.S.; Howells, A.J.; Maxwell, A. The interaction of DNA gyrase with the bacterial toxin CcdB: Evidence for the existence of two gyrase-CcdB complexes. *J. Mol. Biol.* **1999**, *293*, 733–744. [CrossRef] [PubMed]
183. Simic, M.; De Jonge, N.; Loris, R.; Vesnaver, G.; Lah, J. Driving forces of gyrase recognition by the addiction toxin CcdB. *J. Biol. Chem.* **2009**, *284*, 20002–20010. [CrossRef] [PubMed]
184. Smith, A.B.; López-Villarejo, J.; Diago-Navarro, E.; Mitchenall, L.A.; Barendregt, A.; Heck, A.J.; Lemonnier, M.; Maxwell, A.; Díaz-Orejas, R. A common origin for the bacterial toxin-antitoxin systems parD and ccd, suggested by analyses of toxin/target and toxin/antitoxin interactions. *PLoS ONE* **2012**, *7*, e46499. [CrossRef]
185. Ruiz-Echevarría, M.J.; Giménez-Gallego, G.; Sabariegos-Jareño, R.; Díaz-Orejas, R. Kid, a small protein of the parD stability system of plasmid R1, is an inhibitor of DNA replication acting at the initiation of DNA synthesis. *J. Mol. Biol.* **1995**, *247*, 568–577. [CrossRef]
186. Agarwal, S.; Mishra, N.K.; Bhatnagar, S.; Bhatnagar, R. PemK toxin of Bacillus anthracis is a ribonuclease: An insight into its active site, structure, and function. *J. Biol Chem.* **2010**, *285*, 7254–7270. [CrossRef]
187. Masuda, Y.; Miyakawa, K.; Nishimura, Y.; Ohtsubo, E. chpA and chpB, *Escherichia coli* chromosomal homologs of the pem locus responsible for stable maintenance of plasmid R100. *J. Bacteriol.* **1993**, *175*, 6850–6856. [CrossRef]
188. Kamphuis, M.B.; Monti, M.C.; van den Heuvel, R.H.; Santos-Sierra, S.; Folkers, G.E.; Lemonnier, M.; Díaz-Orejas, R.; Heck, A.J.; Boelens, R. Interactions between the toxin Kid of the bacterial parD system and the antitoxins Kis and MazE. *Proteins* **2007**, *67*, 219–231. [CrossRef]
189. Madl, T.; Van Melderen, L.; Mine, N.; Respondek, M.; Oberer, M.; Keller, W.; Khatai, L.; Zangger, K. Structural basis for nucleic acid and toxin recognition of the bacterial antitoxin CcdA. *J. Mol. Biol.* **2006**, *364*, 170–185. [CrossRef]
190. Walling, R.L.; Butler, J.S. Structural Determinants for Antitoxin Identity and Insulation of Cross Talk between Homologous Toxin-Antitoxin Systems. *J. Bacteriol.* **2016**, *198*, 3287–3295. [CrossRef]
191. Yang, M.; Gao, C.; Wang, Y.; Zhang, H.; He, Z.G. Characterization of the interaction and cross-regulation of three Mycobacterium tuberculosis RelBE modules. *PLoS ONE* **2010**, *5*, e10672. [CrossRef]

192. Połom, D.; Boss, L.; Węgrzyn, G.; Hayes, F.; Kędzierska, B. Amino acid residues crucial for specificity of toxin-antitoxin interactions in the homologous Axe-Txe and YefM-YoeB complexes. *FEBS J.* **2013**, *280*, 5906–5918. [CrossRef] [PubMed]
193. Nolle, N.; Schuster, C.F.; Bertram, R. Two paralogous yefM-yoeB loci from Staphylococcus equorum encode functional toxin-antitoxin systems. *Microbiology* **2013**, *159*, 1575–1585. [CrossRef]
194. Ahidjo, B.A.; Kuhnert, D.; McKenzie, J.L.; Machowski, E.E.; Gordhan, B.G.; Arcus, V.; Abrahams, G.L.; Mizrahi, V. VapC toxins from Mycobacterium tuberculosis are ribonucleases that differentially inhibit growth and are neutralized by cognate VapB antitoxins. *PLoS ONE* **2011**, *6*, e21738. [CrossRef] [PubMed]
195. Iqbal, N.; Guérout, A.M.; Krin, E.; Le Roux, F.; Mazel, D. Comprehensive Functional Analysis of the 18 Vibrio cholerae N16961 Toxin-Antitoxin Systems Substantiates Their Role in Stabilizing the Superintegron. *J. Bacteriol.* **2015**, *197*, 2150–2159. [CrossRef] [PubMed]
196. Tandon, H.; Sharma, A.; Wadhwa, S.; Varadarajan, R.; Singh, R.; Srinivasan, N.; Sandhya, S. Bioinformatic and mutational studies of related toxin-antitoxin pairs in Mycobacterium tuberculosis predict and identify key functional residues. *J. Biol. Chem.* **2019**, *294*, 9048–9063. [CrossRef]
197. Ramage, R.H.; Connolly, L.E.; Cox, J.S. Comprehensive functional analysis of Mycobacterium tuberculosis toxin-antitoxin systems: Implications for pathogenesis, stress responses, and evolution. *PLoS Genet.* **2009**, *5*, e1000767. [CrossRef]
198. Jin, G.; Pavelka, M.S., Jr.; Butler, J.S. Structure-function analysis of VapB4 antitoxin identifies critical features of a minimal VapC4 toxin-binding module. *J. Bacteriol.* **2015**, *197*, 1197–1207. [CrossRef]
199. Gucinski, G.C.; Michalska, K.; Garza-Sánchez, F.; Eschenfeldt, W.H.; Stols, L.; Nguyen, J.Y.; Goulding, C.W.; Joachimiak, A.; Hayes, C.S. Convergent Evolution of the Barnase/EndoU/Colicin/RelE (BECR) Fold in Antibacterial tRNase Toxins. *Structure* **2019**, *27*, 1660–1674. [CrossRef]
200. Goulard, C.; Langrand, S.; Carniel, E.; Chauvaux, S. The Yersinia pestis chromosome encodes active addiction toxins. *J. Bacteriol.* **2010**, *192*, 3669–3677. [CrossRef]
201. Sierra, S.S.; Giraldo, R.; Orejas, R.D. Functional interactions between chpB and parD, two homologous conditional killer systems found in the *Escherichia coli* chromosome and in plasmid R1. *FEMS Microbiol. Lett.* **1998**, *168*, 51–58. [CrossRef] [PubMed]
202. Santos-Sierra, S.; Giraldo, R.; Diaz-Orejas, R. Functional interactions between homologous conditional killer systems of plasmid and chromosomal origin. *FEMS Microbiol. Lett.* **1997**, *152*, 51–56. [CrossRef]
203. Zhu, L.; Sharp, J.D.; Kobayashi, H.; Woychik, N.A.; Inouye, M. Noncognate Mycobacterium tuberculosis toxin-antitoxins can physically and functionally interact. *J. Biol. Chem.* **2010**, *285*, 39732–39738. [CrossRef] [PubMed]
204. Pinaud, L.; Ferrari, M.L.; Friedman, R.; Jehmlich, N.; von Bergen, M.; Phalipon, A.; Sansonetti, P.J.; Campbell-Valois, F.X. Identification of novel substrates of Shigella T3SA through analysis of its virulence plasmid-encoded secretome. *PLoS ONE* **2017**, *12*, e0186920. [CrossRef] [PubMed]
205. Berman, H.M.; Westbrook, J.; Feng, Z.; Gilliland, G.; Bhat, T.N.; Weissig, H.; Shindyalov, I.N.; Bourne, P.E. The Protein Data Bank. *Nucleic Acids Res.* **2000**, *28*, 235–242. [CrossRef] [PubMed]
206. Burley, S.K.; Berman, H.M.; Bhikadiya, C.; Bi, C.; Chen, L.; Di Costanzo, L.; Christie, C.; Dalenberg, K.; Duarte, J.M.; Dutta, S.; et al. RCSB Protein Data Bank: Biological macromolecular structures enabling research and education in fundamental biology, biomedicine, biotechnology and energy. *Nucleic Acids Res.* **2019**, *47*, D464–D474. [CrossRef]
207. Li, G.Y.; Zhang, Y.; Inouye, M.; Ikura, M. Inhibitory mechanism of *Escherichia coli* RelE-RelB toxin-antitoxin module involves a helix displacement near an mRNA interferase active site. *J. Biol. Chem.* **2009**, *284*, 14628–14636. [CrossRef]
208. Takagi, H.; Kakuta, Y.; Okada, T.; Yao, M.; Tanaka, I.; Kimura, M. Crystal structure of archaeal toxin-antitoxin RelE-RelB complex with implications for toxin activity and antitoxin effects. *Nat. Struct. Mol. Biol.* **2005**, *12*, 327–331. [CrossRef]
209. Francuski, D.; Saenger, W. Crystal structure of the antitoxin-toxin protein complex RelB-RelE from Methanococcus jannaschii. *J. Mol. Biol.* **2009**, *393*, 898–908. [CrossRef]
210. Garcia-Pino, A.; Christensen-Dalsgaard, M.; Wyns, L.; Yarmolinsky, M.; Magnuson, R.D.; Gerdes, K.; Loris, R. Doc of prophage P1 is inhibited by its antitoxin partner Phd through fold complementation. *J. Biol. Chem.* **2008**, *283*, 30821–30827. [CrossRef]

211. Kamada, K.; Hanaoka, F. Conformational change in the catalytic site of the ribonuclease YoeB toxin by YefM antitoxin. *Mol. Cell* **2005**, *19*, 497–509. [CrossRef] [PubMed]
212. Miallau, L.; Jain, P.; Arbing, M.A.; Cascio, D.; Phan, T.; Ahn, C.J.; Chan, S.; Chernishof, I.; Maxson, M.; Chiang, J.; et al. Comparative proteomics identifies the cell-associated lethality of M. *tuberculosis RelBE-like toxin-antitoxin complexes. Structure* **2013**, *21*, 627–637. [PubMed]
213. Schureck, M.A.; Maehigashi, T.; Miles, S.J.; Marquez, J.; Cho, S.E.; Erdman, R.; Dunham, C.M. Structure of the Proteus vulgaris HigB-(HigA)2-HigB toxin-antitoxin complex. *J. Biol. Chem.* **2014**, *289*, 1060–1070. [CrossRef]
214. Hadži, S.; Garcia-Pino, A.; Haesaerts, S.; Jurėnas, D.; Gerdes, K.; Lah, J.; Loris, R. Ribosome-dependent Vibrio cholerae mRNAse HigB2 is regulated by a beta-strand sliding mechanism. *Nucleic Acids Res.* **2017**, *45*, 4972–4983. [CrossRef]
215. Talavera, A.; Tamman, H.; Ainelo, A.; Konijnenberg, A.; Hadži, S.; Sobott, F.; Garcia-Pino, A.; Hõrak, R.; Loris, R. A dual role in regulation and toxicity for the disordered N-terminus of the toxin GraT. *Nat. Commun.* **2019**, *10*, 972. [CrossRef] [PubMed]
216. Yoon, W.S.; Seok, S.H.; Won, H.S.; Cho, T.; Lee, S.J.; Seo, M.D. Structural changes of antitoxin HigA from Shigella flexneri by binding of its cognate toxin HigB. *Int. J. Biol. Macromol.* **2019**, *130*, 99–108. [CrossRef]
217. Krissinel, E.; Henrick, K. Inference of macromolecular assemblies from crystalline state. *J. Mol. Biol.* **2007**, *372*, 774–797. [CrossRef]
218. Pettersen, E.F.; Goddard, T.D.; Huang, C.C.; Couch, G.S.; Greenblatt, D.M.; Meng, E.C.; Ferrin, T.E. UCSF Chimera—A visualization system for exploratory research and analysis. *J. Comput. Chem.* **2004**, *25*, 1605–1612. [CrossRef]
219. Russell, A.D.; Hatfull, G.F. PhagesDB: The actinobacteriophage database. *Bioinformatics* **2017**, *33*, 784–786. [CrossRef]
220. Dienemann, C.; Bøggild, A.; Winther, K.S.; Gerdes, K.; Brodersen, D.E. Crystal structure of the VapBC toxin-antitoxin complex from Shigella flexneri reveals a hetero-octameric DNA-binding assembly. *J. Mol. Biol.* **2011**, *414*, 713–722. [CrossRef]
221. Lee, I.G.; Lee, S.J.; Chae, S.; Lee, K.Y.; Kim, J.H.; Lee, B.J. Structural and functional studies of the Mycobacterium tuberculosis VapBC30 toxin-antitoxin system: Implications for the design of novel antimicrobial peptides. *Nucleic Acids Res.* **2015**, *43*, 7624–7637. [CrossRef] [PubMed]
222. Korch, B.S.; Contreras, H.; Clark-Curtiss, J.E. Three Mycobacterium tuberculosis Rel toxin-antitoxin modules inhibit mycobacterial growth and are expressed in infected human macrophages. *J. Bacteriol.* **2009**, *191*, 1618–1630. [CrossRef] [PubMed]

Article

Multi-Stress Induction of the *Mycobacterium tuberculosis* MbcTA Bactericidal Toxin-Antitoxin System

Kanchiyaphat Ariyachaokun [1,2,†], Anna D. Grabowska [1,3,†], Claude Gutierrez [1,*] and Olivier Neyrolles [1]

1 Institut de Pharmacologie et de Biologie Structurale, IPBS, Université de Toulouse, CNRS, UPS, 205 route de Narbonne, 31400 Toulouse, France; kanchiyaphat.a@ubu.ac.th (K.A.); dr.anna.grabowska@gmail.com (A.D.G.); olivier.neyrolles@ipbs.fr (O.N.)

2 Department of Biological Science, Ubon Ratchathani University, 85 Sathollmark Rd., Warin Chamrab, Ubon Ratchathani 34190, Thailand

3 Department of Biophysics and Human Physiology, Medical University of Warsaw, 02-091 Warsaw, Poland

* Correspondence: claude.gutierrez@ipbs.fr

† K.A. and A.D.G. contributed equally to this work.

Received: 29 April 2020; Accepted: 14 May 2020; Published: 16 May 2020

Abstract: MbcTA is a type II toxin/antitoxin (TA) system of *Mycobacterium tuberculosis*. The MbcT toxin triggers mycobacterial cell death in vitro and in vivo through the phosphorolysis of the essential metabolite NAD^+ and its bactericidal activity is neutralized by physical interaction with its cognate antitoxin MbcA. Therefore, the MbcTA system appears as a promising target for the development of novel therapies against tuberculosis, through the identification of compounds able to antagonize or destabilize the MbcA antitoxin. Here, the expression of the *mbcAT* operon and its regulation were investigated. A dual fluorescent reporter system was developed, based on an integrative mycobacterial plasmid that encodes a constitutively expressed reporter, serving as an internal standard for monitoring mycobacterial gene expression, and an additional reporter, dependent on the promoter under investigation. This system was used both in *M. tuberculosis* and in the fast growing model species *Mycobacterium smegmatis* to: (i) assess the autoregulation of *mbcAT*; (ii) perform a genetic dissection of the *mbcA* promoter/operator region; and (iii) explore the regulation of *mbcAT* transcription from the *mbcA* promoter (P_{mbcA}) in a variety of stress conditions, including in vivo in mice and in macrophages.

Keywords: tuberculosis; toxin-antitoxin systems; bacterial cell death; NAD^+; stress-response

Key Contribution: A dual fluorescent reporter system, designed to monitor gene expression and regulation in mycobacteria, was constructed and used to address the multi-stress regulation of the MbcTA bactericidal toxin/antitoxin system of *Mycobacterium tuberculosis*.

---

1. Introduction

Mycobacterium tuberculosis (*Mtb*), the etiological agent of tuberculosis (TB), is a strictly human-adapted obligate pathogen of major public health importance (WHO, Global Tuberculosis report, 2019). Indeed, TB was responsible for 1.5 million deaths and about ten million new TB cases were reported in 2018. Of particular concern, an estimated 484,000 new infections were due to *Mtb* strains resistant to one or more frontline antibiotics in 2018, and 80% of these *Mtb* strains were multi-drug resistant (MDR). Therefore, the development of novel therapies against TB is urgently needed. In line with this, we recently started to explore the potential utilization of intrinsic bactericidal toxin/antitoxin (TA) systems as new therapeutic tools [1]. TA systems are genetic modules encoding an endogenous protein toxin, which targets an essential process in the bacterial cell, and a toxin-neutralizing antidote, the so-called antitoxin [2–4]. Under favorable conditions, the antitoxin is stable and prevents auto-intoxication of the TA-producing cell. Although the precise mechanisms remain to be characterized for most TA systems, it has been proposed that when cells are facing environmental stresses, the less stable antitoxin is degraded, leading to an increased production of toxin, resulting in growth arrest of the bacterial cells [5,6]. TA systems are classified into six families, depending on the nature of the antitoxin, i.e., ribonucleic acid (RNA) or protein, and its mechanism of toxin neutralization. For instance, in the most studied family of type II TA systems, the antitoxin is a protein that neutralizes the toxin through direct protein-protein interaction [2,4]. The *Mtb* genome harbors more than 80 TA systems, which were proposed to contribute to virulence and persistence in the infected host [7–11]. Among those, the recently characterized MbcTA system encodes an NAD^+ phosphorylase toxin, able to trigger bacterial cells death through thorough depletion of the NAD^+ pool [1]. In the present study, we investigated the expression and regulation of the *mbcAT* operon. We developed a dual fluorescent reporter system, which we used to dissect the operon promoter/operator region at the genetic level. Using this system, we demonstrate that transcription from P_{mbcA} is induced by a range of stress conditions, reflecting those encountered inside the infected host.

2. Results

2.1. MbcT Toxin Depletes NAD^+ and Is Bactericidal in *Mycobacterium smegmatis*

In order to express the MbcT toxin in *M. smegmatis*, we used an expression vector, pGMCS-TetR-P1-*mbcT*, that can stably integrate into the mycobacterial chromosome, within the *glyV* tRNA gene, through site-specific recombination catalyzed by the L5 mycobacteriophage integrase [12]. In this plasmid, *mbcT* is expressed under the tetracycline-inducible promoter $P_{myc1\ tetO}$ [13], hereafter abbreviated P1. Upon transformation into a *Mtb* strain deleted for the endogenous *mbcAT* operon, pGMCS-TetR-P1-*mbcT* conferred anhydro-tetracycline (Atc)-dependent depletion of the essential metabolite NAD^+, resulting in bacterial cell death [1]. In contrast, growth of a derivative of wild-type *M. smegmatis* strain mc^2 155 carrying this plasmid was not affected by addition of Atc to its growth medium (Figure 1a), presumably because the amount of MbcT produced in such conditions was not sufficient to deplete NAD^+ from the *M. smegmatis* cells (Figure 1d). To verify this hypothesis, we constructed plasmids expressing *mbcT* under the control of stronger Atc-inducible promoters, namely P606Pld and P606 (obtained from D. Schnappinger, Weill Cornell Medical College, NY, USA). Induction of P606Pld-*mbcT* had no effect on mc^2 155 growth (Figure 1b), although it reduced the NAD^+ content to 8% of the amount measured in the uninduced strain carrying pGMCS-TetR-P1-*mbcT*. In contrast, *mbcT* expression from the Atc-induced P606 promoter abolished mycobacterial growth (Figure 1c) and, accordingly, reduced the amount of NAD^+ to less than 2%, compared to that in control cells (Figure 1d). Furthermore, it resulted in an approximately 99% loss of colony forming units (CFU) within few hours following induction (Figure 1e), and assessment of cell viability using the LIVE/DEAD BacLight stains followed by flow cytometry analysis demonstrated that the proportion of propidium iodide-permeable cells, which was very low in uninduced exponentially growing cells (Figure 1e), increased markedly by 8 or 24 h after the addition of Atc (Figure 1f–i). This confirmed the bactericidal effect of the NAD^+ depletion following the induction of P606-*mbcT*. In conclusion, providing that it is driven by a relatively strong promoter, production of MbcT in *M. smegmatis* can recapitulate the bactericidal effect of the toxin previously observed in *Mtb*.

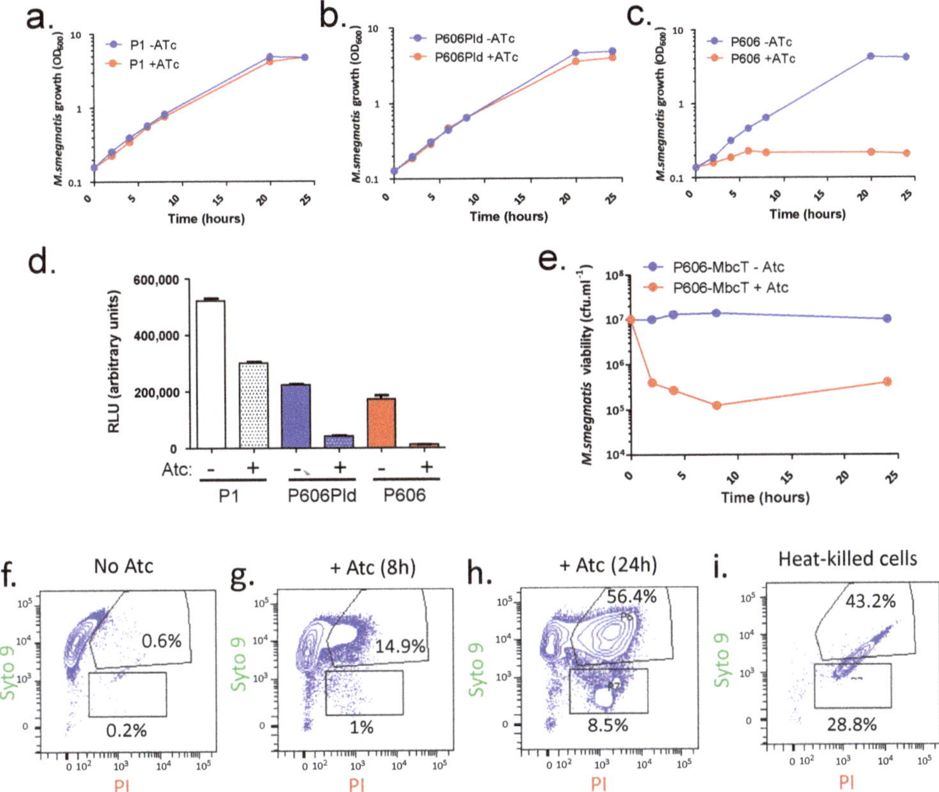

Figure 1. The toxin MbcT depletes NAD^+ and is bactericidal in *Mycobacterium smegmatis*. (**a–c**) Cultures of *M. smegmatis* strain mc^2 155 transformed with plasmids producing MbcT under the control of the indicated promoter were diluted at time 0 in fresh medium supplemented with streptomycin alone (blue) or streptomycin and 200 ng.mL^{-1} anhydro-tetracycline (Atc) (red) and bacterial growth was followed over up to 24 h. (**d**) Samples of the *M. smegmatis* cultures were harvested after 8 h of treatment and relative content of NAD^+ was measured in cell extracts. Values are representative of two independent replicates of the same experiment. (**e**) Samples of the cultures of *M. smegmatis* mc^2 155/pGMCS-TetR-P606-mbcT were harvested at the indicated times, diluted, and plated on L agar. Colonies were counted after 3 days of growth at 37 °C. (**f–h**) Samples of the cultures of mc^2 155/pGMCS-TetR-P606-mbcT were harvested after 8 or 24 h of treatment with Atc, labeled with the LIVE/DEAD BacLight dyes (Syto9 and Propidium Iodide (PI)) and analyzed by flow cytometry (FACS). (**i**) Cells grown without Atc were heat-killed at 98 °C for 30 min before LIVE/DEAD labeling and FACS analysis. Data are representative of two experiments with similar results.

2.2. Construction of a Dual Fluorescent Reporter System to Assess Gene Expression in Mycobacteria

In order to monitor the expression of mycobacterial genes of interest in in vitro cultures or during infection, we developed a dual fluorescent reporter system in which an integrative plasmid carries both (i) a constitutive reporter, serving as an internal standard or probe for the detection of the bacterial cells and (ii) an additional reporter, expressed from the promoter/operator region of the gene under investigation. This reporter system was constructed by multisite gateway recombination [14], using plasmid pDE43-MCS as backbone, and entry vectors carrying either a 5′, a middle or a 3′ module, as depicted in Figure 2a. The 5′ module consists of a fluorescent reporter, which can be either green fluorescent protein (GFP), mCherry, mTurquoise, or mVenus, expressed from the constitutive P1 promoter. The middle entry clone introduces a second fluorescent reporter gene, from the same list but different from the first one, fused to the promoter/operator region of the *mbcAT* operon. Finally, the 3′ entry clone introduces either nothing, using an empty entry clone (pEN23A-MluI), or any gene(s) of interest, which effect on P_{mbcA} is to be tested. Figure 2b depicts an example of such construct, pGMCS-P1-mCherry-P_{mbcA}-GFP-*mbcA*, in which the *mbcA* gene is inserted to assess the effect of MbcA on its own transcription.

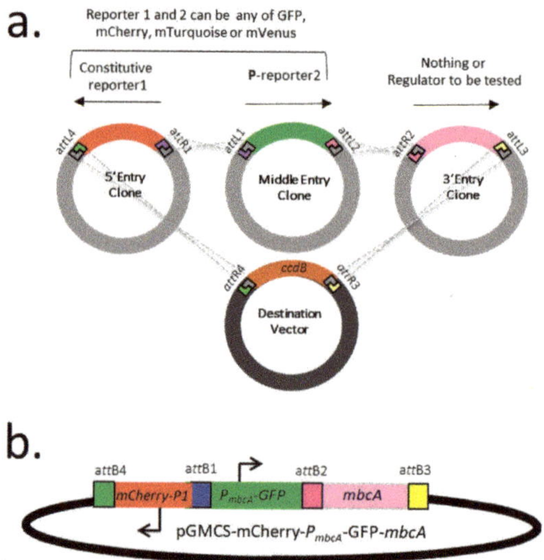

Figure 2. A dual florescent protein reporter system developed for studying gene expression in mycobacteria. (**a**). Gateway cloning strategy used to construct the required plasmids. (**b**). Example of plasmid construction designed to analyze the expression and regulation of *mbcAT*. GFP = green fluorescent protein.

2.3. *Use of a Dual Fluorescent Reporter System to Study the Expression of mbcAT*

2.3.1. Auto-Repression of *mbcAT*

A set of three plasmids, namely pGMCS-P1-mCherry-P_{mbcA}-GFP, pGMCS-P1-mCherry-P_{mbcA}-GFP-*mbcA*, and pGMCS-P1-mCherry-P_{mbcA}-GFP-*mbcAT*, was constructed and transformed into *M. smegmatis* mc^2 155. The resulting strains were grown until exponential phase in complete 7H9 medium and the fluorescence of mCherry and GFP were measured by flow cytometry. The untransformed mc^2 155 strain exhibited a very low background level of fluorescence (Figure 3a). In contrast, the three transformed strains exhibited both red and green fluorescence (Figure 3b–d). Comparison of the mean fluorescence intensities showed a similar level of fluorescence for mCherry in the three constructs, whereas the fluorescence for GFP was approximately 5-fold lower when the MbcA antitoxin was expressed, regardless of MbcT toxin co-expression (Figure 3h). Therefore, we conclude that the transcription from P_{mbcA} is auto-repressed by MbcA, independently of the presence of the toxin. We confirmed these observations in a qualitative assay, by observation of the bacterial cells in fluorescence microscopy (Figure S1). In addition, we also introduced the same set of plasmids into *M. tuberculosis* H37Rv and observed the cells in fluorescence microscopy after fixation with *p*-formaldehyde. As expected, in the wild-type H37Rv, expression of GFP remained very low in all our constructs because of the MbcA repressor produced from the chromosomal copy of the *mbcAT* operon. In contrast, in H37Rv Δ*mbcAT*::Kan^R, the pattern of mCherry and GFP expression was comparable to that observed in *M. smegmatis*, demonstrating that autoregulation occurs similarly in the two mycobacterial species.

Analysis of the MbcA sequence predicts a helix-turn-helix motif between amino-acids 20 and 40, forming a putative DNA binding domain. In order to test whether this motif is important for repression by MbcA, the Arg^{33} and Arg^{37} residues were substituted with glutamic acid by site-directed mutagenesis of pGMCS-P1-mCherry-P_{mbcA}-GFP-*mbcA*, and the resulting plasmids were transformed into *M. smegmatis* mc^2 155. Both modifications abolished the repression by the MbcA variants, leading to an expression level comparable to that observed in the absence of MbcA (Figure 3h). When the $R^{33}E$ substitution was introduced in the plasmid pGMCS-P1-mCherry-P_{mbcA}-GFP-*mbcAT*, the co-expression of the toxin MbcT with the $R^{33}E$ variant of MbcA still resulted in absence of repression.

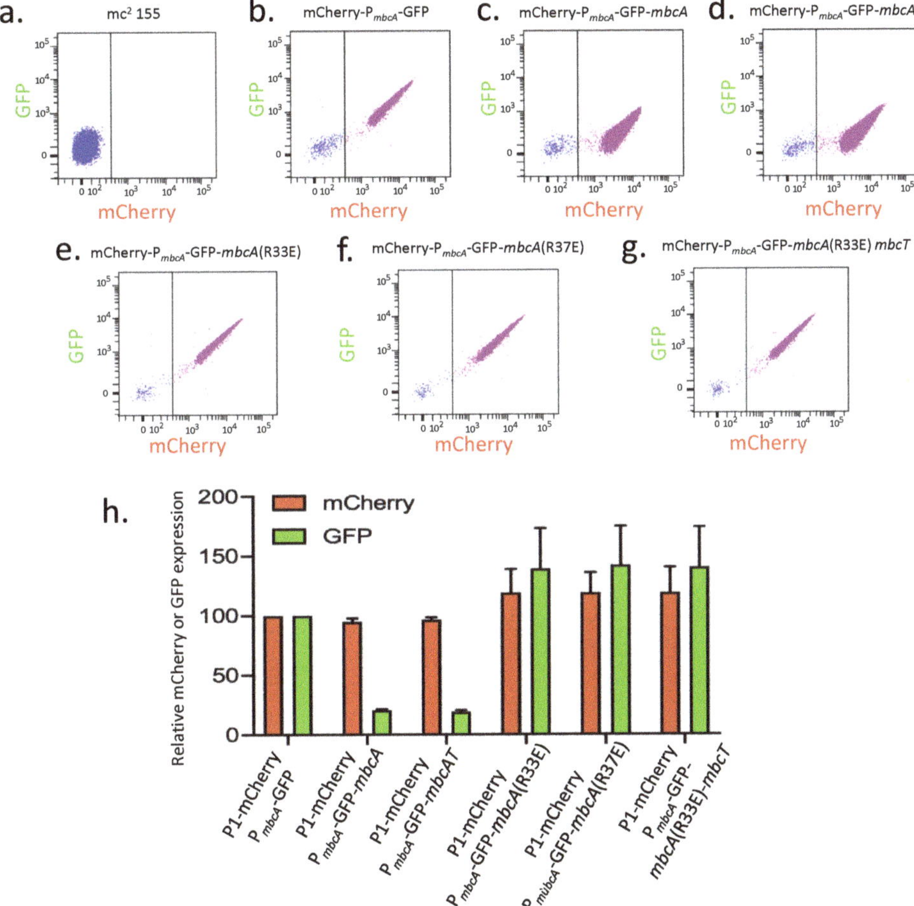

Figure 3. Autorepression of the *mbcAT* operon. (**a**)–(**g**) *M. smegmatis* mc^2 155 wild-type and transformed with the indicated pGMCS-mCherry-P_{mbcA}-GFP derivatives were grown to exponential phase in complete 7H9 medium and analyzed by flow cytometry. Blue and pink points separated by a vertical line indicate the cells negative or positive for mCherry fluorescence, respectively. (**h**) Mean fluorescence intensities (MFI) of the fluorescence positive cells (pink points) relative to that of the strain harboring pGMCS-mCherry-P_{mbcA}-GFP (a) are reported for mCherry (red bars) and for GFP (green bars). Values are the average of two biological replicates with standard deviation.

2.3.2. Genetic Dissection of the *mbcAT* Promoter/Operator Region

In the plasmids described above (Figure 3), the *mbcAT* promoter is located on a 261-bp fragment ($P_{mbcA(261)}$-GFP, labeled in red letters in Figure 4a) amplified with the oligonucleotides Fw1/Rv1 (Table S1). A shorter 103-bp fragment, extending between oligonucleotide Fw2 and Rv1, also contains the entire promoter and operator, because it produced the same pattern of expression ($P_{mbcA(103)}$-GFP, labeled in green letters in Figure 4b). A putative -10 element of σ^A-dependent promoters is present upstream from the *mbcA* transcription start site identified by RNA-seq experiments [15]. We confirmed this promoter location by generating two 239-bp DNA fragments amplified with the oligonucleotides Fw1/Rv2, with the wild-type sequence and Fw1/Rv3, with a mutation changing a consensus T into a non-consensus G in the -10 element. Whereas the first fragment still allowed the expression of GFP ($P_{mbcA(239)}$-GFP, labeled in blue letters in Figure 4b), the T=>G substitution almost completely abolished the production of the downstream encoded GFP ($P_{mbcA(239\text{-}G)}$-GFP, labeled in brown letters in Figure 4b). In addition, we observed that the Fw1/Rv2 fragment was much less sensitive to repression by MbcA. To further analyze the *mbcA* operator, we constructed mutations of 4 (Mut1; changing CAAA into TCGT) or 3 nucleotides (Mut2; changing ACA into GTG), and we observed that Mut1 resulted in a high expression in the presence of MbcA, whereas Mut 2 did not abolish repression by MbcA.

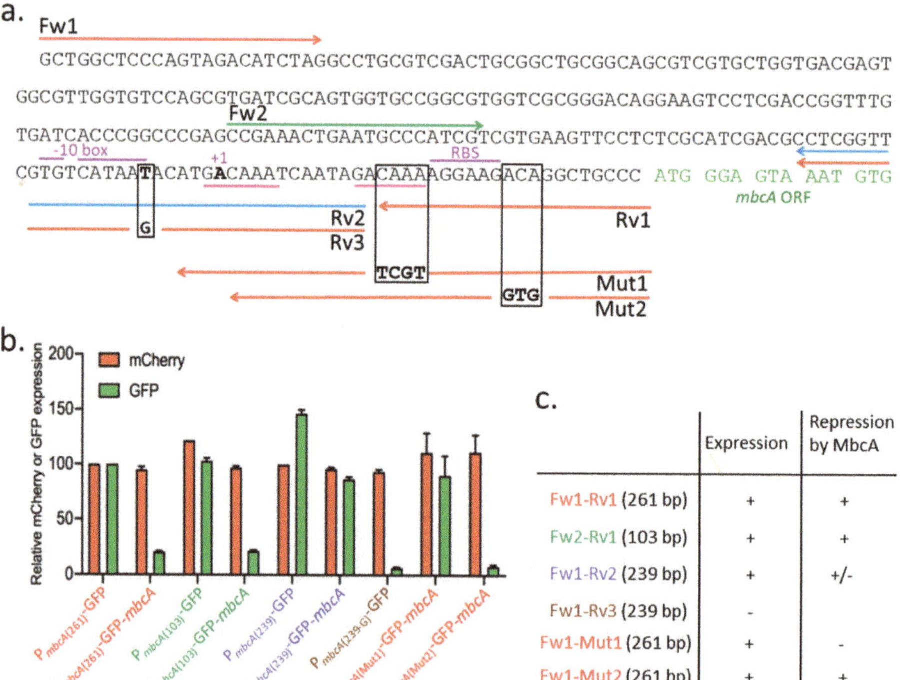

	Expression	Repression by MbcA
Fw1-Rv1 (261 bp)	+	+
Fw2-Rv1 (103 bp)	+	+
Fw1-Rv2 (239 bp)	+	+/-
Fw1-Rv3 (239 bp)	-	
Fw1-Mut1 (261 bp)	+	-
Fw1-Mut2 (261 bp)	+	+

Figure 4. Genetic analysis of the promoter/operator region of the *mbcAT* operon. (**a**) DNA sequence of the *mbcA* promoter/operator region. Arrows indicate oligonucleotides used to amplify various fragments encompassing the promoter region. Black squares surround the nucleotides modified by mutations introduced in the oligonucleotides Rv3 (T=>G), Mut1 (CAAA=>TCGT), and Mut2 (ACA=>GTG). +1 points to the transcription start site. (**b**) Mean fluorescence intensities (MFI) were determined by flow cytometry with strains harboring the different constructs. Red and green bars indicate MFI relative to that of strain harboring pGMCS-mCherry-P_{MbcA}-GFP. Values are the average of two biological replicates with standard deviation. (**c**) Summary of the results obtained with the different constructs.

2.4. Multistress Induction of mbcAT Expression

To assess the regulation of *mbcAT* expression, the plasmid pGMCS-P1-mTurquoise-P_{mbcA}-mVenus-*mbcA* was introduced into wild-type Mtb H37Rv, and the resulting strain was grown to mid-exponential phase and subjected to various stress conditions. As shown in Figure 5, we could observe qualitatively that the expression from P_{mbcA} was increased upon starvation following incubation of the cells in phosphate saline buffer (PBS), or upon treatment with H_2O_2 or with the NO-generating reagent diethylenetriamine/nitric oxide adduct (DETA/NO).

Finally, *Mtb* strain H37Rv/pGMCS-P1-mTurquoise-P_{mbcA}-mVenus-*mbcA* was used to infect mouse or human macrophages, and fluorescence microscopy of fixed infected macrophages demonstrated that transcription from P_{mbcA} is stimulated upon *Mtb* phagocytosis by macrophages (Figure 6).

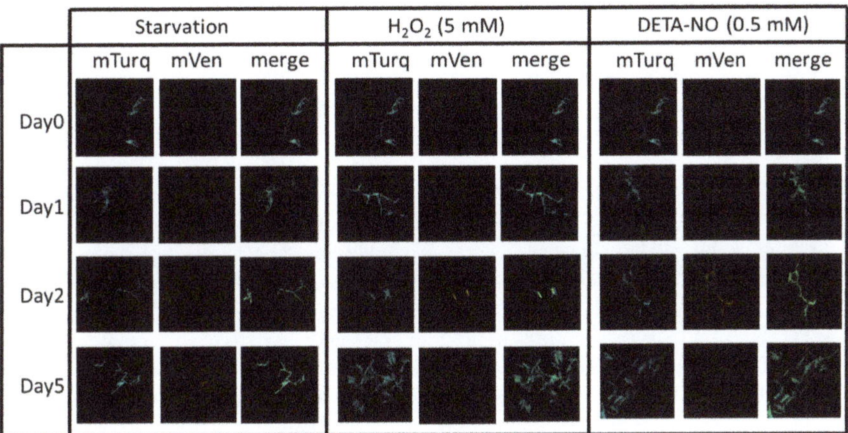

Figure 5. Effect of stress conditions on expression of the *mbcAT* operon in *M. tuberculosis.* Wild-type H37Rv transformed with plasmid pGMCS-P1-mTurq-P_{mbcA}-mVenus-*mbcA* was grown in complete 7H9 medium. At day 0, cells were harvested and resuspended in phosphate saline buffer (PBS) buffer (starvation), or 7H9 medium supplemented with 5 mM H_2O_2 or 0.5 mM DETA/NO. At the indicated times, samples were harvested, fixed with 4% *p*-formaldehyde and observed by fluorescent microscopy using large field Leica DMIRB, magnification 630×.

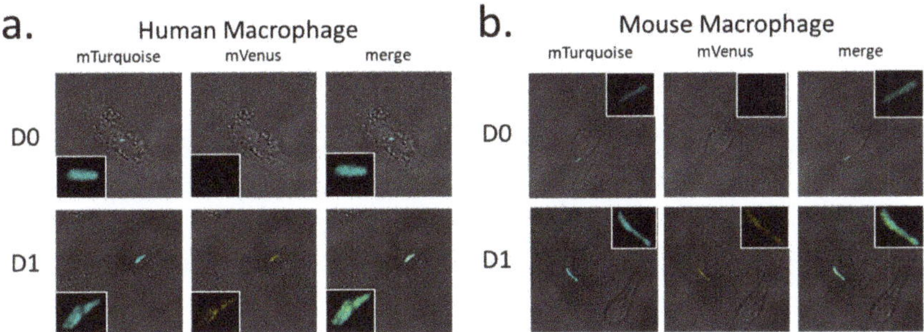

Figure 6. Induction of the *mbcAT* operon expression in *M. tuberculosis* upon infection of macrophages. *Mtb* H37Rv/pGMCS-P1-mTurq-P_{mbcA}-mVenus-*mbcA* was used to infect human (**a**) or mouse (**b**) macrophages at multiplicity of infection of 0.1. Infected cells were harvested at day 0 (D0) or 24 h post-infection (D1), fixed with 4% *p*-formaldehyde and observed by fluorescent microscopy using large field Leica DMIRB, magnification 630×. Images show the merged bright field and either the blue, yellow, or merged fluorescence. Insets show enlarged view of the phagocytized bacteria.

3. Discussion

The dual fluorescent reporter system presented here, constructed using the Gateway cloning technology [12], allows for quantification and/or visualization of bacterial cells, and in parallel for following expression of a gene of interest (Figure 2). As shown here, this system is stable and robust, since it is based on an integrative vector and exhibited only minor variations of the expression of the internal standard reporter (mCherry in Figures 3 and 4). Allowing both quantitative (Figures 3 and 4) and qualitative measurements (Figures 5 and 6), this system can be used to follow gene expression not only in bacterial cultures, or inside infected macrophages, as shown in this work, but also inside host tissues during infection of animal models [16]. Finally, it is also very versatile, since it offers a straightforward way to construct bacterial variants carrying a set of different reporters, as exemplified here with a variety of fluorescent proteins (Figure 2). We believe that it can be of broad interest to investigate gene expression and regulation in mycobacteria.

RNA-seq experiments indicated a potential transcription start site (TSS) of *mbcA* in *M. tuberculosis* (indicated as +1 in Figure 4) and located 34 nucleotides upstream from its translation initiation site (TIS) [15]. Based on bioinformatics analysis, it has been proposed that transcription of the *mbcAT* operon occurs from two overlapping promoters recognized by RNA polymerase holoenzyme using alternative sigma factors [17]. An upstream promoter, with a σ^L-dependent -10 element (GGTTC) located at position -54 to -50 from the TIS, and a second promoter, using a σ^H-dependent -10 element (CGTGTC), located at positions -50 to -45. The σ^L and σ^H alternative sigma factors are both present in *M. smegmatis* [18] and might therefore be involved in *mbcAT* expression in both mycobacterial species used in this study. Our data do not rule out the possibility of *mbcAT* expression from σ^L- or σ^H-dependent promoters, but they demonstrate that yet another -10 element is essential for the expression of *mbcAT* (Figure 4), which sequence **TG**TC**A**T**AA**T fits (nucleotides in bold letters) the consensus of σ^A-dependent promoters with an extended -10 motif (TGNTANNNT [15]). The involvement of this -10 element in *mbcAT* expression has been confirmed by our genetic analysis since its activity was abolished by a mutation changing a consensus base of the -10 element into a non-consensus one, at position -40 with respect to the TIS (Figure 4).

Many type II TA systems are regulated by auto-repression, with the antitoxin binding to an operator site in the vicinity of the operon promoter [2,4]. It is also the case for the *mbcAT* operon (Figures 3 and 4). In solution, MbcA and MbcT form a heterododecameric complex (3 × [MbcT-MbcA]$_2$) arranged around a 3-fold symmetry axis in which MbcA folds into a single structured domain consisting of seven α helices [1]. Helices α2 and α3 form a putative helix-turn-helix motif, and Arg^{33} and Arg^{37}, which are pointing out of helix 3, could be important for the recognition of the operator site. Consistently, substitutions of any of these two arginine residues totally abolished the transcription repression by MbcA (Figure 3). The *mbcA* promoter region harbors a direct repeat of two GACAAA motifs, separated by 12 bp (underlined with a magenta line in Figure 4a). Deletion of the downstream motif on promoter fragment Fw1-Rv2 abolishes the repression by MbcA and mutation Mut1, which disrupts this downstream motif, increases transcription in the presence of MbcA to the level observed in absence of the repressor. In contrast, modification of nucleotides downstream from this motif (mutation Mut2) did not relieve the repression by MbcA (Figure 4b). Although additional biochemical experiments are needed to determine the precise interaction of MbcA with its operator site, these observations suggest that MbcA could bind the GACAAA direct repeats and that both repeats are indispensable to constitute a functional operator.

Several autoregulated type II TA systems are subject to the so-called conditional cooperativity, i.e., a repression mechanism in which the antitoxin-repressor can participate in different types of complexes, alone, or with its cognate toxin, exhibiting different repression efficacies [2,4,19]. Conditional cooperativity has been proposed to contribute to several aspects of the TA systems biology [2]. In such systems, the antitoxin alone binds its operator site with low affinity, resulting in poor repression. When toxin and antitoxin are present in stoichiometric amounts, binding of the toxin to the antitoxin increases its affinity for the operator, leading to a more potent repression of transcription than is observed for the antitoxin alone. Finally, high amounts of toxin can result in a different heteromeric complex exhibiting low affinity for the operator, thereby relieving repression in excess of toxin. As already shown for a number of TA systems [20], our data suggest that the MbcTA system is not subject to conditional cooperativity, since the level of transcription repression by the antitoxin appeared independent on the production of the toxin (Figure 3). We note that in the dodecameric MbcTA complex, the helix-turn-helix motif of MbcA, necessary for the repression (Figure 3), is masked by an interaction with the $\alpha 6$ helix of MbcT [1]. In consequence, the dodecameric complex is most likely unable to participate in repression. In the wild-type H37Rv *M. tuberculosis* strain, the ectopic expression of MbcT under the control of the P1 promoter revealed not to be toxic [1], indicating that expression of the wild-type *mbcAT* operon leads to higher amounts of MbcA antitoxin than of MbcT toxin. Therefore, only a fraction of MbcA must be is involved in MbcT-neutralizing complexes, and it is likely that the remaining excess of MbcA can function as auto-repressor.

Previous work has already reported increased expression of *mbcAT* in stress conditions. The MbcT toxin (Rv1989c) has been identified by proteomic analysis in culture filtrates of PBS-starved *M. tuberculosis* cells, whereas it is not detectable during exponential growth [21]. Transcriptomic studies demonstrated increased amounts of *mbcA* and/or *mbcT* mRNA in stationary compared to exponential phase [15], during the so-called enduring hypoxic response [22], or in *M. tuberculosis* persisters surviving a treatment with D-cycloserine [8]. Our data are in agreement with these observations, and further demonstrate that transcription of *mbcAT* increases upon exposure to oxidative stress caused by H_2O_2 or NO-generating compounds (Figure 5), conditions mimicking those encountered inside macrophages [23–25]. In agreement, we also observed increased transcription from the *mbcAT* promoter within infected macrophages (Figure 6). To date, the physiological role of the MbcTA system has not been firmly established. However, the existence of a gene mimicking the antitoxin MbcA in the genome of several myco-bacteriophages [1] strongly suggests that this system could belong to the family of TA systems used to fight against phage infections [26,27]. The multi-stress induction of *mbcAT*, leading to accumulation of both the toxin and the antitoxin, should not be toxic for the *M. tuberculosis* cells. However, it is reminiscent of the so-called general stress response observed in many bacterial species, which results in the expression of a variety of adaptive responses to stresses not yet encountered by bacterial cells but that contribute to reinforce their resistance if they eventually happen to deal with one of these stresses [28].

The *M. tuberculosis* genome is particularly rich in TA systems [7–11], and, as already proposed for other bacterial pathogens [29,30], it is tempting to speculate that at least some of these systems could be used as novel tools in adjunct therapies with classical antibiotics. However, a number of studies have proposed that activation of TA systems may trigger a non-growing physiological state, or so-called persistence, in which the bacterial cells become tolerant to antibiotics [31–34]. Although the role of TA systems in the development of tolerance is still a matter of debate [35], this raises a serious drawback to their potential therapeutic use. The MbcTA system attracted our attention because it is one of the only two TA systems of *M. tuberculosis* that harbor an essential antitoxin gene, i.e., which cannot be genetically disrupted [36]. We reasoned that this could be the consequence of an atypical and particularly toxic activity of MbcT. Indeed, we have demonstrated that MbcT activity, when expressed in recombinant conditions generating physiological expression levels of the toxin, is bactericidal for *M. tuberculosis* through depletion of NAD^+ [1], a metabolite known to be essential for mycobacteria survival, even in a dormant state [37–39]. In addition, we have shown that production of the MbcT toxin during infection could act in synergy with isoniazid treatment to reduce the bacterial load in mice infected by *M tuberculosis*, supporting the idea that MbcTA could be a novel therapeutic target [1]. The observations reported here strengthen this proposal, given that the induction of *mbcAT* during infection ensures the presence of the toxin inside the infecting *M. tuberculosis* cells, where targeting its activation could help fighting against TB.

4. Materials and Methods

4.1. Bacterial Strains and Cultures

M. tuberculosis strains H37Rv (ATCC27294) and H37Rv Δ*mbcAT*::Kan^R [1] and *M. smegmatis* wild-type strain mc^2 155 (ATCC700084) were routinely grown aerobically at 37 °C in complete 7H9 medium, i.e., Middlebrook 7H9 medium (Difco, Franklin Lakes, NJ, USA) supplemented with 10% albumin-dextrose-catalase (ADC, Difco) and 0.05% Tween 80 (Sigma-Aldrich, Saint-Louis, MI, USA) or on Middlebrook 7H11 agar medium (Difco) supplemented with 10% oleic acid-albumin-dextrose-catalase (OADC, Difco). When required, streptomycin (25 μg mL^{-1}), kanamycin (50 μg mL^{-1}), or anhydrotetracycline (Atc; 200 ng mL^{-1}) were added to the culture media. *E. coli* Stellar strain (Clontech Laboratories, Inc., Mountain View, CA, USA) was grown aerobically at 37 °C in L Broth medium or on L Agar, supplemented with streptomycin (25 μg mL^{-1}).

4.2. Oligonucleotides and Plasmids Used in This Work

Oligonucleotides and plasmids used in this work are listed in Tables S1 and S2, respectively. Plasmids pEN- and pGMCS- were constructed by multisite gateway recombination [14] following the manufacturer's instructions (Life Technologies, Carlsbad, CA, USA). DNA fragments were amplified by PCR, using Phusion DNA polymerase (Thermo-Fisher Scientific, Waltham, MA, USA) with the templates and oligonucleotide pairs indicated in Table S1. Plasmids pEN41- and pEN12- were generated by "BP" cloning, using pDO41A- or pDO221A- as destination vectors, and DNA fragments flanked by *attB4/attB1* or *attB1/attB2*, respectively. Plasmids pGMCS-TetR-P1-*mbcT*, pGMCS-TetR-P606Pld-*mbcT* and pGMCS-TetR-P606-*mbcT* were constructed by "LR" cloning reactions, using plasmid pDE43-MCS as destination vector, plasmid pEN41-TetR as 5′ entry clone, plasmids pEN12-P1, pEN12-P606Pld, or pEN12-P606 as middle entry clone, and plasmid pEN23-*mbcT* as 3′ entry clone. Plasmids pGMCS-P1-reporter1-P_{mbcA}-Reporter2 were constructed by "LR" cloning reactions, using plasmid pDE43-MCS as destination vector, a plasmid pEN41-P1-Reporter as 5′ entry clone, a plasmids pEN12-P_{mbcA}-reporter as middle entry clone, and plasmids pEN23-MluI, pEN23-*mbcA*, or pEN23-*mbcAT* as 3′ entry clone. Site directed mutagenesis of plasmids was performed by amplification by PCR of the plasmid to be mutated with a pair of overlapping oligonucleotides carrying the mutation to be introduced (Table S1). The linear plasmid DNA was purified on agarose gels and circularized using HD In-Fusion cloning reactions (Takara, Kusatsu, Japan). All plasmid constructs were verified by DNA sequencing.

4.3. M. Smegmatis Viability Assays and NAD$^+$ Measurement

Exponentially growing cultures (OD_{600} between 0.05 and 0.2) of *M. smegmatis* mc^2 155 transformed with the desired plasmids were divided in two: half was left in the same medium (uninduced cultures) and the other half was treated with 200 ng mL^{-1} of Atc to induce expression from the tetracycline-inducible promoters. After various time post-induction, samples were harvested and centrifuged to eliminate residual Atc. Cells were resuspended in PBS buffer and dilutions were plated on L Agar supplemented with streptomycin, colonies were counted after 3 days at 37 °C. For labeling with LIVE/DEAD® BacLight dyes (Molecular Probes, Eugene, OR, USA), cells were harvested 8 or 24 h post-Atc induction. Cells were centrifuged, resuspended in PBS buffer and stained as recommended by the manufacturer. Labeled cells were analyzed by fluorescence activated cell sorting using a BD FACS LSRFortessa X20 flow cytometer. Flow cytometry data analysis were performed using FlowJo software (Version 10; Becton, Dickinson and Company, Ashland, OR, USA).

Samples of the above cultures were harvested 8 h post-Atc induction and centrifuged. Cells were resuspended in PBS, adjusting the OD_{600} to 0.2. 0.1 μm-diameter glass beads were added to tubes containing 500 μL of each cell suspensions, and cells were lysed by four 60-s pulses at full speed in a bead-beater device. The samples were centrifuged for 1 min at 20,200× *g* and 50 μL of the lysates were mixed with an equal volume of NAD/NADH-Glo™ Detection Reagent (Promega). Luciferin bioluminescence was measured after 30 min of incubation at room temperature, using a CLARIOstar® plate reader (BMG LABTECH, Champigny s/Marne, France) and normalized to background signal (PBS-only) to monitor the relative amounts of NAD^+ present in the cell extracts.

4.4. Effect of Stress Conditions on P_{mbcA}

Exponential cultures in complete 7H9 medium of *M. tuberculosis* H37Rv transformed with plasmid pGMCS-P1-mTurq-P_{mbcA}-mVenus-*mbcA* were harvested and centrifuged. Cell pellets were and resuspended in PBS buffer (for starvation), or complete 7H9 medium supplemented with 5 mM H_2O_2 or 0.5 mM DETA/NO. At different times, samples were harvested, fixed with 4% *p*-formaldehyde and observed by fluorescent microscopy using large field Leica DMIRB, magnification 630×.

4.5. Human and Mice Macrophage Cultures and Infection

Human monocytes were obtained from healthy blood donors by Etablissement Français du Sang, EFS, Toulouse, France). Written informed consents were obtained from the donors before sample collection (under EFS Contract n°21PLER2017-0035 valid until July 02 2020, which was approved by the French Ministry of Science and Technology, agreement nr. AC2009-921, following articles L1243-4 and R1243-61 of the French Public Health Code). Monocytes were prepared following a previously published procedure [40]. Cells were purified using CD14 microbead positive selection and MACS separation columns (Miltenyi Biotec, Bergisch Gladbach, Germany), according to manufacturer's instructions. For differentiation of monocyte-derived macrophages, monocytes were allowed to adhere to glass coverslips (VWR international, Radnor, PA, USA) in 6-well plates (Thermo-Fisher Scientific), at 1.5×10^6 cells/well, for 1 h at 37 °C in pre-warmed RPMI-1640 medium (GIBCO, Thermo-Fisher Scientific, Waltham, MA, USA). The medium was then supplemented with the following additives: 10% Fetal Bovine Serum (Sigma-Aldrich), 1 mM sodium pyruvate (GIBCO), 50 μM β-mercaptoethanol (GIBCO), and 20 ng mL^{-1} human Macrophage Colony-Stimulating Factor (Miltenyi Biotec, Bergisch Gladbach, Germany). Cells were allowed to differentiate for seven days at 37 °C under 5% CO_2 atmosphere. Murine bone-marrow derived macrophages were prepared as described [41]. Bone-marrow cells were flushed out of the femurs and tibias of 6- to 8-week-old female C57BL/6 mice and cultured in Petri dishes (2×10^6 cells per dish) in pre-warmed RPMI-1640 GlutaMax medium (GIBCO or Life Technologies, Carlsbad, CA, USA), supplemented with 10% FBS (Sigma-Aldrich), 1 mM sodium pyruvate (GIBCO or Life Technologies), 50 μM β-mercaptoethanol (GIBCO or Life Technologies), and 20 ng mL^{-1} M-CSF (PeproTech, Rocky Hill, NJ, USA) at 37 °C under 5% CO_2 atmosphere. Medium was refreshed every three days of culture.

For infection, *M. tuberculosis* strain H37Rv/pGMCS-P1-mTurq-P_{mbcA}-mVenus-*mbcA* was grown to exponential phase in complete 7H9 medium. Mycobacterial clumps were disaggregated by at least 20 passages through a 25G needle, and macrophages were infected at MOI of 0.1 in complete RPMI medium for 4 h at 37 °C under 5% CO_2 atmosphere. Cells were then washed with RPMI and further incubated at 37 °C in RPMI supplemented medium. Infected cells were harvested at day 0 or 24 h post-infection, fixed with 4% *p*-formaldehyde and observed by fluorescent microscopy using large field Leica DMIRB, at a magnification of 630X.

Supplementary Materials: The following are available online at http://www.mdpi.com/2072-6651/12/5/329/s1, Figure S1: Auto-repression of the *mbcAT* operon expression in *M. smegmatis* and *M. tuberculosis*, Table S1: Primers used for plasmid constructions, Table S2: Plasmids used in this work.

Author Contributions: Conceptualization, K.A., A.D.G., and C.G., and O.N.; funding acquisition, C.G. and O.N.; methodology and investigation, K.A., A.D.G., and C.G.; writing—original draft preparation, C.G. and A.D.G.; writing—review and editing, C.G., A.D.G., K.A., and O.N. All authors have read and agreed to the published version of the manuscript.

Funding: This research was supported by the European Union (TBVAC2020), Agence Nationale de la Recherche/Programme d'Investissements d'Avenir (ANR-11-EQUIPEX-0003; ANR-13-BSV8-0010-01), MSDAVENIR, Fondation pour la Recherche Médicale (DEQ20160334902 and post-doctoral fellowship to A.D.G.), the Bettencourt Schueller Foundation and the Thai Ministery of Science and Technology (fellowship to K.A.).

Acknowledgments: We thank D. Schnappinger (Weill Cornell Medical College; New York, USA) for the gift of plasmids for Gateway cloning; Jérôme Rech and Jean-Yves Bouet (LMGM-CNRS; Toulouse, France) for the gift of plasmid carrying the mTurquoise and mVenus reporter; and A. Peixoto and E. Näser (IPBS Genotoul-TRI imaging facility) for help with microscopy and flow cytometry analysis.

Conflicts of Interest: The authors declare no conflict of interest.

References

1. Freire, D.M.; Gutierrez, C.; Garza-Garcia, A.; Grabowska, A.D.; Sala, A.J.; Ariyachaokun, K.; Panikova, T.; Beckham, K.S.H.; Colom, A.; Pogenberg, V.; et al. An NAD(+) Phosphorylase Toxin Triggers Mycobacterium tuberculosis Cell Death. *Mol. Cell* **2019**, *73*, 1282–1291.e8. [CrossRef]
2. Harms, A.; Brodersen, D.E.; Mitarai, N.; Gerdes, K. Toxins, Targets, and Triggers: An Overview of Toxin-Antitoxin Biology. *Mol. Cell* **2018**, *70*, 768–784. [CrossRef] [PubMed]
3. Lobato-Marquez, D.; Diaz-Orejas, R.; Garcia-Del Portillo, F. Toxin-antitoxins and bacterial virulence. *FEMS Microbiol. Rev.* **2016**, *40*, 592–609. [CrossRef] [PubMed]
4. Page, R.; Peti, W. Toxin-antitoxin systems in bacterial growth arrest and persistence. *Nat. Chem. Biol.* **2016**, *12*, 208–214. [CrossRef] [PubMed]
5. Deter, H.S.; Jensen, R.V.; Mather, W.H.; Butzin, N.C. Mechanisms for Differential Protein Production in Toxin-Antitoxin Systems. *Toxins* **2017**, *9*, 211. [CrossRef] [PubMed]
6. Song, S.; Wood, T.K. Toxin/Antitoxin System Paradigms: Toxins Bound to Antitoxins Are Not Likely Activated by Preferential Antitoxin Degradation. *Adv. Biosyst.* **2020**, *4*, e1900290. [CrossRef] [PubMed]
7. Akarsu, H.; Bordes, P.; Mansour, M.; Bigot, D.J.; Genevaux, P.; Falquet, L. TASmania: A bacterial Toxin-Antitoxin Systems database. *PLoS Comput. Biol.* **2019**, *15*, e1006946. [CrossRef]
8. Keren, I.; Minami, S.; Rubin, E.; Lewis, K. Characterization and transcriptome analysis of Mycobacterium tuberculosis persisters. *MBio* **2011**, *2*, e00100–e00111. [CrossRef]
9. Ramage, H.R.; Connolly, L.E.; Cox, J.S. Comprehensive functional analysis of Mycobacterium tuberculosis toxin-antitoxin systems: Implications for pathogenesis, stress responses, and evolution. *PLoS Genet.* **2009**, *5*, e1000767. [CrossRef]
10. Sala, A.; Bordes, P.; Genevaux, P. Multiple toxin-antitoxin systems in Mycobacterium tuberculosis. *Toxins* **2014**, *6*, 1002–1020. [CrossRef]
11. Slayden, R.A.; Dawson, C.C.; Cummings, J.E. Toxin-antitoxin systems and regulatory mechanisms in Mycobacterium tuberculosis. *Pathog. Dis.* **2018**, *76*, fty039. [CrossRef] [PubMed]
12. Lee, M.H.; Pascopella, L.; Jacobs, W.R., Jr.; Hatfull, G.F. Site-specific integration of mycobacteriophage L5: Integration-proficient vectors for Mycobacterium smegmatis, Mycobacterium tuberculosis, and bacille Calmette-Guerin. *Proc. Natl. Acad. Sci. USA* **1991**, *88*, 3111–3115. [CrossRef] [PubMed]
13. Ehrt, S.; Guo, X.V.; Hickey, C.M.; Ryou, M.; Monteleone, M.; Riley, L.W.; Schnappinger, D. Controlling gene expression in mycobacteria with anhydrotetracycline and Tet repressor. *Nucleic Acids Res.* **2005**, *33*, e21. [CrossRef] [PubMed]
14. Schnappinger, D.; Ehrt, S. Regulated Expression Systems for Mycobacteria and Their Applications. *Microbiol. Spectr.* **2014**, *2*, MGM2-0018-2013. [CrossRef] [PubMed]
15. Cortes, T.; Schubert, O.T.; Rose, G.; Arnvig, K.B.; Comas, I.; Aebersold, R.; Young, D.B. Genome-wide mapping of transcriptional start sites defines an extensive leaderless transcriptome in Mycobacterium tuberculosis. *Cell Rep.* **2013**, *5*, 1121–1131. [CrossRef] [PubMed]
16. Levillain, F.; Poquet, Y.; Mallet, L.; Mazeres, S.; Marceau, M.; Brosch, R.; Bange, F.C.; Supply, P.; Magalon, A.; Neyrolles, O. Horizontal acquisition of a hypoxia-responsive molybdenum cofactor biosynthesis pathway contributed to Mycobacterium tuberculosis pathoadaptation. *PLoS Pathog.* **2017**, *13*, e1006752. [CrossRef]
17. Thakur, Z.; Saini, V.; Arya, P.; Kumar, A.; Mehta, P.K. Computational insights into promoter architecture of toxin-antitoxin systems of Mycobacterium tuberculosis. *Gene* **2018**, *641*, 161–171. [CrossRef]
18. Waagmeester, A.; Thompson, J.; Reyrat, J.M. Identifying sigma factors in Mycobacterium smegmatis by comparative genomic analysis. *Trends Microbiol.* **2005**, *13*, 505–509. [CrossRef]

19. Garcia-Pino, A.; Balasubramanian, S.; Wyns, L.; Gazit, E.; De Greve, H.; Magnuson, R.D.; Charlier, D.; Van Nuland, N.A.; Loris, R. Allostery and intrinsic disorder mediate transcription regulation by conditional cooperativity. *Cell* **2010**, *142*, 101–111. [CrossRef]
20. Song, S.; Wood, T.K. Post-segregational Killing and Phage Inhibition Are Not Mediated by Cell Death Through Toxin/Antitoxin Systems. *Front. Microbiol.* **2018**, *9*, 814. [CrossRef]
21. Albrethsen, J.; Agner, J.; Piersma, S.R.; Hojrup, P.; Pham, T.V.; Weldingh, K.; Jimenez, C.R.; Andersen, P.; Rosenkrands, I. Proteomic profiling of Mycobacterium tuberculosis identifies nutrient-starvation-responsive toxin-antitoxin systems. *Mol. Cell. Proteom.* **2013**, *12*, 1180–1191. [CrossRef] [PubMed]
22. Rustad, T.R.; Harrell, M.I.; Liao, R.; Sherman, D.R. The enduring hypoxic response of Mycobacterium tuberculosis. *PLoS ONE* **2008**, *3*, e1502. [CrossRef] [PubMed]
23. Flannagan, R.S.; Cosio, G.; Grinstein, S. Antimicrobial mechanisms of phagocytes and bacterial evasion strategies. *Nat. Rev. Microbiol.* **2009**, *7*, 355–366. [CrossRef] [PubMed]
24. Russell, D.G. Mycobacterium tuberculosis and the intimate discourse of a chronic infection. *Immunol. Rev.* **2011**, *240*, 252–268. [CrossRef] [PubMed]
25. Tsunawaki, S.; Nathan, C.F. Enzymatic basis of macrophage activation. Kinetic analysis of superoxide production in lysates of resident and activated mouse peritoneal macrophages and granulocytes. *J. Biol. Chem.* **1984**, *259*, 4305–4312. [PubMed]
26. Dy, R.L.; Przybilski, R.; Semeijn, K.; Salmond, G.P.; Fineran, P.C. A widespread bacteriophage abortive infection system functions through a Type IV toxin-antitoxin mechanism. *Nucleic Acids Res.* **2014**, *42*, 4590–4605. [CrossRef]
27. Pecota, D.C.; Wood, T.K. Exclusion of T4 phage by the hok/sok killer locus from plasmid R1. *J. Bacteriol.* **1996**, *178*, 2044–2050. [CrossRef]
28. Gottesman, S. Trouble is coming: Signaling pathways that regulate general stress responses in bacteria. *J. Biol. Chem.* **2019**, *294*, 11685–11700. [CrossRef]
29. Lee, K.Y.; Lee, B.J. Structure, Biology, and Therapeutic Application of Toxin-Antitoxin Systems in Pathogenic Bacteria. *Toxins* **2016**, *8*, 305. [CrossRef]
30. Williams, J.J.; Hergenrother, P.J. Artificial activation of toxin-antitoxin systems as an antibacterial strategy. *Trends Microbiol.* **2012**, *20*, 291–298. [CrossRef]
31. Helaine, S.; Cheverton, A.M.; Watson, K.G.; Faure, L.M.; Matthews, S.A.; Holden, D.W. Internalization of Salmonella by macrophages induces formation of nonreplicating persisters. *Science* **2014**, *343*, 204–208. [CrossRef]
32. Keren, I.; Kaldalu, N.; Spoering, A.; Wang, Y.; Lewis, K. Persister cells and tolerance to antimicrobials. *FEMS Microbiol. Lett.* **2004**, *230*, 13–18. [CrossRef]
33. Ronneau, S.; Helaine, S. Clarifying the Link between Toxin-Antitoxin Modules and Bacterial Persistence. *J. Mol. Biol.* **2019**, *431*, 3462–3471. [CrossRef] [PubMed]
34. Wang, X.; Wood, T.K. Toxin-antitoxin systems influence biofilm and persister cell formation and the general stress response. *Appl. Environ. Microbiol.* **2011**, *77*, 5577–5583. [CrossRef] [PubMed]
35. Fraikin, N.; Goormaghtigh, F.; Van Melderen, L. Type II Toxin-Antitoxin Systems: Evolution and Revolutions. *J. Bacteriol.* **2020**, *202*, e00763-19. [CrossRef]
36. DeJesus, M.A.; Gerrick, E.R.; Xu, W.; Park, S.W.; Long, J.E.; Boutte, C.C.; Rubin, E.J.; Schnappinger, D.; Ehrt, S.; Fortune, S.M.; et al. Comprehensive Essentiality Analysis of the Mycobacterium tuberculosis Genome via Saturating Transposon Mutagenesis. *MBio* **2017**, *8*, e02133-16. [CrossRef] [PubMed]
37. Kim, J.H.; O'Brien, K.M.; Sharma, R.; Boshoff, H.I.; Rehren, G.; Chakraborty, S.; Wallach, J.B.; Monteleone, M.; Wilson, D.J.; Aldrich, C.C.; et al. A genetic strategy to identify targets for the development of drugs that prevent bacterial persistence. *Proc. Natl. Acad Sci. USA* **2013**, *110*, 19095–19100. [CrossRef]
38. Rodionova, I.A.; Schuster, B.M.; Guinn, K.M.; Sorci, L.; Scott, D.A.; Li, X.; Kheterpal, I.; Shoen, C.; Cynamon, M.; Locher, C.; et al. Metabolic and bactericidal effects of targeted suppression of NadD and NadE enzymes in mycobacteria. *MBio* **2014**, *5*, e00747-13. [CrossRef]
39. Vilcheze, C.; Weinrick, B.; Wong, K.W.; Chen, B.; Jacobs, W.R., Jr. NAD+ auxotrophy is bacteriocidal for the tubercle bacilli. *Mol. Microbiol.* **2010**, *76*, 365–377. [CrossRef]

40. Troegeler, A.; Lastrucci, C.; Duval, C.; Tanne, A.; Cougoule, C.; Maridonneau-Parini, I.; Neyrolles, O.; Lugo-Villarino, G. An efficient siRNA-mediated gene silencing in primary human monocytes, dendritic cells and macrophages. *Immunol. Cell Biol.* **2014**, *92*, 699–708. [CrossRef]
41. Benard, A.; Sakwa, I.; Schierloh, P.; Colom, A.; Mercier, I.; Tailleux, L.; Jouneau, L.; Boudinot, P.; Al-Saati, T.; Lang, R.; et al. B Cells Producing Type I IFN Modulate Macrophage Polarization in Tuberculosis. *Am. J. Respir. Crit. Care Med.* **2018**, *197*, 801–813. [CrossRef] [PubMed]

Article

Mechanisms of Tolerance and Resistance to Chlorhexidine in Clinical Strains of *Klebsiella pneumoniae* Producers of Carbapenemase: Role of New Type II Toxin-Antitoxin System, PemIK

Ines Bleriot [1,2], Lucia Blasco [1,2], Mercedes Delgado-Valverde [3], Ana Gual-de-Torrella [3], Anton Ambroa [1], Laura Fernandez-Garcia [1,2], Maria Lopez [1,4], Jesus Oteo-Iglesias [2,4,5], Thomas K. Wood [6], Alvaro Pascual [2,3,4], German Bou [1,2,4], Felipe Fernandez-Cuenca [2,3,4,†] and Maria Tomas [1,2,4,*,†]

1 Microbiology Department-Research Institute Biomedical A Coruña (INIBIC), Hospital A Coruña (CHUAC), University of A Coruña (UDC), 15006 A Coruña, Spain; bleriot.ines@gmail.com (I.B.); luciablasco@gmail.com (L.B.); anton17@mundo-r.com (A.A.); laugemis@gmail.com (L.F.-G.); maria.lopez.diaz@sergas.es (M.L.); German.Bou.Arevalo@sergas.es (G.B.)
2 Study Group on Mechanisms of Action and Resistance to Antimicrobials (GEMARA) the Behalf of the Spanish Society of Infectious Diseases and Clinical Microbiology (SEIMC), 28003 Madrid, Spain; jesus.oteo@isciii.es(J.-O.I.); apascual@us.es (A.P.); felipefc@us.es (F.F.-C.)
3 Clinical Unit for Infectious Diseases, Department of Microbiology and Medicine, Microbiology and Preventive Medicine, Hospital Universitario Virgen Macarena, University of Seville, Biomedicine Insititute of Seville (IBIS), 41009 Seville, Spain; mercdss@gmail.com (M.D.-V.); gualdetorrella.ana@gmail.com (A.G.-d.-T.)
4 Spanish Network for Research in Infectious Diseases (REIPI), 41071 Seville, Spain
5 Reference and Research Laboratory for Antibiotic Resistance and Health Care Infections, National Centre for Microbiology, Institute of Health Carlos III, 28222 Majadahonda, Spain
6 Department of Chemical Engineering, Pennsylvania State University, University Park, PA 16801, USA; tuw14@psu.edu
* Correspondence: MA.del.Mar.Tomas.Carmona@sergas.es; Tel.: +34-981-176-399; Fax: +34-981-178-273
† These authors equally contributed to this work.

Received: 23 July 2020; Accepted: 29 August 2020; Published: 2 September 2020

Abstract: Although the failure of antibiotic treatment is normally attributed to resistance, tolerance and persistence display a significant role in the lack of response to antibiotics. Due to the fact that several nosocomial pathogens show a high level of tolerance and/or resistance to chlorhexidine, in this study we analyzed the molecular mechanisms associated with chlorhexidine adaptation in two clinical strains of *Klebsiella pneumoniae* by phenotypic and transcriptomic studies. These two strains belong to ST258-KPC3 (high-risk clone carrying β-lactamase KPC3) and ST846-OXA48 (low-risk clone carrying β-lactamase OXA48). Our results showed that the *K. pneumoniae* ST258-KPC3CA and ST846-OXA48CA strains exhibited a different behavior under chlorhexidine (CHLX) pressure, adapting to this biocide through resistance and tolerance mechanisms, respectively. Furthermore, the appearance of cross-resistance to colistin was observed in the ST846-OXA48CA strain (tolerant to CHLX), using the broth microdilution method. Interestingly, this ST846-OXA48CA isolate contained a plasmid that encodes a novel type II toxin/antitoxin (TA) system, PemI/PemK. We characterized this PemI/PemK TA system by cloning both genes into the IPTG-inducible pCA24N plasmid, and found their role in persistence and biofilm formation. Accordingly, the ST846-OXA48CA strain showed a persistence biphasic curve in the presence of a chlorhexidine-imipenem combination, and these results were confirmed by the enzymatic assay (WST-1).

Keywords: tolerance; persistence; cross-resistance; toxin-antitoxin system; PemI/PemK; *Klebsiella pneumoniae*

Key Contribution: Combination of resistance, tolerance and persistence mechanisms seen in clinical isolates under biocide and antimicrobial stress. PemI/PemK system TA system discovered in the plasmid carrying β-lactamase OXA48.

---

1. Introduction

The increase in antimicrobial resistance due to the emergence of multi-drug resistant (MDR) pathogens is one of the world's greatest public health challenges, as it can lead to an era without effective antibiotics [1]. Recently, the World Health Organization (WHO) published a list of "priority pathogens", which includes those microorganisms considered a serious threat to human health. Some members of this list are carbapenem-resistant pathogens and are known under the acronym of ESKAPE, including among other species, *Klebsiella pneumoniae* [1–3]. *K. pneumoniae* is a Gram-negative, opportunistic bacteria pathogen associated with a wide range of diseases such as urinary tract infections, pneumoniae, septicemia, wounds, and soft tissue infections [4]. Carbapenem resistance is increasing rapidly worldwide, particularly among *K. pneumoniae*. The main carbapenem-resistance mechanism is acquisition of plasmid-encoded carbapenemases, which may belong to the molecular class A (i.e., KPC-type), B (i.e., imipenem (IMP)-type, VIM-type, NDM-type) and D (i.e., OXA-48-type). The high-risk clones of *K. pneumoniae*, in contrast to low-risk clones, have an extraordinary ability to persist and spread in the nosocomial environment, disseminating these carbapenemases and therefore being involved in nosocomial outbreaks [4].

Nevertheless, much less attention has been paid to the presence or occurrence of resistance to antiseptics and biocides, such as chlorhexidine (CHLX) [5], widely used in hospital settings. CHLX is a symmetric bis-biguanide molecule comprising two chloroguanide chains that are connected by a central hexamethylene chain, and carry two positive charges at physiological pH. CHLX is sparingly soluble in water, and thereby normally formulated with either acetate or gluconate to form water-soluble salts [6]. The antimicrobial effect of this compound is based on damaging the bacterial membrane, leading to the subsequent leakage of cytoplasmatic material. Therefore, mechanisms conferring resistance toward CHLX include multidrug efflux pumps and cell membrane changes [5]. Moreover, CHLX adaptation has been associated with the emergence of stable resistance to the last-resort antibiotic colistin (polymyxin E) [7–9].

In general, the failure of antibiotic treatments has been associated with resistance mechanisms. However, it has recently been noted that other mechanisms such as tolerance and persistence were also involved [10]. The recovery of persistent cells is one of the main causes of prolonged and recurrent infections, that can lead to the complete failure of antibiotic treatments [11]. In this context, it is important to distinguish between resistant, tolerant, and persistent bacteria [12]. The term resistance is generally used to describe the inherited ability of a bacterial population to grow in the presence of high concentrations of antibiotics, regardless of the duration of treatment [12], due to active defense mechanisms associated with mutations [10]. Whereas, the term tolerance is used to describe the ability, inherited or not, of a bacterial population to survive the transient exposure of high concentrations of antibiotics without causing changes in minimum inhibitory concentrations (MICs), due to the deceleration of essential biological processes [10–12]. It is important to emphasize that despite the slow-growth rate, tolerant bacteria keep a metabolically active state. In contrast to resistance and tolerance, persistence is characterized by the ability, not inherited, of a bacterial subpopulation (around 0.001–1%) [10] to resist antibiotics by growth arrest due to the inactivation of their metabolism and their non-replicative state, thus it is due to their dormant state. Persistent bacteria exhibit transient levels of tolerance to antibiotics that do not affect their MICs, so once the drug pressure is removed and their metabolism is reactivated, they can rapidly re-grow. Nowadays, it is known that multiple molecular mechanisms are involved in the formation of persistent bacteria such as the stringent response molecule (p)ppGpp, stress response, SOS response, quorum sensing, toxin-antitoxin (TA) systems, efflux pumps, the ROS response and energy metabolism, among others [11].

The involvement of TA systems in cell physiology, specifically in: (i) biofilm formation by regulating fimbriae [13,14], (ii) bacterial persistence, by generating slowly-growing cells tolerant to antibiotics and environmental changes [15–19], (iii) plasmid maintenance [20,21], (iv) general stress response [22], and (v) phage inhibition [23–25] is becoming clearer [18]. A TA system is a module of two genes encoding a stable toxin and an unstable antitoxin. Under normal growth conditions the antitoxin inhibits the toxin, but under stress conditions the antitoxin is degraded, leaving the toxin free to inhibit the basic cellular processes like DNA replication or protein synthesis, and also promoting plasmid maintenance, slow growth and latency [18]. These systems are widely distributed and found in the bacterial chromosome, plasmids, and bacteriophages [2].

In this context, this study provides a better comprehension of the molecular mechanisms associated with chlorhexidine adaptation (CA) in two clinical strains of *K. pneumoniae*, both of which produce carbapenemase: ST258-KPC3 (high-risk clone carrying β-lactamase KPC3) and ST846-OXA48 (low-risk clone carrying β-lactamase OXA48), from a phenotypic and transcriptomic point of view. It should be noted that international high-risk clones of *K. pneumoniae* are among the most common nosocomial pathogens. The success of these clones is due to their facility to spread their plasmids, which carry a considerable variety of antimicrobial resistance genes [26,27]. Thus, the study of this type of clones is of great clinical relevance. Moreover, this study aims to characterize a new toxin-antitoxin system (PemIK) located in a plasmid inside the ST846-OXA48CA strain, and to examine the possible role of this system in persistence and in biofilm formation.

2. Results

2.1. Results Subsection

2.1.1. Time-Killing Curve in the Presence of CHLX (10 × MIC)

The time-killing curves of the strains ST258-KPC3CA and ST846-OXA48CA in the presence of CHLX (10 × MIC) showed two different growth patterns (Figure 1). The strain ST258-KPC3CA showed a slight reduction in its bacterial population, occurring in the first two hours of CHLX exposure, decreasing from 6 LogCFU/mL (1.47×10^6 CFU/mL) to 4 LogCFU/mL (7.75×10^4 CFU/mL) at 2 h (Figure 1A). This slight reduction in the bacterial population occurs during the activation of defense mechanisms, such as efflux pumps, which can reduce the effective concentration of the drug in the cell. In contrast, the ST846-OXA48CA strain, dramatically reduced its bacterial population during the first four hours of CHLX exposure, decreasing from of a bacterial population of 6 LogCFU/mL (2.43×10^6 CFU/mL) to 2 LogCFU/mL (1.05×10^2 CFU/mL) at 4 h (Figure 1B). After this period, both bacterial strains grew again reaching respectively a bacterial population of 6 LogCFU/mL (6.63×10^6 CFU/mL) and 5 LogCFU/mL (2.75×10^5 CFU/mL) at 48 h. The curves have the characteristics of a resistant strain in the first case (ST258-KPC3CA) and a tolerant strain in the second one (ST846-OXA48CA), according to the definition of each mechanisms. Resistance is the ability of bacterial population to grow at a similar rate in the presence of an environmental stress, while the tolerance is the ability of a bacterial population to withstand the stress.

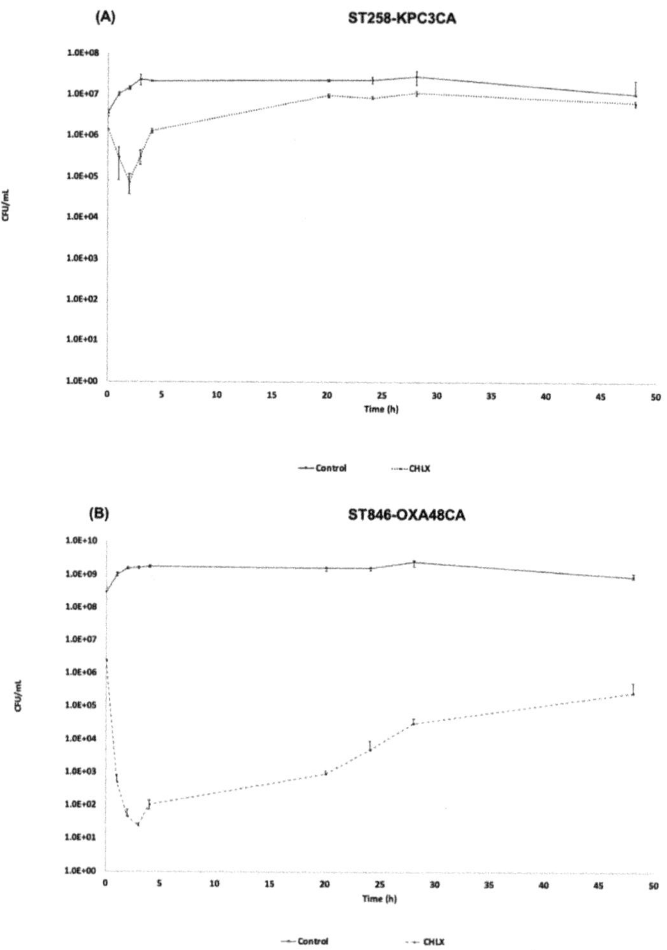

Figure 1. Time-killing curve in the presence of chlorhexidine (CHLX) (10 × minimum inhibitory concentration (MIC)) in *K. pneumoniae* chlorhexidine adaptation (CA) strains ST258-KPC3CA (**A**) and ST846-OXA48CA (**B**). The same strains without being exposed to biocide pressure are used as controls. The errors bar represents the standard deviation of the three replicates experiment.

2.1.2. Transcriptomic Study

All the transcriptomic results are deposited in the NCBI database as a GenBank BioProject (Code number: PRJNA609262) and GEO series (Code number: GSE147316). The transcriptomic profile from the ST258-KPC3CA isolate indicated a probable CHLX resistant profile. Indeed, this strain has a higher number of overexpressed genes (Log2fold change > 1.5), especially for transporters and efflux pumps such as the methyl viologen resistance gene *smvA* (Log2fold change: 3.635), which is involved in the cationic biocide resistance. However, strain ST846-OXA48CA showed what it was as a CHLX tolerant profile, with repressed genes (Log2FoldChange < 1.5) for efflux pumps, TA systems, SOS response, and ppGpp mechanisms (Table 1). This strain also showed high levels of expression of genes, *pmrD*, and *pmrK* (Log2fold change 2.360 and 1.570, respectively), characteristics of the colistin resistance. Therefore, these transcriptomic results corroborated the results obtained by the time-killing curves, showing activation of molecular mechanisms of resistance and tolerance molecular mechanisms in response to CHLX in ST258-KPC3CA and ST846-OXA48CA strains, respectively.

Table 1. Gene expression in response to CHLX in the strain ST258-KPC3CA and ST846-OXA48CA.

Mechanism	Gene [a]	Description	ST258-KPC3CA		ST846-OXA48CA	
			Log2 FoldChange	ID Gene	Log2 FoldChange	ID Gene
Transporter	smvA	Methyl viologen protein (cationic biocide resistance)	3.635	HGAILKPD_00917	1.209	EMNICGIE_00134
	actP	Acetate permease ActP (cation/acetate symporter)	2.724	HGAILKPD_00571	0.649	EMNICGIE_02128
	csbX	MFS superfamily	2.549	HGAILKPD_04325	0.618	EMNICGIE_04796
	lldP	L-lactate permease	2.486	HGAILKPD_04496	0.085	EMNICGIE_00243
	cysW	Ferric iron ABC transporter	2.181	HGAILKPD_02877	−0.266	EMNICGIE_02673
	potA	ABC transporter	1.749	HGAILKPD_02785	−0.351	EMNICGIE_03352
	pmrD	Signal transduction protein PmrD (colistin resistance)	-	-	2.360	EMNICGIE_04427
	pmrK	Polymyxin resistance protein PmrK (colistin resistance)	-	-	1.570	EMNICGIE_02839
ATP metabolism	atpD	ATP synthase beta chain	−0.209	HGAILKPD_02375	−0.232	EMNICGIE_00435
TA systems	ortT	Orphan toxin OrtT	0.731	HGAILKPD_02791	0.727	EMNICGIE_00095
	pemI	Programmed cell death antitoxin PemI	-	-	−0.100	EMNICGIE_05097
	pemK	Programmed cell death toxin PemK	-	-	−0.302	EMNICGIE_05098
(p)ppGpp	gppA	Guanosine-5′-triphosphate,3′-diphosphate pyrophosphatase	−0.765	HGAILKPD_02586	−0.399	EMNICGIE_03280
ROS response	cydA	Cytochrome d ubiquinol oxidase subunit I	1.318	HGAILKPD_03209	0.441	EMNICGIE_03423
	cybB	Cytochrome b561	0.456	HGAILKPD_02756	0.196	EMNICGIE_00060
SOS system	yedK	Putative SOS response-associated peptidase YedK	1.117	HGAILKPD_04848	0.003	EMNICGIE_04152
	yebG	DNA damage-inducible gene in SOS regulon	0.722	HGAILKPD_02193	0.455	EMNICGIE_04716

[a] All gene expression showed have a p-value < 0.05, and overexpression and repression were considered from as a Log2fold change of at least 1.5 and 0.5, respectively. (-) Not detected. The rows shaded differently reveal those genes of high relevance in this study, the cationic biocide resistance gene, the colistin resistance genes, and the novel TA system PemI/PemK. Background colour indicates the genes of most interest from this study.

2.1.3. Antimicrobial Susceptibly Testing

The antimicrobial susceptibility test was done for wild-type ST258-KPC3 and ST846-OXA48 wild-type and the two CA strains. According to adaptation to CHLX, an increase in MICs of CHLX was observed in both CA strains. However, no differences in the minimum inhibitory concentration (MIC) values were observed for the other antibiotics tested, except for the colistin, in that in the ST846-OXA48CA strain showed an increase in the MIC value of 32-fold, that corresponds to a resistance value (Table 2).

Table 2. MIC values (µg/mL) of different antibiotics for ST258-KPC3, ST258-KPC3CA, ST846-OXA48, and ST846-OXA48CA.

	MIC (µg/mL)														
Strain	**CHLX**	**CIP**	**TGC**	**TOB**	**IMP**	**MRP**	**GEN**	**CAZ**	**TZP**	**SAM**	**NET**	**DOX**	**AMK**	**MIN**	**CST**
ST258-KPC3	9.8	>32	2	64	4	8	4	>32	>32	1024	128	2	16	4	0.25
ST258-KPC3CA	39.1	>32	2	64	4	8	4	>32	>32	1024	128	2	16	4	0.25
ST846-OXA48	19.5	8	8	32	16	>32	32	>32	>32	128	16	8	2	8	0.5
ST846-OXA48CA	78.2	8	8	32	16	>32	32	>32	>32	128	16	8	2	8	16

CHLX, Chlorhexidine; CIP, Ciprofloxacin; TGC, Tigecycline; TOB, Tobramycin; IMP, Imipenem; MRP, Meropenem; GEN, Gentamycin; CAZ, Ceftazidime; TZP, Piperacillin-tazobactam; SAM, Sulbactam; NET, Netilmicin; AMK, Amikacin; MIN, Minociclin; CST, Colistin.

2.1.4. Characterization of the New TA System, PemI/PemK, Present in a Plasmid in the Strain ST846-OXA48CA

A PemI/PemK TA system, whose closest relative is the type II TA toxin-antitoxin system PemK/MazF family toxin belonging to Enterobacteriacea (Query: 78%; Identity: 99.35%; Code number: WP_077688581.1) and which have not been previously described in *K. pneumoniae*, it was identified by transcriptomic analysis (Table 1) as encoded by a plasmid of *K. pneumoniae* ST846-OXA48CA (Figure S1). This plasmid harbors several genes, such as *repA, dsbC, trbA, trbC, lusR*, CPBP metalloprotease, *umuD, umuC*, restriction endonuclease, *IS1, ssb, mobC, nikA, dotD*/TraH family lipoprotein, secretion systems type IV, *traO, traP, traQ, traW, traX, dotA*/*traY*, and *repC*, in addition to another TA system, the RelE/RelB TA system. This PemI/PemK TA system is composed of a 258 bp antitoxin gene (*pemI*) and a 333 bp toxin gene (*pemK*). To confirm that this system is a TA system, *pemI/pemK* and *pemK* genes alone were cloned into the overexpression vector pCA24N, widely used in the literature to overexpress TA systems [28,29], and transformed into the cured plasmid strain ST846-OXA48CA CP (i.e., lacking plasmids and therefore the plasmid that encodes the PemI/PemK TA system). The toxicity of this TA system was tested by growth curves overexpressing both *pemI/pemK* and *pemK* (Figure 2). Overexpression of *pemK* in ST846-OXA48CA CP/pCA24N (*pemK*) inhibited bacterial growth, while overexpression of the *pemI/pemK* system (ST846-OXA48CA CP/pCA24N (*pemIK*)) in the strain led to normal bacterial growth, which was slightly impaired compared to the empty plasmid. Therefore, the plasmid-based PemI/PemK TA system found in *K. pneumoniae* ST846-OXA48CA is functional.

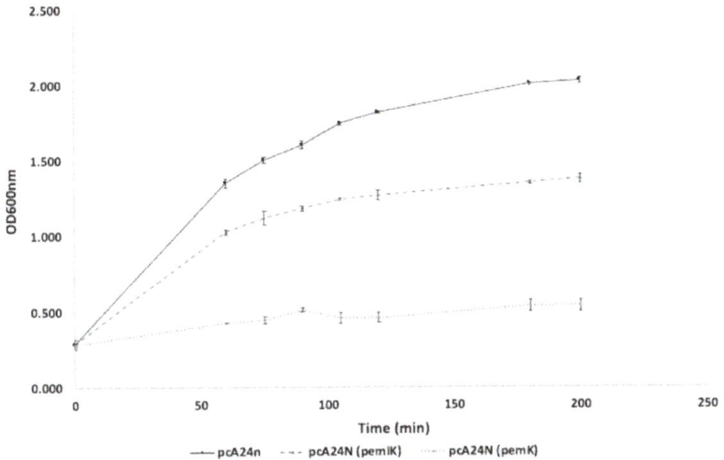

Figure 2. Growth curves of ST846-OXA48CA CP containing pCA24N plasmids with *pemIK* (dark grey, dashed line) and *pemK* (light grey, dotted line) in the presence of 1 mM IPTG. The strain ST846-OXA48CA CP is used as a control as it carries the empty plasmid pCA24N (black). The errors bar represents the standard deviation of the three experimental replicates.

2.1.5. Biofilm Formation Assay

Since TA systems have been associated with the arrest of bacterial growth and the formation of biofilms, we studied the effect of the PemI/PemK TA system and the PemK toxin on biofilm formation (Figure 3). Production of PemK toxin resulted in a significant decrease in biofilm formation compared to the control (ST846-OXA48CA CP/pCA24N) (p-value < 0.001). Moreover, production of PemI/PemK restored a similar phenotype as the control, lacking a significant difference in biofilm formation (p-value < 0.05). Therefore, the PemI/PemK TA system influences *K. pneumoniae* biofilm formation.

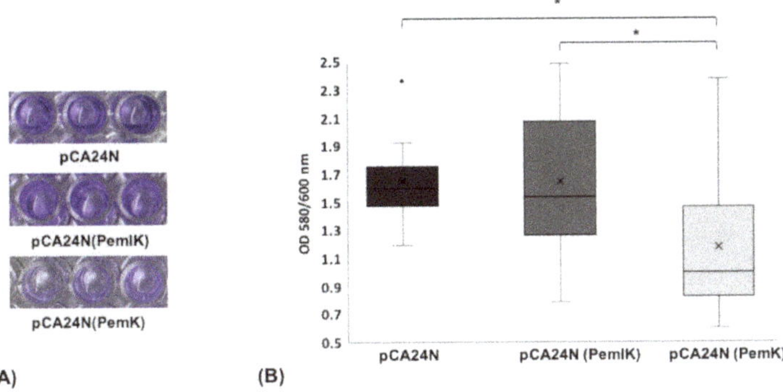

Figure 3. Biofilm formation assay. (**A**) Biofilm strained with 10% crystal violet was dissolved in 30% acetic acid. (**B**) Box and whisker plot of the optical density of biofilm produced by the strains ST846-OXA48CA CP/pCA24N, ST846-OXA48CA CP/pCA24N (PemIK), and ST846-OXA48CA CP/pCA24N (PemK). The biofilm formation was expressed as the ratio between OD580/600 nm, in order to normalize the data. Boxes indicate the lower and upper quartile. Horizontal lines in each box represents the median value of biofilm formation. The mean biofilm formation for each strain is indicated by a +. Vertical lines extending from each box represent the minimum and maximum biofilm formation. *, p-value < 0.05. All experiments were performed in triplicates.

2.1.6. Time-Killing Curve in the Presence of Imipenem or in Combination with Chlorhexidine for ST846-OXA48CA and ST846-OXA48CA CP

The time-killing curves of *K. pneumoniae* ST846-OXA48CA tolerant to CHLX (Figure 4A) were performed in the presence of imipenem (IMP) (50 × MIC) alone or in combination with CHLX (10 × MIC). A drastic reduction in the number of CFU was observed during the first four hours in the presence of IMP (50 × MIC), decreasing from 7 LogCFU/mL (9.35×10^7 CFU/mL) to 3 LogCFU/mL (1.50×10^3 CFU/mL). However, bacterial regrowth occurred after four hours, reaching similar levels of CFU/mL as the control (8 LogCFU/mL (6.83×10^8 CFU/mL) vs. 9 LogCFU/mL (1.73×10^9 CFU/mL)) at 28 h and exceeding it at 48 h (9 LogCFU/mL (1.25×10^9 CFU/mL) vs. 9 LogCFU/mL (1.07×10^9 CFU/mL)). In the case of the combination of IMP and CHLX, a greater CFU reduction than IMP alone was observed, with no CFU detected after 4 h. Nevertheless, a regrowth of the bacterial population was observed after 28 h. Thus, ST846-OXA48CA in presence of the combination of IMP and CHLX showed a characteristic behavior of the persistent subpopulation.

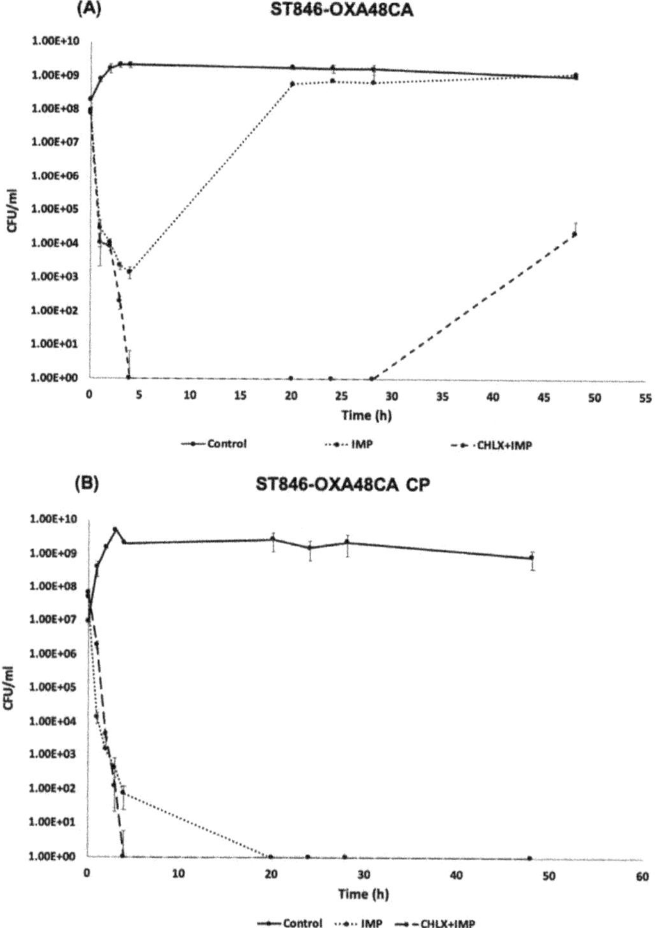

Figure 4. Time-killing curve in the presence of IMP (50 × MIC) and in presence of the combination of IMP (50 × MIC) and CHLX (10 × MIC) for the strains of *K. pneumoniae* ST846-OXA48CA (**A**) and ST846-OXA48CA CP (**B**). The controls are the strains without exposure to any stress (IMP, IMP + CHLX). The error bars represent the standard deviation of the three replicates of the experiment.

In the case of the ST846-OXA48CA CP strain (Figure 4B), in which the plasmid was removed by means of a curing agent, 3% sodium dodecyl sulfate (SDS) (10% *w*/*v* pH = 7.4), a drastic reduction in the number of CFU, was observed both with IMP alone as for in the combination of IMP and CHLX. In fact, in the case of IMP alone no CFUs were recovered at 20 h while, in the case of the combination, no CFUs were recovered at 4 h (Figure 4). Finally, the culture was considered dead as no regrowth was observed throughout the rest of the assay. Thus, the ST846-OXA48CA CP strain, unlike the ST846-OXA48CA strain, showed more sensitive behavior curve pattern in the presence of IMP alone and in the presence of the combination of drug. These results may suggest that the absence of the plasmid containing both the β-lactamase OXA48 and the TA system PemI/PemK could be a factor responsible for the absence of regrowth. Moreover, the lack of the TA system PemI/PemK could be implicated in the non-appearance of a persistent subpopulation in presence of the combination of IMP and CHLX, contrary to what happens in the strain ST846-OXA48 CA.

2.1.7. Enzymatic Analysis Using the Cell Proliferation Reagent WST-1

The results of the time-killing curves in the presence of the IMP and CHLX combination was confirmed by enzymatic analysis using the cell proliferation reagent WST-1 (Figure 5), which measures the omnipresent reducing agents NADH and NADPH as biochemical markers to evaluate the metabolic activity of the cell [30]. Indeed, ST846-OXA48CA lacks metabolic activity/cell proliferation at 24 h ($OD_{480\ nm} < 0.01$), whereas it presents a significant increase at 48 h ($OD_{480\ nm} > 0.4$; p-value < 0.0001), confirming regrowth in the bacterial culture. In contrast, the ST846-OXA48CA CP strain, despite showing significant differences (p-value < 0.002) between 24 and 48 h in terms of metabolic activity/cell proliferation, is considered as a dead culture since its $OD_{480\ nm}$ is less than 0.1.

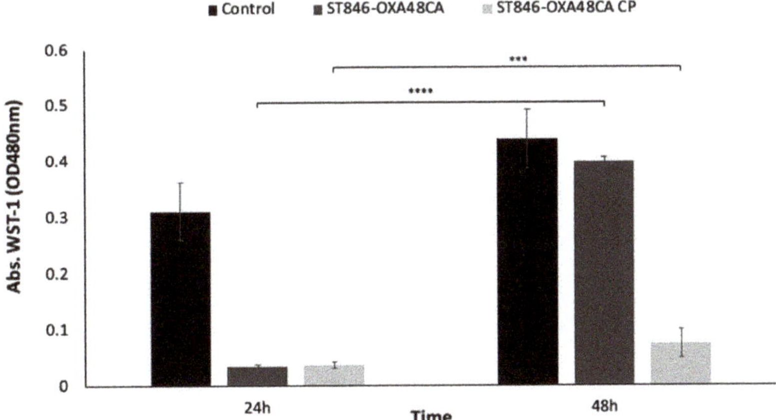

Figure 5. Enzymatic activity by the colorimetric assay (WST-1 based) of the strain *K. pneumoniae* ST846-OXA48CA and ST846-OXA48CA CP in the presence of the combination of IMP (50 × MIC) and CHLX (10 × MIC). The growth control is ST846-OXA48CA strain without antibiotic pressure. ***, p-value < 0.001 and ****, p-value < 0.0001. The errors bars represent the standard deviation of the three experiment replicates.

3. Discussion

Due to the emergence of MDR pathogens over the past few decades, public health officials faces new challenges, such as the alarming increase in antimicrobial resistance, as well as the emerging link between resistance strategies used by bacteria against antibiotics and biocides [18]. This last problem is even more worrisome due to the routinely and uncontrolled use of antiseptics and biocides in clinical practice [9]. One example of this is CHLX, a bis-biguanide antiseptic of cationic nature that has bactericidal activity through membrane disruption [31]. For these reasons it of great interest to decipher the molecular mechanisms involved in the adaptation to CHLX in clinical strains of *K. pneumoniae*, producers of carbapenemases.

In order to determine at the molecular level the effect of the adaptation to CHLX in a strains of *K. pneumoniae*, we performed a phenotypic study in the presence of CHLX (10 × MIC), which showed that the adaptation to CHLX led to the activation of two different molecular mechanisms in the clinical strains of *K. pneumoniae* ST258-KPC3CA and ST846-OXA48CA. In effect, the ST258-KPC3CA strain presented a growth curve typical of resistant bacteria, where a slight reduction in the bacterial population occurs during the time of activation of defense mechanisms [32] (e.g., efflux pumps, TA systems, quorum network), followed by a regrowth period similar to the control. In contrast, ST846-OXA48CA had a characteristic growth curve of tolerant bacteria, where the strain undergoes a drastic reduction or arrest of growth during the first four hours of exposure to the bactericide [31]. These results were corroborated by the transcriptomic study where the transcriptomic profile of the ST258-KPC3CA strain revealed the overexpression of a larger number of genes compared to ST846-OXA48CA strain, especially those related to the overexpression of efflux pumps, which are generally considered as a basic molecular mechanism associated with resistance [33]. In our study, the efflux pump gene that was overexpressed in ST258-KPC3CA was the methyl viologen resistance gene *smvA*. SmvA is responsible for resistance to CHLX in *K. pneumoniae* strains previously adapted to CHLX, due to the interruption of the *smvR* gene (*smvA* repressor) [9].

The antimicrobial susceptibility test revealed the appearance of cross-resistance to colistin (polymyxin E), in the ST846-OXA48CA strain tolerant to CHLX, increasing its MIC value 32-fold. These results were also corroborated by the transcriptomic studies, in which a high level of expression of the colistin resistance genes, *pmrD* and *pmrK* could be observed. This phenomenon of cross-resistance was also described previously, where five out of six strains of *K. pneumoniae* adapted to CHLX also presented resistance to colistin, increasing their MIC values from 2–4 mg/L to 64 mg/L [9]. This phenotype is due to the common biochemical characteristics of CHLX and colistin: in fact, both are cationic compounds with hydrophobic functions [6]. As for the other antibiotics tested, no change in the MIC value was observed in the strain adapted to CHLX compared to the wild-types.

In the previous study of Fernández-García et al. (2018), a combination of β-lactam antibiotic IMP with the CHLX biocide caused in some strains of *Acinetobacter baumannii* the formation of a subpopulation of persistent bacteria [12]. This phenomenon could be observed in the ST846-OXA48CA strain in the presence of the same combination. However, we have seen that the combination of this β-lactam antibiotic and the biocide did not lead to the appearance of a persistent subpopulation in ST846-OXA48 CA CP, cured of the plasmid, but rather led to the death of the bacterial culture after 4 h. This suggests that the presence of the TA systems carried by the plasmid, PemI/PemK and RelE/RelB, could be responsible of the emergence of a persistent subpopulation. As it has long been shown that TA systems are involved in the formation of persistent bacterial subpopulations [19,34,35]; indeed, TA systems are genetic elements composed of a toxin, which inhibits bacterial growth by interfering with essential cellular processes, and an antitoxin, which is able to neutralize the effect of the toxin in normal growth conditions [11,18,36]. The *pemI/pemK* genetic module present in the plasmid of *K. pneumoniae* ST846-OXA48CA, has never been described in *K. pneumoniae.* This TA system was characterized and overexpression assays confirmed that this module corresponds to a TA system. The overexpression of the *pemK* gene led to the inhibition of bacterial growth; however, overexpression of the *pemIK* module led to normal bacterial growth. The same phenomenon was observed in *Bacillus anthracis* where the overexpression of *pemK* in the pHCMC05 vector was severely toxic to the growth of *B. anthracis* cells [37].

In recent decades, many studies have shown that TA systems are associated with the formation of biofilms [22,38]. Biofilms are characterized by a dense multicellular community of microorganisms, constituted after the attachment of bacteria to a biotic or an abiotic surface [39]. In fact, the first TA system linked to biofilm formation was the MqsR/MqsA system of *Escherichia coli* [13]. The toxins have been described as modulators of biofilm formation [13,40]. In this study, we have seen that the overexpression of the PemK toxin contributed to a significant decrease in biofilm formation. The effect that we observed for the toxin is corroborated by the study of García-Contreras et al. (2008), [41], where the *hha*-deletion mutant significantly increased biofilm formation. In addition, the complemented *hha* mutant showed a consistently and dramatically inhibition of biofilm formation. Besides, Ma et al. (2019), described that the disruption of MazF toxin, in *Staphylococcus aureus*, led to an increase in biofilm formation in an *ica*-cluster dependent way, as the disruption of *mazF* produced an increase in the level of expression of *icaA*, *icaB*, and *icaC* genes [42]. Furthermore, in the study of Kim et al. (2009), the authors described that the overexpression of five toxins conduced to a decrease in biofilm formation at 8 h, although at 24 h they observed an increase in biofilm [14].

4. Conclusions

This is the first study that describes the different effects of the adaptation to CHLX in two clinical strains of *K. pneumoniae*, producers of carbapenemase, that become resistant (ST258-KPC3CA) and tolerant (ST846-OXA48CA) to CHLX. This adaptation has lead, in the case of ST846-OXA48CA strain, to the development of cross-resistance to colistin, an antibiotic of last resort in hospital infections.

Furthermore, this study is the first one to describe the relationship between the mechanisms of bacterial persistence and the combination of a β-lactam antibiotic (IMP) and a biocide (CHLX) in the clinical isolate of *K. pneumoniae* ST846-OXA48CA. Finally, a new PemI/PemK TA system was identified in a plasmid of the ST846-OXA48CA strain. Its subsequent characterization demonstrated it participates in the development of persister cells as well as the establishment of biofilms.

5. Materials and Methods

5.1. Bacterial Strains and CHLX Adaptation

Two clinical strains of *K. pneumoniae*, producers of carbapenemases harbored in plasmids, ST258-KPC3 (high-risk clone) and ST846-OXA48 (low-risk clone) were used in this study. These clinical strains belonging to different sequence types (ST) and are from urine and sputum samples, respectively. The attribution of the ST was carried out in the study of Esteban-Cantos et al. (2017) according to the scheme of the Pasteur Institute (http://bigsdb.web.pasteur.fr/Klebsiella/) [43].

In order to obtain the CHLX adapted strains, the clinical strains of *K. pneumoniae* clinical ST258-KPC3 and ST846-OXA48, were exposed for two weeks to $\frac{1}{4}$ of their MIC of CHLX (9.8 and 19.5 μg/mL), respectively in liquid media with aeration. The antibiotic was replaced every 24 h. After these two weeks exposure, the two CA strains were obtained with MIC values of 39.1 and 78.2 μg/mL, respectively.

Furthermore, to study the role of the plasmid (Figure S1) encoding the PemI/PemK TA system in the ST846-OXA48CA strain plasmids were removed from the strain. The cured plasmid ST846-OXA48CA CP strain of *K. pneumoniae* was generated following the protocol of El-Mansi et al. (2000) [44], where 3% sodium dodecyl sulfate (SDS) (10% *w*/*v* pH = 7.4) was used as a curing agent. To check the effective loss of the plasmid, PCR was carried out using the verification primers and, to corroborate the results, plasmid extractions were also performed and subsequently loaded on a 1% agarose gel.

All the bacterial strains and plasmids used in this study shown in the Table 3.

Table 3. Description of the bacterial strains and plasmids used in this study.

Strain or Plasmid	Main Characteristics	Source or Reference
ST258-KPC3	*K. pneumoniae* high-risk clone carrying β-lactamase KPC3	This study
ST258-KPC3 CA	*K. pneumoniae* high-risk clone carrying β-lactamase KPC3 adapted to CHLX	This study

Table 3. *Cont.*

ST846-OXA48	*K. pneumoniae* low-risk clone carrying β-lactamase OXA48	This study
ST846-OXA48 CA	*K. pneumoniae* low-risk clone carrying β-lactamase OXA48 adapted to CHLX	This study
ST846-OXA48 CA CP	*K. pneumoniae* low-risk clone carrying β-lactamase OXA48 adapted to CHLX and cured plasmid strain	This study
pCA24N	Expression plasmid Cm^R, LacIq	[45]
pCA24N (*pemIK*)	Expression plasmid pCA24N with the TA systems *pemIK*	This study
pCA24N (*pemK*)	Expression plasmid pCA24N with the TA systems *pemIK*	This study

5.2. Time-Killing Curve

The different time-killing curves were performed according to Hofsteenge et al. (2016) [8] in low-nutrient Luria-Bertani (LN-LB) broth (2 g/L tryptone, 1 g/L yeast extract and 5 g/L NaCl). The culture was incubated at 37 °C with shaking (180 rpm) until it reached the optical density at 600 nm ($OD_{600\ nm}$) of 0.6. At that moment, CHLX digluconate (10 × MIC) (Sigma-Aldrich, Darmstadt, Germany), IMP (50 × MIC) (Sigma-Aldrich, Darmstadt, Germany), or the combination of IMP (50 × MIC) and CHLX (10 × MIC), were added. Bacterial concentrations (CFU/mL) were determined at 0, 1, 2, 3, 4, 20, 24, 28, and 48 h by serial dilutions and plating on LB agar (10 g/L tryptone, 5 g/L yeast extract, 5 g/L NaCl, and 20 g/L agar). All experiments were performed in triplicates.

The MIC value for CHLX and IMP of the strains ST258-KPC3CA and ST846-OXA48CA are shown in Table 2 (MIC value of CHLX: 39.1 μg/mL and 78.2 μg/mL, respectively; and value of MIC of IMP of strain ST846-OXA48CA is 16 μg/mL). Therefore, 391 μg/mL and 782 μg/mL of CHLX was added in the time-killing curves in presence of CHLX digluconate (10 × MIC) in the strains ST258-KPC3CA and ST846-OXA48CA, respectively. In the case of the strains ST846-OXA48CA and ST846-OXA48CA CP, where the time-killing curves were performed in presence of (50 × MIC) IMP alone or in combination with (10 × MIC) CHLX, 800 μg/mL of IMP was added.

5.3. Transcriptomic Study

The strain ST258-KPC3, ST258-KPC3CA, ST846-OXA48, and ST846-OXA48CA were cultured on solid LB plates, with or without CHLX according to the requirement of the strains and incubated at 37 °C for 24 h. One colony was removed and inoculated in liquid LB medium, with (1/16 × MIC) CHLX for the CA strain and incubated overnight at 37 °C with shaking (180 rpm). The inoculum was diluted (1:100) in LB media with (1/4 × MIC) CHLX for CA strains and without CHLX for wild-type strains. Then, when the culture reached a logarithmic growth phase ($OD_{600\ nm}$: 0.6), the RNA was extracted using the High Pure RNA Isolation kit (Roche, Mannheim, Germany), following the kit instructions, and the extract was treated with DNase (Roche, Mannheim, Germany). The RNA was subsequently quantified in a Nanodrop ND-100 spectrophometer (NanoDrop Technologies, Waltham, MA, USA) and was analyzed with an Agilent 2100 Bioanalyzer with RNA 6000 Nano reagents and RNA Nano Chips (Agilent Technologies, Santa Clara, CA, USA) to determine the quality and the integrity of the samples. All extractions were carried out for quadruple. After RNA extraction of the four replicates of each strain, rRNA and tRNA were removed using the Ribo-Zero rRNA removal kits (bacteria) (Illumina, San diego, CA, USA). The rRNA depletion was checked using Agilent RNA ScreenTape Assay (TapeStation 4200 Agilent, Santa Clara, CA, USA) and the RNA was quantified by Qubit TM RNA HS Assay Kit (Thermo Fisher Scientific, Waltham, MA, USA). Then, the transcriptomic libraries were performed using the Ion RNAseq Kit v2 in combination with the Ion Xpress TM RNA-Seq Barcode 01-16 Kit. Sequencing was carried out by emulsion PCR with Ion Sphere Particles (IPs). The enrichment of the library carrying IPs and the subsequent loading of the chip was developed in the Ion Chef automated system using the Ion PI Hi-Q Chef kit (Thermo Fisher Scientific, Waltham, MA, USA) and Ion PI chips v3 (Thermo Fisher Scientific, Waltham, MA, USA). Chip analysis was carried out on the Ion Proton sequencer (Thermo Fisher Scientific, Waltham, MA, USA). The generated data was analyzed using the specific software of the Torrent Suite 5.6.2 platform (Thermo Fisher Scientific, Waltham, MA, USA). The resulting readings (approx. 60 million) were exported in a FASTQ file using the FileExporter 4.6.0.0 (Thermo Fisher Scientific, Waltham, MA, USA) plugin. The alignment of the sequences against their respective controls was performed with the STAR software. Subsequently, the aligned readings were counted using the htseq-count software (HTSeq 0.6.1.p2). All data was normalized using DESeq2.

5.4. Antimicrobial Susceptibility Test

MIC values of chlorhexidine, ciprofloxacin, tigecycline, tobramycin, imipenem, meropenem, gentamycin, piperacillin-tazobactam, ceftazidime, sulbactam, netilmicin, doxycycline, amikacin, minocycline, and colistin were determined by broth microdilution according to Clinical and Laboratory Standards Institute (CLSI) 2018 [46].

5.5. Construction of pCA24N (pemIK) and pCA24N (pemK)

The plasmids pCA24N (*pemIK*) and pCA24N (*pemK*) were constructed by amplifying the *pemI/pemK* module and the *pemK* gene in the expression vector pCA24N (cmR; LacIq) [45] inducible by IPTG (Fisher Scientific). The insertion of these genes was performed at the BseRI and NotI restriction sites under the control of T5-lac promotor (Table 4). Final constructions were verified by DNA sequencing.

Table 4. Oligonucleotide used for cloning and sequencing.

Primer Name	Sequences	Sense	Reference
	Oligonucleotide for clonation		
PemI_Fow(BseRI)	GAGGAGAAATTAACTATCATGCATACCACTCGACTG	5′-3′	This study
PemK_Fow(BseRI)	GAGGAGAAATTAACTATCATGGAAAGAGGGGAAATC	5′-3′	This study
PemK_Rev(NotI)	ATAAGAATGCGGCCGCCGCTCAGGTCAGGATGGTGGC	5′-3′	This study
	Oligonucleotide for sequencing		
pCA24N Up	GCCCTTTCGTCTTCAC	5′-3′	This study
pCA24N Down	GAACTCCATCTGGATTTGTT	5′-3′	This study

5.6. Toxicity Assay

The toxicity assay was performed as previously described Wood T.L. and Wood T.K. (2016) [47] with certain modifications. Overnight cultures of *K. pneumoniae* strains ST846-OXA48CA CP/pCA24N, ST846-OXA48CA CP/pCA24N (*pemIK*) and ST846-OXA48CA CP/pCA24N (*pemK*) were inoculated into 25 mL of LB broth medium with chloramphenicol (60 µg/mL; Sigma Aldrich, Darmstadt, Germany) to maintain the plasmid. IPTG (1 mM) was added when $OD_{600\ nm}$ was 0.3. For 200 min, the $OD_{600\ nm}$ was measured to determine growth evolution. All experiments were performed in triplicates.

5.7. Biofilm Formation Assay

The biofilm formation assay was performed in a 96-well polystyrene plate for 30 h. Briefly, cells of ST846-OXA48CA CP/pcA24N, ST846-OXA48CA CP/pCA24N (*pemIK*) and ST846-OXA48CA CP/pCA24N (*pemK*) were inoculated at $OD_{600\ nm}$ was 0.05 and incubated at 37 °C without shaking. After 6 h, IPTG at 1 mM was added and cells were incubated at 37 °C without shaking. 0.4 mM of IPTG was added every 8 h to avoid its degradation and to be able to observe the action of either the TA system or the toxin, as they are under an IPTG inducible promoter. Then, the cell density ($OD_{600\ nm}$)

and total biofilm ($OD_{580\ nm}$) were measured by using 10% crystal violet staining, and quantified in a NanoQuant plate reader. Normalized biofilm was calculated by dividing the total biofilm by the bacterial growth for each strain. All experiments were performed in triplicates.

5.8. Enzymatic Assay Using the Cell Proliferation Reagent WST-1

The cell proliferation/metabolic activity of the ST846-OXA48CA and ST846-OXA48CA CP strains in presence of the combination of IMP (50 × MIC) and CHLX (10 × MIC) was analyzed using a colorimetric enzymatic assay based on the water soluble tetrazolium salt (WST-1) reagent and electron mediators (Roche, Mannheim, Germany). Tetrazolium salts have become some of the most widely used tools in cell biology for measuring the metabolic activity of cells ranging from mammalian to microbial origin [48,49]. Briefly, the cultures of ST846-OXA48 and ST846-OXA48CA CP were incubated at 37 °C with shaking (180 rpm) until $OD_{600\ nm}$ was 0.6. At that moment the combination of IMP and CHLX, was added. After 24 and 48 h of antibiotic exposition and two washing, the culture cell ($OD_{600\ nm}$ = 0.1) was put in 96-well polystyrene plate (Corning Incorporated, NY, USA) and 10 μL of the reagent was added. After 1 h of incubation at 37 °C without shaking and 10 min with shaking (180 rpm), the optical density was measured at $OD_{480\ nm}$. The ST846-OXA48CA strain without antibiotic addition was taken as a control. The $OD_{480\ nm}$ of the medium culture (LN-LB) in the presence of WST-1 reagent was used to normalize all data. All experiments were performed in triplicates.

5.9. Statistical Analysis

Statistical analysis was based on the number of populations and comparisons. A Student's *t*-test was used to compare two populations. All statistical analysis were performed using the GraphPad (Prism 8) software.

Supplementary Materials: The following are available online at http://www.mdpi.com/2072-6651/12/9/566/s1, Figure S1: Schema representative of the most relevant genes of the *K. pneumoniae* plasmid P2 (CP048940.1). This plasmid has more than 90% of homology by BLAST nucleotide (https://blast.ncbi.nlm.nih.gov/Blast.cgi) with the plasmid harbouring the pemI/pemK TA systems in the strain ST846-OXA48CA. This figure was created with the software Geneious version 2020.2 created by Biomatters. Available from https://www.geneious.com.

Author Contributions: Conceptualization of the study was carried out by, I.B., L.B., M.D.-V., A.G.-d.-T., A.A., L.F.-G., M.L., Methodology, validation, formal analysis and writing, T.K.W., J.O.-I., A.P., G.B., F.F.-C., M.T., Supervision, analysis, writing and funding. All authors have read and agreed to the published version of the manuscript.

Funding: This study was funded by grants PI16/01163 and PI19/00878 awarded to M. Tomás within the State Plan for R+D+I 2013–2016 (National Plan for Scientific Research, Technological Development and Innovation 2008–2011) and co-financed by the ISCIII-Deputy General Directorate for Evaluation and Promotion of Research - European Regional Development Fund "A way of Making Europe" and Instituto de Salud Carlos III FEDER, Spanish Network for the Research in Infectious Diseases (REIPI, RD16/0016/0001, RD16/CIII/0004/0002 and RD16/0016/0006) and by the Study Group on Mechanisms of Action and Resistance to Antimicrobials, GEMARA (SEIMC, http://www.seimc.org/).

Acknowledgments: Spanish Network for the Research in Infectious Diseases (REIPI, RD16/0016/0001, RD16/CIII/0004/0002 and RD16/0016/0006) and by the Study Group on Mechanisms of Action and Resistance to Antimicrobials, GEMARA (SEIMC, http://www.seimc.org/).

Conflicts of Interest: The authors declare no conflicts of interest.

References

1. Mulani, M.S.; Kamble, E.E.; Kumkar, S.N.; Tawre, M.S.; Pardesi, K.R. Emerging Strategies to Combat ESKAPE Pathogens in the Era of Antimicrobial Resistance: A Review. *Front. Microbiol.* **2019**, *10*, 539. [CrossRef] [PubMed]
2. Yang, Q.E.; Walsh, T.R. Toxin-antitoxin systems and their role in disseminating and maintaining antimicrobial resistance. *FEMS Microbiol. Rev.* **2017**, *41*, 343–353. [CrossRef] [PubMed]
3. Santajit, S.; Indrawattana, N. Mechanisms of Antimicrobial Resistance in ESKAPE Pathogens. *BioMed Res. Int.* **2016**, *2016*, 2475067. [CrossRef]

4. Cadavid, E.; Robledo, S.M.; Quiñones, W.; Echeverri, F. Induction of Biofilm Formation in. *Antibiotics* **2018**, *7*, 103. [CrossRef]
5. Cieplik, F.; Jakubovics, N.S.; Buchalla, W.; Maisch, T.; Hellwig, E.; Al-Ahmad, A. Resistance Toward Chlorhexidine in Oral Bacteria—Is There Cause for Concern? *Front. Microbiol.* **2019**, *10*, 587. [CrossRef]
6. Hashemi, M.M.; Holden, B.S.; Coburn, J.; Taylor, M.F.; Weber, S.; Hilton, B.; Zaugg, A.L.; McEwan, C.; Carson, R.; Andersen, J.L.; et al. Proteomic Analysis of Resistance of Gram-Negative Bacteria to Chlorhexidine and Impacts on Susceptibility to Colistin, Antimicrobial Peptides, and Ceragenins. *Front. Microbiol.* **2019**, *10*, 210. [CrossRef]
7. Kampf, G. Acquired resistance to chlorhexidine—Is it time to establish an 'antiseptic stewardship' initiative? *J. Hosp. Infect.* **2016**, *94*, 213–227. [CrossRef]
8. Venter, H.; Henningsen, M.L.; Begg, S.L. Antimicrobial resistance in healthcare, agriculture and the environment: The biochemistry behind the headlines. *Essays Biochem.* **2017**, *61*, 1–10. [CrossRef]
9. Wand, M.E.; Bock, L.J.; Bonney, L.C.; Sutton, J.M. Mechanisms of Increased Resistance to Chlorhexidine and Cross-Resistance to Colistin following Exposure of Klebsiella pneumoniae Clinical Isolates to Chlorhexidine. *Antimicrob. Agents Chemother.* **2017**, *61*. [CrossRef]
10. Trastoy, R.; Manso, T.; Fernández-García, L.; Blasco, L.; Ambroa, A.; Pérez Del Molino, M.L.; Bou, G.; García-Contreras, R.; Wood, T.K.; Tomás, M. Mechanisms of Bacterial Tolerance and Persistence in the Gastrointestinal and Respiratory Environments. *Clin. Microbiol. Rev.* **2018**, *31*. [CrossRef]
11. Harms, A.; Maisonneuve, E.; Gerdes, K. Mechanisms of bacterial persistence during stress and antibiotic exposure. *Science* **2016**, *354*. [CrossRef] [PubMed]
12. Fernández-García, L.; Fernandez-Cuenca, F.; Blasco, L.; López-Rojas, R.; Ambroa, A.; Lopez, M.; Pascual, Á.; Bou, G.; Tomás, M. Relationship between Tolerance and Persistence Mechanisms in Acinetobacter baumannii Strains with AbkAB Toxin-Antitoxin System. *Antimicrob. Agents Chemother.* **2018**, *62*. [CrossRef] [PubMed]
13. Ren, D.; Bedzyk, L.A.; Thomas, S.M.; Ye, R.W.; Wood, T.K. Gene expression in Escherichia coli biofilms. *Appl. Microbiol. Biotechnol.* **2004**, *64*, 515–524. [CrossRef] [PubMed]
14. Kim, Y.; Wang, X.; Ma, Q.; Zhang, X.S.; Wood, T.K. Toxin-antitoxin systems in Escherichia coli influence biofilm formation through YjgK (TabA) and fimbriae. *J. Bacteriol.* **2009**, *191*, 1258–1267. [CrossRef]
15. Moyed, H.S.; Bertrand, K.P. hipA, a newly recognized gene of Escherichia coli K-12 that affects frequency of persistence after inhibition of murein synthesis. *J. Bacteriol.* **1983**, *155*, 768–775. [CrossRef]
16. Korch, S.B.; Henderson, T.A.; Hill, T.M. Characterization of the hipA7 allele of Escherichia coli and evidence that high persistence is governed by (p)ppGpp synthesis. *Mol. Microbiol.* **2003**, *50*, 1199–1213. [CrossRef]
17. Bokinsky, G.; Baidoo, E.E.; Akella, S.; Burd, H.; Weaver, D.; Alonso-Gutierrez, J.; García-Martín, H.; Lee, T.S.; Keasling, J.D. HipA-triggered growth arrest and β-lactam tolerance in Escherichia coli are mediated by RelA-dependent ppGpp synthesis. *J. Bacteriol.* **2013**, *195*, 3173–3182. [CrossRef]
18. Page, R.; Peti, W. Toxin-antitoxin systems in bacterial growth arrest and persistence. *Nat. Chem. Biol.* **2016**, *12*, 208–214. [CrossRef]
19. Wood, T.K.; Knabel, S.J.; Kwan, B.W. Bacterial persister cell formation and dormancy. *Appl. Environ. Microbiol.* **2013**, *79*, 7116–7121. [CrossRef]
20. Ogura, T.; Hiraga, S. Mini-F plasmid genes that couple host cell division to plasmid proliferation. *Proc. Natl. Acad. Sci. USA* **1983**, *80*, 4784–4788. [CrossRef]
21. Kamruzzaman, M.; Iredell, J. A ParDE-family toxin antitoxin system in major resistance plasmids of Enterobacteriaceae confers antibiotic and heat tolerance. *Sci. Rep.* **2019**, *9*, 9872. [CrossRef] [PubMed]
22. Wang, X.; Wood, T.K. Toxin-antitoxin systems influence biofilm and persister cell formation and the general stress response. *Appl. Environ. Microbiol.* **2011**, *77*, 5577–5583. [CrossRef] [PubMed]
23. Pecota, D.C.; Wood, T.K. Exclusion of T4 phage by the hok/sok killer locus from plasmid R1. *J. Bacteriol.* **1996**, *178*, 2044–2050. [CrossRef] [PubMed]
24. Hazan, R.; Engelberg-Kulka, H. Escherichia coli mazEF-mediated cell death as a defense mechanism that inhibits the spread of phage P1. *Mol. Genet. Genom.* **2004**, *272*, 227–234. [CrossRef]
25. Song, S.; Wood, T.K. Post-segregational Killing and Phage Inhibition Are Not Mediated by Cell Death Through Toxin/Antitoxin Systems. *Front. Microbiol.* **2018**, *9*, 814. [CrossRef]
26. Mathers, A.J.; Peirano, G.; Pitout, J.D. The role of epidemic resistance plasmids and international high-risk clones in the spread of multidrug-resistant Enterobacteriaceae. *Clin. Microbiol. Rev.* **2015**, *28*, 565–591. [CrossRef]

27. Domokos, J.; Damjanova, I.; Kristof, K.; Ligeti, B.; Kocsis, B.; Szabo, D. Multiple Benefits of Plasmid-Mediated Quinolone Resistance Determinants in. *Front. Microbiol.* **2019**, *10*, 157. [CrossRef]
28. Wang, X.; Lord, D.M.; Cheng, H.Y.; Osbourne, D.O.; Hong, S.H.; Sanchez-Torres, V.; Quiroga, C.; Zheng, K.; Herrmann, T.; Peti, W.; et al. A new type V toxin-antitoxin system where mRNA for toxin GhoT is cleaved by antitoxin GhoS. *Nat. Chem. Biol.* **2012**, *8*, 855–861. [CrossRef]
29. Fernandez-Garcia, L.; Kim, J.S.; Tomas, M.; Wood, T.K. Toxins of toxin/antitoxin systems are inactivated primarily through promoter mutations. *J. Appl. Microbiol.* **2019**, *127*, 1859–1868. [CrossRef]
30. Präbst, K.; Engelhardt, H.; Ringgeler, S.; Hübner, H. Basic Colorimetric Proliferation Assays: MTT, WST, and Resazurin. *Methods Mol. Biol.* **2017**, *1601*, 1–17. [CrossRef]
31. Gilbert, P.; Moore, L.E. Cationic antiseptics: Diversity of action under a common epithet. *J. Appl. Microbiol.* **2005**, *99*, 703–715. [CrossRef] [PubMed]
32. Munita, J.M.; Arias, C.A. Mechanisms of Antibiotic Resistance. *Microbiol. Spectr.* **2016**, *4*. [CrossRef] [PubMed]
33. Brauner, A.; Fridman, O.; Gefen, O.; Balaban, N.Q. Distinguishing between resistance, tolerance and persistence to antibiotic treatment. *Nat. Rev. Microbiol.* **2016**, *14*, 320–330. [CrossRef] [PubMed]
34. Keren, I.; Shah, D.; Spoering, A.; Kaldalu, N.; Lewis, K. Specialized persister cells and the mechanism of multidrug tolerance in Escherichia coli. *J. Bacteriol.* **2004**, *186*, 8172–8180. [CrossRef] [PubMed]
35. Tashiro, Y.; Kawata, K.; Taniuchi, A.; Kakinuma, K.; May, T.; Okabe, S. RelE-mediated dormancy is enhanced at high cell density in Escherichia coli. *J. Bacteriol.* **2012**, *194*, 1169–1176. [CrossRef]
36. Lewis, K. Persister cells. *Annu. Rev. Microbiol.* **2010**, *64*, 357–372. [CrossRef]
37. Agarwal, S.; Bhatnagar, R. Identification and characterization of a novel toxin-antitoxin module from Bacillus anthracis. *FEBS Lett.* **2007**, *581*, 1727–1734. [CrossRef]
38. Kędzierska, B.; Hayes, F. Emerging Roles of Toxin-Antitoxin Modules in Bacterial Pathogenesis. *Molecules* **2016**, *21*, 790. [CrossRef]
39. Martínez, L.C.; Vadyvaloo, V. Mechanisms of post-transcriptional gene regulation in bacterial biofilms. *Front. Cell. Infect. Microbiol.* **2014**, *4*, 38. [CrossRef]
40. González Barrios, A.F.; Zuo, R.; Hashimoto, Y.; Yang, L.; Bentley, W.E.; Wood, T.K. Autoinducer 2 controls biofilm formation in Escherichia coli through a novel motility quorum-sensing regulator (MqsR, B3022). *J. Bacteriol.* **2006**, *188*, 305–316. [CrossRef]
41. García-Contreras, R.; Zhang, X.S.; Kim, Y.; Wood, T.K. Protein translation and cell death: The role of rare tRNAs in biofilm formation and in activating dormant phage killer genes. *PLoS ONE* **2008**, *3*, e2394. [CrossRef] [PubMed]
42. Ma, D.; Mandell, J.B.; Donegan, N.P.; Cheung, A.L.; Ma, W.; Rothenberger, S.; Shanks, R.M.Q.; Richardson, A.R.; Urish, K.L. The Toxin-Antitoxin MazEF Drives Staphylococcus aureus Biofilm Formation, Antibiotic Tolerance, and Chronic Infection. *mBio* **2019**, *10*. [CrossRef] [PubMed]
43. Esteban-Cantos, A.; Aracil, B.; Bautista, V.; Ortega, A.; Lara, N.; Saez, D.; Fernández-Romero, S.; Pérez-Vázquez, M.; Navarro, F.; Grundmann, H.; et al. The Carbapenemase-Producing Klebsiella pneumoniae Population Is Distinct and More Clonal than the Carbapenem-Susceptible Population. *Antimicrob. Agents Chemother.* **2017**, *61*. [CrossRef]
44. El-Mansi, M.; Anderson, K.J.; Inche, C.A.; Knowles, L.K.; Platt, D.J. Isolation and curing of the *Klebsiella pneumonia* large indigenous plasmid using sodium dodecyl sulphate. *Res. Microbiol.* **2000**, *151*, 201–208. [CrossRef]
45. Kitagawa, M.; Ara, T.; Arifuzzaman, M.; Ioka-Nakamichi, T.; Inamoto, E.; Toyonaga, H.; Mori, H. Complete set of ORF clones of Escherichia coli ASKA library (a complete set of *E. coli* K-12 ORF archive): Unique resources for biological research. *DNA Res.* **2005**, *12*, 291–299. [CrossRef] [PubMed]
46. Clinical and Laboratory Standard Institute. *Performance Standards for Antimicrobial Sucesceptibilty Testing*, 28th ed.; CLSI suplement M100; Clinical and Laboratory Standard Institute: Wayne, PA, USA, 2018.
47. Wood, T.L.; Wood, T.K. The HigB/HigA toxin/antitoxin system of Pseudomonas aeruginosa influences the virulence factors pyochelin, pyocyanin, and biofilm formation. *Microbiologyopen* **2016**, *5*, 499–511. [CrossRef]

48. Berridge, M.V.; Herst, P.M.; Tan, A.S. Tetrazolium dyes as tools in cell biology: New insights into their cellular reduction. *Biotechnol. Annu. Rev.* **2005**, *11*, 127–152. [CrossRef]
49. Tsukatani, T.; Suenaga, H.; Higuchi, T.; Akao, T.; Ishiyama, M.; Ezoe, K.; Matsumoto, K. Colorimetric cell proliferation assay for microorganisms in microtiter plate using water-soluble tetrazolium salts. *J. Microbiol. Methods* **2008**, *75*, 109–116. [CrossRef]

 *toxins* 

Article

The Toxin-Antitoxin Systems of the Opportunistic Pathogen *Stenotrophomonas maltophilia* of Environmental and Clinical Origin

Laurita Klimkaitė, Julija Armalytė *, Jūratė Skerniškytė and Edita Sužiedėlienė *

Institute of Biosciences, Life Sciences Center, Vilnius University, LT-1025 Vilnius, Lithuania; laurita.klimkaite@gmail.com (L.K.); jurate.skerniskyte@gmc.vu.lt (J.S.)
* Correspondence: julija.armalyte@gf.vu.lt (J.A.); edita.suziedeliene@cr.vu.lt (E.S.)

Received: 31 July 2020; Accepted: 25 September 2020; Published: 1 October 2020

Abstract: *Stenotrophomonas maltophilia* is a ubiquitous environmental bacterium that has recently emerged as a multidrug-resistant opportunistic pathogen causing bloodstream, respiratory, and urinary tract infections. The connection between the commensal environmental *S. maltophilia* and the opportunistic pathogen strains is still under investigation. Bacterial toxin–antitoxin (TA) systems have been previously associated with pathogenic traits, such as biofilm formation and resistance to antibiotics, which are important in clinical settings. The same species of the bacterium can possess various sets of TAs, possibly influencing their overall stress response. While the TA systems of other important opportunistic pathogens have been researched, nothing is known about the TA systems of *S. maltophilia*. Here, we report the identification and characterization of *S. maltophilia* type II TA systems and their prevalence in the isolates of clinical and environmental origins. We found 49 putative TA systems by bioinformatic analysis in *S. maltophilia* genomes. Despite their even spread in sequenced *S. maltophilia* genomes, we observed that *relBE*, *hicAB*, and previously undescribed COG3832-ArsR operons were present solely in clinical *S. maltophilia* isolates collected in Lithuania, while *hipBA* was more frequent in the environmental ones. The kill-rescue experiments in *Escherichia coli* proved *higBA*, *hicAB*, and *relBE* systems to be functional TA modules. Together with different TA profiles, the clinical *S. maltophilia* isolates exhibited stronger biofilm formation, increased antibiotic, and serum resistance compared to environmental isolates. Such tendencies suggest that certain TA systems could be used as indicators of virulence traits.

Keywords: *Stenotrophomonas maltophilia*; opportunistic pathogen; clinical origin; environmental origin; toxin-antitoxin system; biofilm; antibiotic resistance

Key Contribution: Previously undescribed *Stenotrophomonas maltophilia* type II TA systems were identified and their abundance in clinical and environmental *S. maltophilia* isolates evaluated. A difference in TA content and virulence-related characteristics between *S. maltophilia* isolates of clinical and environmental origins was observed.

---

1. Introduction

Stenotrophomonas maltophilia is a bacterium of class *Gammaproteobacteria*, ubiquitously found in the natural environment (soil, plants, animals, foods, or aqueous habitats) [1]. Recently, *S. maltophilia* came into attention as a globally emerging multidrug-resistant opportunistic pathogen causing nosocomial and community-acquired infections to immunocompromised patients [2]. The mortality rate of *S. maltophilia* infection can exceed 35% [3], with the World Health Organization (WHO) listing *S. maltophilia* as one of the leading multidrug-resistant pathogens [4]. The most known *S. maltophilia* features contributing to its successful adaptation in clinical environments include intrinsic antibiotic resistance mechanisms and its ability to form biofilms [5], although other less characterized factors (synthesis of extracellular enzymes, bacterial motility, and quorum sensing) are also known to influence its pathogenesis [6]. *S. maltophilia* is also known as a highly diverse species, since numerous studies have revealed genotypic diversity between clinical *S. maltophilia* isolates [7–9], even when analyzing samples from the same hospital [10,11]. Moreover, no significant genotypic or phenotypic differences between clinical and environment *S. maltophilia* isolates could be identified [12]. There is still an open question whether all the members of this species can become virulent and is there a specific marker to distinguish clinical and environmental *S. maltophilia* isolates.

The ability of pathogens to adapt to the clinical environment and to develop resistance to treatment can often be attributed to their ability to quickly exchange genetic information [13]. Gaining antibiotic resistance genes or virulence factors increases the chances of survival and gives rise to nosocomial opportunistic pathogens [14]. The ability of the bacterial species to mobilize their DNA could, therefore, be considered as a virulence factor in itself.

Toxin–antitoxin (TA) systems are abundant genetic elements mostly found in prokaryotes to encode a toxic protein, which inhibits cell growth and an antitoxin molecule (protein or RNA), which protects cells from toxin effects [15]. Human pathogenic bacteria have been shown to possess multiple sets of TAs, the type II TA systems, consisting of toxin and antitoxin proteins, being the most represented module and encoded by both plasmids and chromosomes [16]. Once the bacteria acquires DNA encoding the TA system, it becomes difficult to lose it due to the toxin's higher stability and eventual suppression of the cell growth in case of loss of the genes [17]. Apart from being addictive, other functions have been proposed for the TA systems, such as a role in virulent phenotype and alleviating bacterial survival under stress conditions [18]. Moreover, the connection between TA systems and virulence traits, such as biofilm formation [19,20] and host colonization [18], was demonstrated. Interestingly, the same species can have various sets of TA systems [21], indicating their ability to be transferred horizontally. Analysis of the prevalence of TA systems in the genomes of 12 dangerous epidemic bacteria has shown that pathogens encoded a higher number of TA systems and reduced genomes compared to their closest non-pathogenic relatives [22]. This suggests that TAs might participate in the manifestation of the evolved pathogenic phenotype. The pathogenic *Mycobacterium tuberculosis* represents the prominent example, which genome encodes a remarkably high number of TA systems compared with nonpathogenic *Mycobacterium* strains [23]. Altogether, a possible TA systems role in bacteria pathogenicity is possible.

In this study, we aimed to identify and characterize previously undescribed *S. maltophilia* type II TA systems, as well as to evaluate and compare the abundance of these TA modules in clinical and environmental *S. maltophilia* isolates. Additionally, we accessed virulence-related features of *S. maltophilia*, including antibiotic and serum resistance, biofilm, pellicle, and capsule formation in order to find out if the correlation between the presence of TA systems and virulence-associated traits is apparent in *Stenotrophomonas*. We observed a difference in TA content and virulence-related characteristics between *S. maltophilia* isolates of clinical and environmental origins.

2. Results

2.1. Detection of *S. maltophilia* TA Systems in Genomes

To date, there are several tools available to detect TA systems in bacterial genomes: TADB 2.0 database tool TA-finder [24], RASTA-Bacteria [25], and TASmania [26]. However, only TA-finder was useful in our case, since RASTA-Bacteria limited the size of the analyzed DNA fragments, while TASmania relied on the Ensembl database, which at the time of this search did not have *S. maltophilia* genomes. On the date of our analysis, 21 fully assembled genomes were available to analyze with TA-finder. If the TA pair contained the same predicted domains, they were considered homologous; if they matched with a protein of a different domain, they were considered a distinct TA pair (e.g., RelE-Xre and RelE-RHH were considered as two distinct TAs). The presence of each predicted TA system was checked against all the genomes using BLAST. The final results for 49 detected TA systems are presented in Figure 1. Interestingly, some of the TA system genes were not identified in the genomes by TA-finder, but their presence was later detected by BLAST. After additional analysis, we found that the gene sequences in those genomes contained mutations that disrupted the reading frame, which could be due to sequencing errors or genuine mutations. The possible pseudogenes belonging to TAs are indicated in Figure 1.

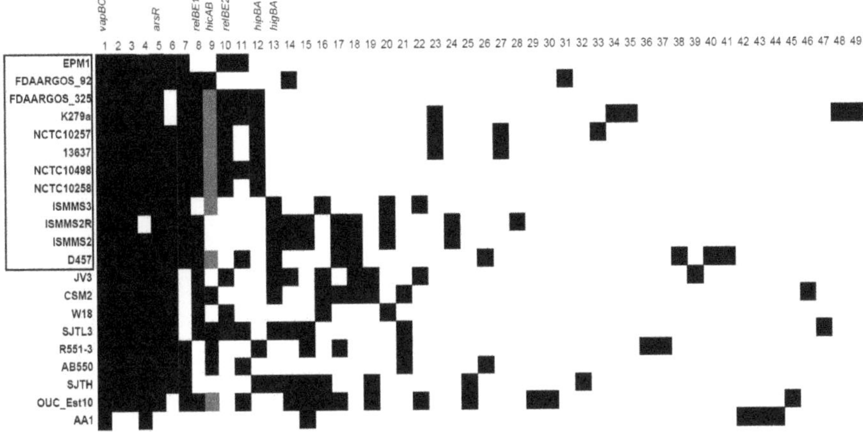

Figure 1. The predicted toxin–antitoxin (TA) systems in *S. maltophilia* genomes. The genomes with names bracketed in black rectangle were isolated from clinical sources. Black squares indicate the presence of a TA system, whereas grey squares indicate the presence of a pseudogene version. The information about the presented TA systems (1–49) is described in Table S1. TA systems analyzed in this work are indicated above.

As can be seen from Figure 1, there were no clear TA distribution patterns characteristic to annotated genomes of *S. maltophilia* from clinical or environmental origins. The average number of predicted TA systems per genome did not significantly differ between the *S. maltophilia* of both origins (Figure 2a).

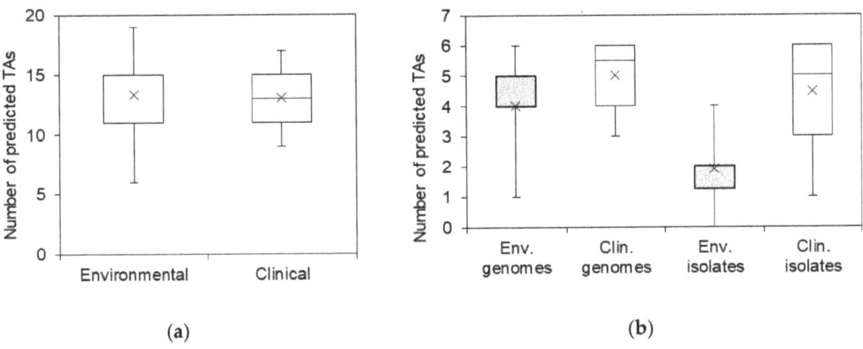

Figure 2. The prevalence of predicted TA systems in sequenced *S. maltophilia* genomes and *Stenotrophomonas* spp. isolates. (**a**) The distribution of total predicted TA systems in annotated *S. maltophilia* genomes; (**b**) the distribution of selected most prevalent TA systems in annotated genomes of *S. maltophilia* and in isolates of clinical and environmental origin. Boxes indicate upper and lower quartiles, whiskers indicate minimum and maximum values, and crosses indicate mean values.

Not all gene pairs predicted as TA systems are active TA modules [23,27,28]. Therefore, we aimed to check if the prevailing predicted TA systems of *S. maltophilia* were functional. For functional analysis, the TA systems that were present in at least 1/3 of the analyzed genomes were chosen. To increase the likelihood of the selected operons acting as TA systems, we chose TAs where both genes were no longer than 1 kb [25] (with the exception of *hipBA* system, where the toxin gene is known to be larger) [29]. The distance between the toxin and antitoxin genes was set to be 50 bp or less, which increased the possibility that the genes belonged to an operon and were expressed from a single transcript. The genes, which were clearly annotated for a function and not related to TAs (e.g., antibiotic resistance genes, membrane transporters), were excluded from further analysis. The final set of seven TAs, selected for functional characterization, is presented in Table 1.

Table 1. The characteristics of TA systems prevalent in annotated genomes of *S. maltophilia*.

Predicted TA Family	Protein Domains (T-A) [1]	Found in Genomes (n = 21)	Conservativity of Operon (%) [2]	Prevalence in Other Bacteria [3]	Remarks
vapBC	PIN-AbrB	21	84	*Stenotrophomonas* spp., *Xantomonas* spp.	
-	COG3832- ArsR	20	90	*Stenotrophomonas* spp. and other genera	
relBE	RelE-Xre	15	73	*Stenotrophomonas* spp.	T before A
relBE	RelE-RHH	10	92	*Stenotrophomonas* spp.	
relBE (higBA)	HigB-Xre	8	78	*Stenotrophomonas* spp., *Pseudomonas* spp.	T before A
hicAB	HicA-HicB	14	79	*Stenotrophomonas* spp.	T before A
hipBA	HipA-Xre	8	70	*Stenotrophomonas* spp. (T common to *Burkholderia* spp.)	T is 440 a.a.

[1] Predicted domains (domain-like); [2] in *Stenotrophomonas* spp.; [3] the prevalence of selected TA systems was identified by BLAST analysis against domain *Bacteria*; T: toxin, A: antitoxin.

The majority of the selected systems belonged to well-known type II TA families. There were three TA systems belonging to the *relBE* family: *relBE1* (containing RelE-Xre domains), *relBE2* (RelE-RHH domains), and *higBA* (HigB-Xre domains). However, one of the most common predicted TAs in *S. maltophilia*, containing COG3832-ArsR domains, was the least described. It was predicted as a TA system bioinformatically more than 10 years ago [30], yet only once was it analyzed thoroughly [23].

The predicted TA systems were conservative not only in *S. maltophilia* but in *Stenotrophomonas* spp. with high sequence conservativity (>70%) (Table 1). Therefore, we decided to test the presence and spread of these TAs in both *S. maltophilia* and *Stenotrophomonas* spp. isolates from our collection.

2.2. The Prevalence of TA Systems in Stenotrophomonas spp. of Clinical and Environmental Origin

To detect selected TA systems, a set of primers was designed after aligning all available sequences for a TA system in a *Stenotrophomonas* genus and identifying the most conserved regions (Table S2). Next, the collection of *Stenotrophomonas* spp. of clinical (n = 21) and environmental (n = 14) origins (Figure 3) was screened for the presence of TAs. All isolates of clinical origin were attributed to *S. maltophilia* species, while six of the environmental isolates belonged to *S. maltophilia*. The rest of the environmental isolates were attributed to either *S. rhizophila* or *Stenotrophomonas* sp.

All of the examined TA systems were detected in at least four isolates (Figure 3). *vapC* and *hipBA* modules were the most prevalent (31 and 21 isolates, respectively), and were evenly spread in both clinical and environmental bacteria. However, *hipBA* was more common in isolates of environmental origin. Three TAs (*relBE1*, *relBE2*, and COG3832-ArsR) were prevalent in isolates of clinical origin yet were absent in environmental isolates. *hicAB* system was detected in the majority of clinical isolates and in only two isolates of environmental origin. Surprisingly, we observed a clear difference in TA profiles between *Stenotrophomonas* isolates of clinical and environmental origins, contradicting a similar distribution of these TAs in the sequenced genomes. We found that certain TAs were more likely to be present in the clinical isolates of *S. maltophilia*. Moreover, clinical isolates of *S. maltophilia* had more TA systems per isolate, compared to the environmental isolates (Figure 2b).

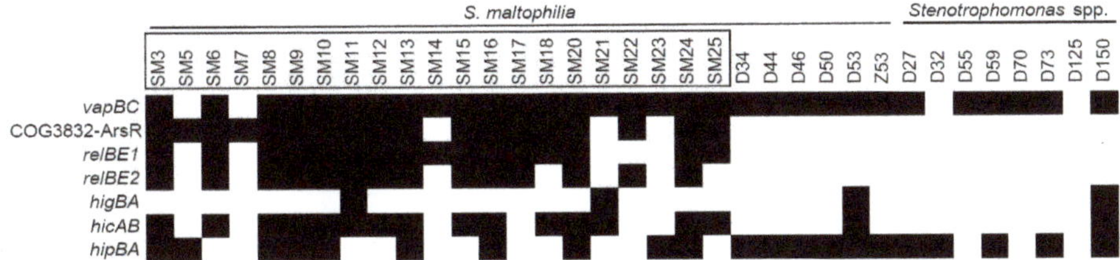

Figure 3. The prevalence of predicted TA systems in *Stenotrophomonas* spp. of clinical and environmental origins. The isolates with their names bracketed in black rectangles were derived from clinical sources. The presence of TAs, indicated as black boxes, was detected by PCR as described in the Materials and Methods section.

2.3. Functionality of *S. maltophilia* TA Systems

To test if the predicted TA systems were functional, we cloned toxin and antitoxin genes into separate inducible vectors to perform kill-rescue experiments in *E. coli*, as described in Materials and methods. The predicted toxin and antitoxin expression was induced with arabinose, and IPTG, respectively.

Only *S. maltophilia relBE1*, *relBE2* and *higBA* TA systems, belonging to *relBE* family, were functional in kill-rescue experiments in *E. coli*, as evident from the growth inhibition upon induction of the toxin and growth restoration, when antitoxin was concomitantly induced (Figure 4a–c). The predicted toxins of *hipBA*, *vapBC*, *hicAB*, and COG3832-ArsR TA systems did not display any toxicity in the kill-rescue assay (Figure 4d–f).

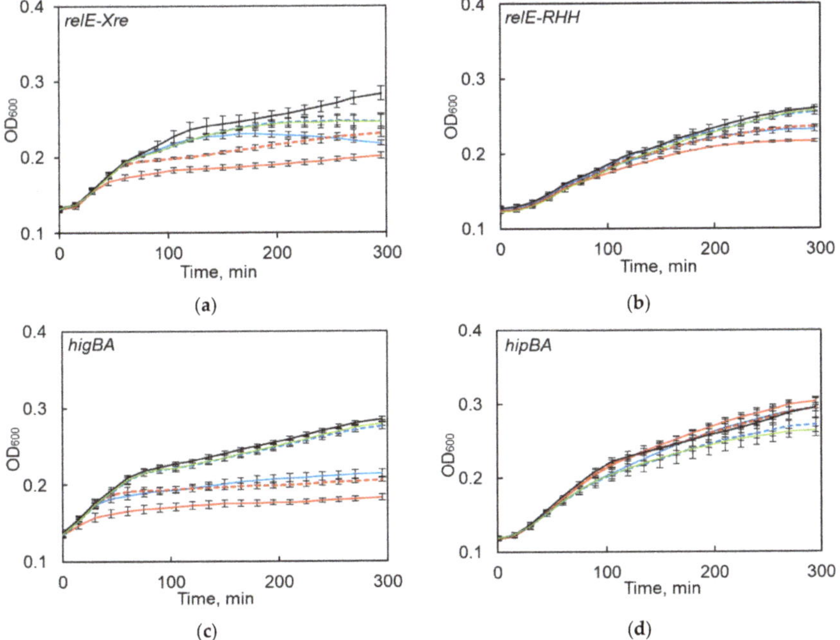

Figure 4. *Cont.*

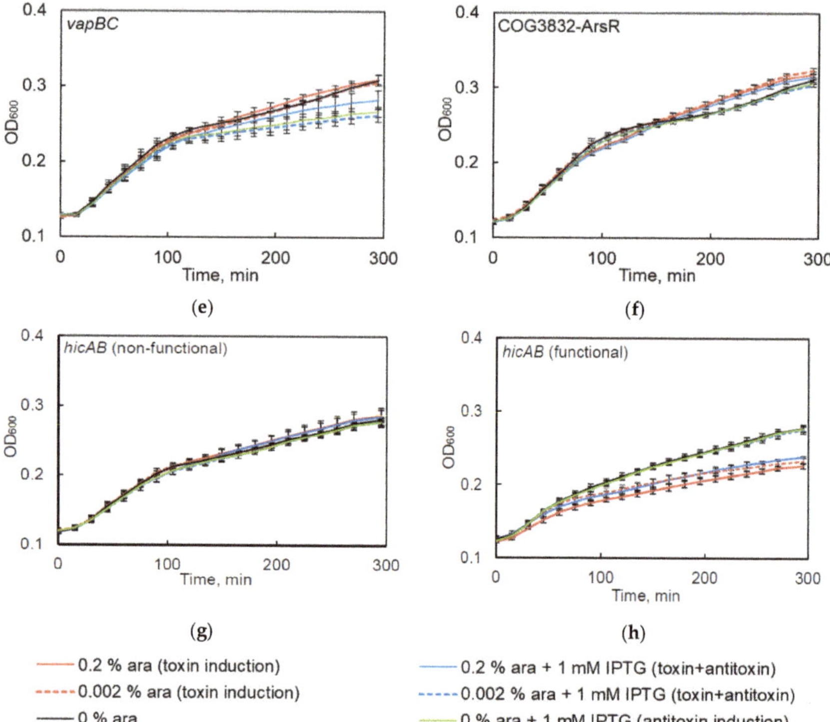

Figure 4. The functionality of predicted *S. maltophilia* TA systems as tested by the kill-rescue assay in *E. coli*. (**a–c**) the growth inhibition and restoration of functional TA systems; (**d–f**) the kill-rescue assays for the predicted TA that appear to be non-functional in *E. coli* system; (**g**,**h**) representative growth curves for a functional and a non-functional version of *hicAB*. The toxin and antitoxin genes were cloned into separate inducible plasmids, the toxin was induced by adding arabinose (ara), and its activity counteracted by inducing antitoxins with IPTG. The bacterial growth was measured as OD_{600}, as described in the Materials and Methods section. The experiments were repeated at least three times; error bars indicate standard deviation.

We previously observed that one of the tested TAs (i.e., *hicAB*) was often present in the sequenced *S. maltophilia* genomes as a pseudogene (Figure 1). Therefore, when we first observed *hicAB* to be non-functional in the kill-rescue assay (Figure 4g), we decided to clone it from several other *S. maltophilia* isolates and to use the kill-rescue assay. Out of eight cloned versions of *hicAB*, three were non-functional, and five systems displayed a distinct kill-rescue phenotype. The growth of *E. coli* after the induction of a representative functional *S. maltophilia hicAB* is presented in Figure 4h. Sequencing of the cloned non-functional versions of *hicAB* revealed the same frameshift mutation in the toxin gene.

According to our results, *S. maltophilia* COG3832-ArsR, *hipBA*, and *vapBC* predicted TA systems were non-functional in *E. coli*. To find out, if this was due to the differences of the host or TAs, which were indeed non-functional, we created an expression system for *S. maltophilia* by using a broad-host range plasmid pBAD1 with an arabinose controlled promoter and cloned the predicted toxins. *S. maltophilia relE* toxin of the *relE-Xre* TA system was proven to be functional in *E. coli* and was thus used as a control. Further, *relE* toxin showed toxicity upon induction (not shown), while COG3832-ArsR, *hipBA*, and *vapC* toxins did not show any toxicity in *S. maltophilia* under the same induction conditions (not shown).

2.4. Characterization of *Stenotrophomonas* spp. Isolates

Despite the similar distribution of predicted TAs observed in the *S. maltophilia* sequenced genomes of clinical and environmental origins (Figure 1), the most prevalent TA modules in bacterial isolates showed characteristic profiles regarding their origin, with clear prevalence for some TAs in clinical isolates (Figure 3). Since the majority of the clinical *S. maltophilia* isolates were obtained in the same hospital, we wanted to find out if there was a clonality between the analyzed isolates. The genotyping of *Stenotrophomonas* spp. isolates was undertaken by a random amplified polymorphic DNA (RAPD) assay with two sets of random primers as described in the Materials and Methods section. Each isolate significantly differed and was not grouped into clusters according to their origin (Figure 5). This thereby confirmed that the *S. maltophilia* of clinical origin were sporadic cases.

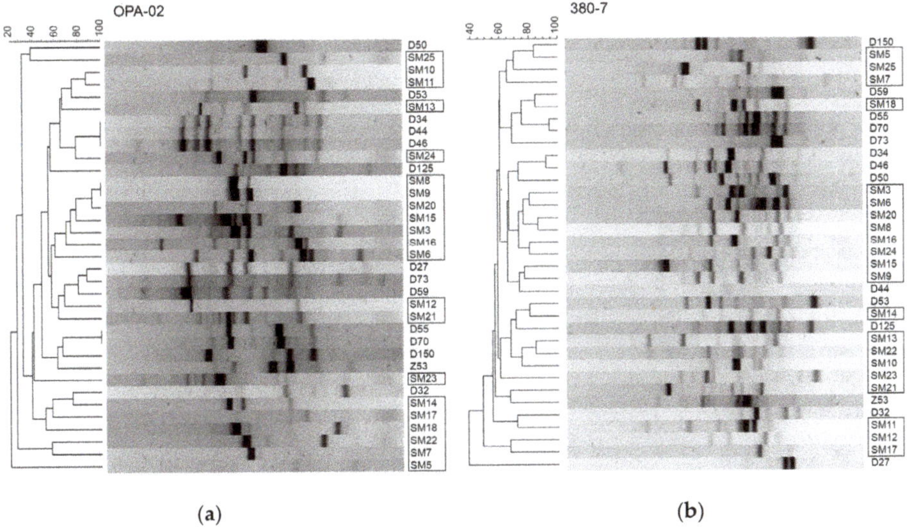

Figure 5. The genotyping of *Stenotrophomonas* spp. isolates. Random amplified polymorphic DNA (RAPD) assay was performed as described in the Materials and Methods section using OPA-02 (**a**) and 380-7 (**b**) primers (Table S2). The isolates bracketed in black rectangles were isolated from clinical sources.

The genotyping of *Stenotrophomonas* spp. from clinical and environmental origins showed that isolates could not be grouped into clusters based on their origin of isolation. However, uneven distribution of TAs (i.e., the types of TAs and total TA number per isolate) among the isolates indicated that there were certain traits that distinguished the isolates of different origins. Therefore, we decided to compare other virulence-related traits of the isolates and see if the presence of TAs could be considered as a marker for virulence of the strain.

One of the characteristics that makes *S. maltophilia* a dangerous opportunistic pathogen is its innate resistance to antibiotics [31]. We tested the resistance of the isolates against a set of antibiotics (trimethoprime, chloramfenicol, gentamicine, ceftazidime, meropenem, and tetracyclin), including trimethoprime-sulfametoxazole and ciprofloxacine, which are commonly used against *S. maltophilia* infections. Resistance and sensitivity values were determined according EUCAST clinical breakpoints for *S. maltophilia* or *Pseudomonas aeruginosa*, if such data for *S. maltophilia* was not available. We observed significantly higher resistance profiles among the isolates of the clinical origin (Figure 6). Tetracyclin, ceftazidime, meropenem, and gentamicin were effective against most environmental isolates, while the majority of the clinical ones showed resistance.

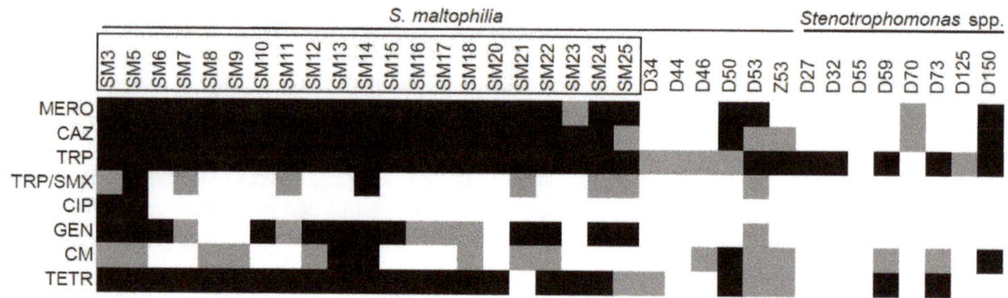

Figure 6. The resistance to antibiotics of *Stenotrophomonas* spp. of clinical and environmental origin. The isolates with names bracketed in black rectangles were isolated from clinical sources. Black squares indicate resistance; grey squares indicate intermediate resistance to antibiotic. The antibiotics are as indicated: meropenem (MERO), ceftazidime (CAZ), trimethoprime (TRP), trimethoprime-sulfametoxazole (TRP/SMX), ciprofloxacine (CIP), gentamicine (GEN), chloramphenicol (CM), and tetracyclin (TETR).

The next virulence-related trait we assessed was biofilm formation, a complex community structure composed of adhesive bacteria and matrix components. Various bacteria can form biofilms on medical devices, such as catheters or intubation tubes, since it increases pathogen survival during stress (e.g., antibiotic pressure, desiccation, or laminar flow) [32]. Previous studies have shown the ability of *S. maltophilia* to form biofilms [33]. We first checked the ability of the isolates to form biofilms at 37 °C, as described in the Materials and Methods section, since this was an expected temperature inside the infected host. The majority of *S. maltophilia* from clinical origins formed strong biofilms (Figure 7, Figure S1). Only two environmental isolates displayed weak biofilm growth at 37 °C (Figure 7). However, the environmental isolates showed severely reduced growth at 37 °C, indicating that their inability to form biofilms could be due to suboptimal growth conditions. Therefore, we tested the ability of all the isolates to form biofilms at 28 °C, a temperature that resembled environmental (outside the host) conditions. No change in biofilm formation was observed for the clinical strains; however, several of the environmental strains increased their biofilm formation under permissive temperature. Thus, the results demonstrated a clear differentiation of the strains of different in ability to form biofilm at both 37 °C and 28 °C.

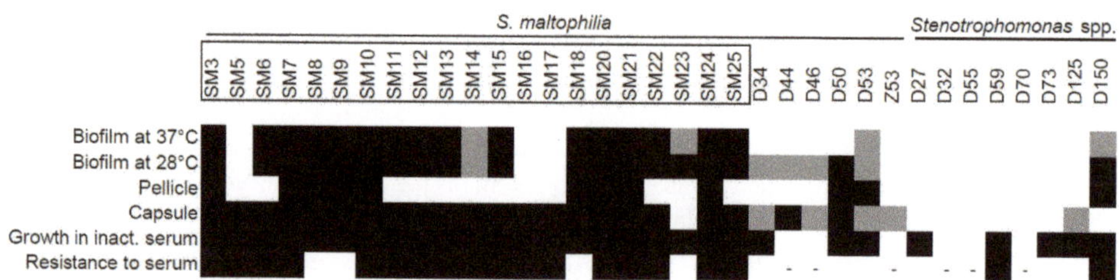

Figure 7. Biofilm formation and capsule-related characteristics of *Stenotrophomonas* spp. isolates. The biofilm, pellicle, capsule-forming phenotype, growth in inactivated serum and resistance to serum were determined as described in the Material and Methods section. The isolates with names bracketed in black rectangles were isolated from clinical sources. Black squares indicate the ability to form biofilm, pellicle, the presence of capsular polysaccharides (CPS), or the ability to grow in serum; grey squares indicate the ability to form weak biofilms or weak capsule-forming phenotype. "-" indicates that the isolates were not exposed to non-inactivated serum due to their inability to survive in inactivated serum.

We then analyzed the ability of the isolates to form pellicle, a type of biofilm, which formed at the air–liquid interface. This specific structure is associated with the increased virulence of bacteria [34]. As can be seen in Figure 7, only the isolates with biofilm forming capacity were able to form pellicle, though this structure was not common to all such isolates and was more frequent in the clinical strains.

The ability to form capsule is considered as an important virulence trait in bacteria, together with resistance to antibiotics, biofilm formation, adhesiveness, desiccation, and other features [35–37]. The role of capsular exopolysaccharides in *S. maltophilia* virulence is still unknown. Therefore, we were interested to discover if bacterial isolates could produce polysaccharides that were indicative of capsule formation. The capsular polysaccharides were extracted and analyzed as described in the Materials and Methods section. The vast majority of clinical *S. maltophilia* isolates (95%) produced high molecular mass polysaccharides, which were associated with capsular polysaccharides (CPS). This phenotype was demonstrated by only a third (36%) of environmental *Stenotrophomonas* spp. isolates (Figure 7, Figure S2). However, *S. maltophilia* environmental isolates tended to produce capsule-associated polysaccharides more often than *Stenotrophomonas* of other species.

The ability to produce exopolysaccharide capsule is indicative of a strain's ability to avoid killing by a host's immune system [36,38]. We aimed to test if isolates differed in their ability to survive serum-mediated killing. First, we tested the ability of isolates to grow in a heat-treated serum with an inactivated complement system. All of the clinical isolates were able to grow, yet only half of the environmental isolates grew in the presence of inactivated serum (Figure 7). Next, we assessed the growth of *Stenotrophomonas* isolates in the presence of the active serum. The majority of clinical *S. maltophilia* strains were resistant to complement components when grown in an active serum, whereas only two of the environmental isolates showed resistance, thereby demonstrating an inability of environmental strains to resist a host immune system and cause an infection.

3. Discussion

In this study, we aimed to investigate the prevalence and profiles of TA systems in a newly emerging opportunistic pathogen, *S. maltophilia,* as well as to compare the TA presence with the manifestation of the virulence features of clinical and environmental isolates.

While the role of plasmid-borne TA systems is known as plasmid maintenance [39], the biological function of chromosome encoded TA systems is far less clear. It has been shown that chromosome encoded TA systems help stabilize genome integrity by preventing large scale genome reductions, which keeps genes needed for survival in unfavorable conditions [40,41] and causes growth arrest that helps bacteria survive environmental stress [39]. The most widespread type II TA system families affect the translation apparatus of a cell by cleaving or otherwise disrupting the functionality of mRNA (*relBE* (including *higBA*) [42,43], *hicAB* [44], and *mazEF* [45]) or tRNA (*vapBC* [46] and *hipBA* [29]). It has also been reported that TA systems function as virulence factors. Thus, a *higBA* TA system in *Pseudomonas aeruginosa* influences different virulence factor expression and biofilm formation [20]. *Mycobacterium tuberculosis MazEF* type TA systems play a role in pathogen intracellular survival inside the host [47]. A *Staphylococcus aureus SavRS* TA system functions as a regulator of virulence traits [48]. Moreover, previous studies have shown that the number of TA systems is enlarged in pathogenic strains compared with closely related nonpathogenic bacteria (*M. tuberculosis* encodes 88 putative TA systems [49], while non-harmful *M. smegmatis* encodes only three TA modules [50]). In this study, we investigated whether the same trend was characteristic to TA systems in the opportunistic pathogen *S. maltophilia*. The data observed on *vapBC* and *relBE* families, which are two major type II TA families [51,52], were in line with the observations on the prevalence of TA systems.

The amplicon detection-based screen of the TA systems in bacterial isolates revealed unexpected results, since *relBE1*, *relBE2,* and COG3832-ArsR TA systems were found solely in clinical *S. maltophilia* isolates and were not detected in environmental samples. Moreover, some preferences were observed for *hicAB* and *hipBA* TAs. The dissimilarity between observations of bioinformatic TA analysis in genomes available in the databases and our detection might be explained by the different geographic locations of bacterial sources. The bacteria analyzed in our study were isolated from soil in Lithuania, while the origin of *S. maltophilia* genomes in the databases varied greatly. However, the genotyping assay proved that the analyzed isolates did not belong to the same clonal group and showed high genetic diversity. Similar results were previously reported for *S. maltophilia* [53,54], supporting our findings about genotypic diversity among clinical and environmental *S. maltophilia* isolates from Lithuania.

In the present study, the putative COG3832-ArsR TA system was found in nearly all analyzed *S. maltophilia* genomes. COG3832-ArsR operon was previously proposed to be a TA system on the basis of comparative prokaryotic genomic analysis [30]. This module was found among 10 of the most induced TA systems in drug tolerant persister cells for *M. tuberculosis* [23], though functional analysis of this operon was not done in any TA system research. However, *S. maltophilia* COG3832-ArsR did not show toxicity in the kill-rescue assay and therefore its attribution to TA systems remains questionable. Altogether, functional analysis revealed that four of the seven analyzed *S. maltophilia* TA systems were functionally active TA operons in *E. coli*. The confirmed TAs belonged to previously described classes. To exclude the possibility of host incompatibility, we transferred the inactive toxins into *S. maltophilia*, yet no toxicity was observed. Interestingly, we identified two types of the same *S. maltophilia hicAB* TA systems via bioinformatic analysis in bacterial isolates. Previous studies have reported that TA system genes may contain mutations, which can result in nonfunctional pseudogene formation [55]. Loss of function mutations in virulence-associated genes can indicate genomic plasticity and adaptation to different environments [56], suggesting that *hicAB* TA systems may be favorable or, otherwise, unfavorable when living under particular conditions.

In our study, only the chromosome sequences of *S. maltophilia* were used for TA system prediction, however, the detected TA systems might be located on both chromosomes and mobile elements. TA systems are known to be transmitted via horizontal gene transfer [21,57]. It was shown that in *S. maltophilia,* trimethoprim/sulfamethoxazole resistance genes were found in class 1 integron and plasmid [58], indicating the ability *S. maltophilia* to acquire genetic information through horizontal gene transfer. Therefore, it is possible that some of the analyzed TA systems can spread via mobile elements, yet further studies are needed to confirm this possibility.

In this study, we demonstrated that clinical and environmental *Stenotrophomonas* isolates possessed not only distinct TA profiles but other important virulence-related traits. *S. maltophilia* is notorious for its multidrug resistance associated with intrinsic resistance factors, such as low membrane permeability, multidrug resistance efflux pumps, antibiotic-modifying enzymes, and ability to acquire resistance via mutations or acquisition of resistance genes through horizontal gene transfer [31]. Our study revealed a high antibiotic resistance of clinical *S. maltophilia* isolates, all of which displayed resistance to at least five out of eight antimicrobials. Environmental *Stenotrophomonas* isolates were more sensitive to antibiotics compared to bacteria of clinical origins, including meropenem (though *S. maltophilia* is known to have metallo-β-lactamases, providing resistance to various β-lactam antibiotics including carbapenems) [59,60]. The data on the antibiotic resistance level of clinical and environmental *S. maltophilia* isolates are contradictory [60–62], although it has been proposed that some mechanisms of antibiotic resistance occur in the environment and could remain in clinical isolates [63].

Biofilm formation is one of the best characterized *S. maltophilia* virulence traits. *S. maltophilia* forms biofilms on glass and different types of plastic [33], including medical devices (indwelling venous catheters [64], prosthetic valves [65], and lens implants [66]). Biofilm formation on biotic surfaces (e.g., human epithelial respiratory cells) has also been reported [67,68]. In this study, we showed that clinical and environmental *S. maltophilia* isolates differed in biofilm formation abilities, which was in line with previous studies [9,69]. Pellicle is a type of biofilm, formed at the air–liquid interface [70] and protecting bacterial population from environmental stresses. Further, it can increase survival rates by increasing accessibility to oxygen [71]. Less than a half of both clinical and environmental isolates were able to form pellicle, indicating that this phenotype may not be directly linked to virulence traits in *S. maltophilia*.

Although the ability of *S. maltophilia* to form capsule has not been described, studies of capsule functions in other bacterial pathogens showed its crucial role in the protection from unfavorable environmental conditions, including antibiotic treatment, host immune response, and resistance to disinfectants and heat [72–75]. Based on the separation of exopolysaccharides on SDS-PAGE gel, in some of the strains we observed the fraction of high molecular mass polysaccharides, as well as fractions resembling lipopolysaccharides [76]. Typically, such fractions represent capsular exopolysaccharides [37]. We demonstrated that CPS associated high molecular mass exopolysaccharides were detected in nearly all clinical isolates, while only 36% of environmental *Stenotrophomonas* spp. isolates displayed this phenotype. However, the majority of environmental isolates belonging to the *S. maltophilia* species did synthesize extracellular polysaccharides. Therefore, further studies are needed to investigate the importance of capsules in environmental isolates. Nevertheless, the presence of the capsule phenotypes among clinical *S. maltophilia* isolates supports the idea that capsule is beneficial to helping unfavorable conditions and host immune responses survive [38]. The hospital-related strains were also distinguished by increased resistance to serum-induced stress, especially when compared to strains of environmental origin. The ability of isolates to grow at 37 °C (temperature of the host) and resist the complement system of the active serum indicates the increased infectivity potential of these strains. Some of the environmental isolates did not grow even in the heat-inactivated serum, which indicates the inability of these strains to multiply inside the host even before the encounter with the immune system. This could hint at the inability of environmental isolates to persist and grow due to improper nutrients and blood serum composition.

Based on the phenotypes of clinical and environmental *S. maltophilia* and genotyping results, we can assume that the environment could be the primary source of opportunistic *S. maltophilia* infections. However, further development of virulence potential might include an exchange of genetic information and switching on potential virulence genes.

4. Conclusions

Both clinical and environmental *S. maltophilia* genomes contain a variety of type II TA systems, the most common of which belong to *vapBC*, *relBE*, *hipBA*, *hicAB*, and putative COG3832-ArsR families. While bioinformatics assay did not show differences in TA distribution between clinical and environmental TAs, the screening of our collection of isolates did. *relBE* and COG3832-ArsR operons were present solely in clinical *S. maltophilia* isolates, while *hipBA* was more frequent in the environmental isolates. Together with different TA composition, the clinical *S. maltophilia* isolates exhibited clearly stronger biofilm formation, as well as increased antibiotic and serum resistance compared to environmental isolates.

5. Materials and Methods

5.1. Bioinformatic Assays

In total, 21 complete *S. maltophilia* genomes from NCBI database (date of accession 2018-10-13) (Table S3) were analyzed using the TADB 2.0 database tool TA-finder [24]. To identify TA systems in *S. maltophilia* genomes, a basic local alignment search tool (BLAST) [77] was used. The Clustal Omega tool [78] was used for sequence alignments.

5.2. The Bacteria Used in the Study and Growth Conditions

Plasmids and strains used in this study are listed in Table 2. Clinical and environmental *Stenotrophomonas* spp. isolates are listed in Table 3. *Stenotrophomonas* spp. and *E. coli* were grown in an LB medium, unless indicated otherwise. *E. coli* and clinical *S. maltophilia* isolates were grown at 37 °C. Environmental *S. maltophilia* and *Stenotrophomonas* spp. isolates were grown at 28 °C, unless indicated otherwise. DNA was transformed to *E. coli* by heat shock [79] and electroporation was used for

S. maltophilia [80]. Antibiotics (ampicillin 100 mg/L, chloramphenicol 30 mg/L, and kanamycin 60 mg/L) were used to maintain plasmids when needed.

Table 2. Strains and plasmids used in this study.

Strain/Plasmid	Description	Reference
E. coli JM107	*endA1 glnV44 thi-1 relA1 gyrA96 Δ(lac-proAB) [F′ traD36 proAB⁺ lacI^q lacZΔM15] hsdR17(R_K^- m_K^+) λ^-*	[81]
E. coli BW25113 F′	*proA⁺B⁺lacIqΔlacZ)M15 zzf::mini-Tn10* (KanR)	[82]
pBAD30	Expression plasmid	[83]
pUHE 25-2(cat)	Expression plasmid	[82]
pBAD1_gfp	*gfp* gene cloned into pBAD-MCS-1 vector	[84]

Table 3. Clinical and environmental *Stenotrophomonas* spp. isolates used in this study.

Isolate	Genus/Species	Origin	Source [1]	Isolation Year
D27	*S. maltophilia/rhizophila*	Environmental	Soil (Conventional wheat)	2016
D32	*Stenotrophomonas* sp.	Environmental	Soil (Conventional wheat)	2016
D34	*S. maltophilia*	Environmental	Soil (Conventional wheat)	2016
D44	*S. maltophilia*	Environmental	Soil (Conventional wheat)	2016
D46	*S. maltophilia*	Environmental	Soil (Conventional wheat)	2016
D50	*S. maltophilia*	Environmental	Soil (Conventional wheat)	2016
D53	*S. maltophilia*	Environmental	Soil (Conventional wheat)	2016
D55	*Stenotrophomonas* sp.	Environmental	Soil (Organic rapeseed)	2016
D59	*Stenotrophomonas* sp.	Environmental	Soil (Organic rapeseed)	2016
D70	*S. rhizophila*	Environmental	Soil (Organic rapeseed)	2016
D73	*Stenotrophomonas* sp.	Environmental	Soil (Organic rapeseed)	2016
D125	*Stenotrophomonas* sp.	Environmental	Soil (Organic maize)	2016
D150	*Stenotrophomonas* sp.	Environmental	Soil (Conventional maize)	2016
Z53	*S. maltophilia*	Environmental	Fish (Carp)	2016
SM3	*S. maltophilia*	Clinical	NCI	2017
SM5	*S. maltophilia*	Clinical	VCCH LM	2018
SM6	*S. maltophilia*	Clinical	VCCH LM	2018
SM7	*S. maltophilia*	Clinical	VCCH LM	2018
SM8	*S. maltophilia*	Clinical	NCI	2019
SM9	*S. maltophilia*	Clinical	NCI	2019
SM10	*S. maltophilia*	Clinical	VCCH LM	2019
SM11	*S. maltophilia*	Clinical	VUH SK PD	2019
SM12	*S. maltophilia*	Clinical	VUH SK PD	2019
SM13	*S. maltophilia*	Clinical	VUH SK PD	2019
SM14	*S. maltophilia*	Clinical	VCCH LM	2019
SM15	*S.maltophilia*	Clinical	KCPHC	2019
SM16	*S. maltophilia*	Clinical	VCCH LM	2019
SM17	*S. maltophilia*	Clinical	NCI	2019
SM18	*S. maltophilia*	Clinical	NCI	2019
SM20	*S. maltophilia*	Clinical	NCI	2019
SM21	*S. maltophilia*	Clinical	NCI	2019
SM22	*S. maltophilia*	Clinical	NCI	2019
SM23	*S. maltophilia*	Clinical	NCI	2019
SM24	*S. maltophilia*	Clinical	NCI	2019
SM25	*S. maltophilia*	Clinical	NCI	2019

[1] NCI—National Cancer Institute (Lithuania); VCCH LM—Vilnius City Clinical Hospital, Laboratory of Microbiology; VUH SK PD—Vilnius University Hospital Santaros Klinikos, Pediatrics Department, Division of Infectious Diseases; KCPHC—Kaunas City Public Health Center. The isolates from soil were isolated and described previously [85].

5.3. Detection of TA Systems in Stenotrophomonas

For the detection of *S. maltophilia* TA systems, primers targeting conservative gene fragments were designed (Table S2) and PCR was performed using DreamTaq polymerase (Thermo Fisher Scientific).

5.4. Cloning of TA System Genes

All enzymes used for cloning were from Thermo Fisher Scientific and were used according to the manufacturer's recommendations. For gene cloning, a high-fidelity Phusion polymerase was used with the primers listed in Table S2. A *hicAB* TA system was cloned using SM6, SM9, SM11, SM12, SM13, SM16, SM20, and SM24 isolate DNA as the template. For *relBE2* and *higBA* the isolate SM11 was used, for *relBE1, hipBA, vapBC,* and COG3832-ArsR the isolate SM9 DNA was used as a template. For TA functional analysis in *E. coli*, all toxin genes were cloned into pBAD30 plasmid; antitoxin genes were cloned into pUHcat plasmid. Both vectors were hydrolyzed with SphI, then overhangs were blunted with T4 polymerase. Final restriction was performed with HindIII. To clone the toxins into a broad host range pBAD1 vector, pBAD1_GFP plasmid was used. First, the pBAD1_GFP vector was hydrolyzed with XbaI and other steps were performed as described above.

5.5. Kill-Rescue Assay

The kill-rescue assay was performed as described previously [86]. Briefly, *E. coli* BW25113 F′ strains harboring plasmids pBAD_tox and pUHEcat_antitox were grown overnight, diluted 1:500 into fresh medium supplemented with 0.2% glucose, and grown to its early exponential phase (optical density at 600 nm (OD_{600}) = 0.12). The expression of toxin and/or antitoxin was then induced with arabinose (0.002% and 0.2%) and/or IPTG (1 mM), respectively. Growth was measured as OD_{600} in a Tecan Infinite M200 Pro plate reader at 37 °C with shaking.

5.6. Random Amplified Polymorphic DNA (RAPD) Assay

For the RAPD assay, DNA amplification was performed at the same time on the same thermocycler in a final reaction volume of 25 µL, containing 2.5 µL of 10× polymerase reaction buffer with KCl: 0.5 µL 10 mM dNTP, 1.25 µL 50 mM $MgCl_2$, 2 µL 10 pmol/µL OPA-02 or 380-7 primer, 0.15 µL DreamTaq polymerase, and 5 µL of bacteria lysate.

Cycling conditions are as follows: initial denaturation at 95 °C for 2 min followed by 39 cycles of denaturation at 95 °C for 30 s, annealing at 36 °C/34 °C for 1 min (OPA-02 and 380-7 primers, respectively), extension at 72 °C for 2 min and final extension at 72 °C for 5 min.

PCR products were loaded on a 1% (*w/v*) agarose gel with 0.5 mg/mL of ethidium bromide (1 × TAE buffer, 7 V/cm voltage, 1.5 h). Electrophoresis gels were visualized with Bio Rad Molecular Imager and analyzed using GelCompar II software (Applied Maths) with the Dice coefficient set at 1% and band tolerance set at 1.5% using the UPGMA method.

5.7. The Evaluation of Antibiotic Resistance

Antimicrobial resistance analysis was performed by growing *Stenotrophomonas* isolates on agarose LB plates with selected concentrations of antimicrobial agents for 16 h at 28 °C. Control growth was considered as growth of isolates on the LB plates without antimicrobial agents. The isolate was considered as resistant to tested antibiotics only if growth visually looked the same as the control. In the case of significantly weaker growth, the isolate was referred to as an intermediate resistant and in the absence of growth, the isolate was considered as sensitive.

5.8. Biofilm Formation Assay

Biofilm formation experiments were performed as described elsewhere [38]. Briefly, overnight *Stenotrophomonas* spp. cultures (grown in TSB medium (Oxoid) at 37 °C for 16 h) were inoculated into the 150 µL TSB medium containing wells of 96 U-form polystyrene plate (30-fold dilution), then incubated stationary at 28 °C or 37 °C for 24 h. The OD_{600} of planktonic culture was measured and wells were washed 3 times with PBS buffer. Adherent cells were stained with 0.1% crystal violet dye for 15 min, then washed 5 times with the PBS buffer. Dye was dissolved in 96% ethanol for OD_{580}

measurements. The OD_{580}/OD_{600} ratio was calculated to normalize the number of biofilm forming cells to the total cell number.

5.9. Pellicle Formation Assay

The pellicle formation was performed as previously described [87]. Briefly, overnight cultures *Stenotrophomonas* spp. (grown 16 h in TSB medium at 37 °C) were inoculated into the TSB medium containing wells of a flat-bottom 12 well polystyrene microplate in a total volume of 3 mL (1000-fold dilution). The cultures were incubated stationary at 28 °C for 30 h. Results were evaluated qualitatively by monitoring the presence/absence of bacterial biomass formed in the air–liquid contact.

5.10. Extraction of Capsular Polysaccharides

The bacteria were grown on LB agar plates for 16 h and suspended in the PBS buffer. Capsular polysaccharide extraction and analysis were performed as described earlier [38]. Briefly, polysaccharides were released by vortexing for 30 s. After centrifugation at 9000× *g* for 10 min, polysaccharides were precipitated in 75% ice-cold ethanol. The pellet was resuspended in SDS sample buffer and boiled for 5 min. Samples were loaded on the 12% SDS-PAGE gels. After electrophoresis, gels were stained overnight with 0.1% (*w*/*v*) Alcian Blue.

5.11. Serum Resistance

Serum resistance was evaluated by measuring bacterial growth in LB medium, heat inactivated FBS (fetal bovine serum (Gibco, 12657029)), and active FBS. FBS was inactivated by incubation at 56 °C for 30 min with constant shaking. Overnight cultures were inoculated at ×1000 dilution to LB medium and 80% FBS (20% of LB medium) or inactive 80% FBS. Growth curves were measured at 37 °C with periodic shaking every 20 min using a Tecan Infinite M200 Pro microplate reader.

Supplementary Materials: The following are available online at http://www.mdpi.com/2072-6651/12/10/635/s1, Figure S1: Biofilm formation of clinical and environmental *Stenotrophomonas* spp. isolates, Figure S2: The polysaccharide production of *Stenotrophomonas* spp. isolates of clinical and environmental origin, Table S1: TA systems detected by TA-finder in annotated *S. maltophilia* genomes, Table S2: Primers used in this study, Table S3: *S. maltophilia* genomes used in this study.

Author Contributions: J.A. and E.S. conceived and designed the experiments; L.K. and J.S. performed the experiments; L.K., J.S. and J.A. analyzed the data; J.A. and L.K. wrote the paper. All authors have read and agreed to the published version of the manuscript.

Funding: This research received no external funding.

Acknowledgments: We would like to thank Modestas Ružauskas for *Stenotrophomonas* spp., Jolanta Miciulevičienė, R. Plančiūnienė and K. Sužiedėlis for the clinical *S. maltophilia* isolates.

Conflicts of Interest: The authors declare no conflict of interest.

References

1. Brooke, J.S. Stenotrophomonas maltophilia: An Emerging Global Opportunistic Pathogen. *Clin. Microbiol. Rev.* **2012**, *25*, 2–41. [CrossRef]
2. Ryan, R.P.; Monchy, S.; Cardinale, M.; Taghavi, S.; Crossman, L.; Avison, M.B.; Berg, G.; van der Lelie, D.; Dow, J.M. The versatility and adaptation of bacteria from the genus Stenotrophomonas. *Nat. Rev. Microbiol.* **2009**, *7*, 514–525. [CrossRef]
3. Falagas, M.E.; Kastoris, A.C.; Vouloumanou, E.K.; Rafailidis, P.I.; Kapaskelis, A.M.; Dimopoulos, G. Attributable mortality of Stenotrophomonas maltophilia infections: A systematic review of the literature. *Future Microbiol.* **2009**, *4*, 1103–1109. [CrossRef] [PubMed]
4. WHO | World Health Organization. Available online: https://www.who.int/drugresistance/AMR_Importance/en/ (accessed on 28 July 2020).

5. Pompilio, A.; Savini, V.; Fiscarelli, E.; Gherardi, G.; Di Bonaventura, G. Clonal Diversity, Biofilm Formation, and Antimicrobial Resistance among Stenotrophomonas maltophilia Strains from Cystic Fibrosis and Non-Cystic Fibrosis Patients. *Antibiotics* **2020**, *9*, 15. [CrossRef] [PubMed]
6. Trifonova, A.; Strateva, T. *Stenotrophomonas maltophilia*—A low-grade pathogen with numerous virulence factors. *Infect. Dis.* **2019**, *51*, 168–178. [CrossRef] [PubMed]
7. Adamek, M.; Overhage, J.; Bathe, S.; Winter, J.; Fischer, R.; Schwartz, T. Genotyping of Environmental and Clinical Stenotrophomonas maltophilia Isolates and their Pathogenic Potential. *PLoS ONE* **2011**, *6*, e27615. [CrossRef]
8. Bostanghadiri, N.; Ghalavand, Z.; Fallah, F.; Yadegar, A.; Ardebili, A.; Tarashi, S.; Pournajaf, A.; Mardaneh, J.; Shams, S.; Hashemi, A. Characterization of Phenotypic and Genotypic Diversity of Stenotrophomonas maltophilia Strains Isolated From Selected Hospitals in Iran. *Front. Microbiol.* **2019**, *10*. [CrossRef]
9. Pompilio, A.; Pomponio, S.; Crocetta, V.; Gherardi, G.; Verginelli, F.; Fiscarelli, E.; Dicuonzo, G.; Savini, V.; D'Antonio, D.; Di Bonaventura, G. Phenotypic and genotypic characterization of Stenotrophomonas maltophiliaisolates from patients with cystic fibrosis: Genome diversity, biofilm formation, and virulence. *BMC Microbiol.* **2011**, *11*, 159. [CrossRef]
10. Madi, H.; Lukić, J.; Vasiljević, Z.; Biočanin, M.; Kojić, M.; Jovčić, B.; Lozo, J. Genotypic and Phenotypic Characterization of Stenotrophomonas maltophilia Strains from a Pediatric Tertiary Care Hospital in Serbia. *PLoS ONE* **2016**, *11*, e0165660. [CrossRef]
11. Valdezate, S.; Vindel, A.; Martin-Davila, P.; Del Saz, B.S.; Baquero, F.; Canton, R. High Genetic Diversity among Stenotrophomonas maltophilia Strains Despite Their Originating at a Single Hospital. *J. Clin. Microbiol.* **2004**, *42*, 693–699. [CrossRef]
12. Lira, F.; Berg, G.; Martínez, J.L. Double-Face Meets the Bacterial World: The Opportunistic Pathogen Stenotrophomonas maltophilia. *Front. Microbiol.* **2017**, *8*. [CrossRef] [PubMed]
13. Dzidic, S.; Bedeković, V. Horizontal gene transfer-emerging multidrug resistance in hospital bacteria. *Acta Pharmacol. Sin.* **2003**, *24*, 519–526. [PubMed]
14. Brown, S.P.; Cornforth, D.M.; Mideo, N. Evolution of virulence in opportunistic pathogens: Generalism, plasticity, and control. *Trends Microbiol.* **2012**, *20*, 336–342. [CrossRef] [PubMed]
15. Harms, A.; Brodersen, D.E.; Mitarai, N.; Gerdes, K. Toxins, Targets, and Triggers: An Overview of Toxin-Antitoxin Biology. *Mol. Cell* **2018**, *70*, 768–784. [CrossRef]
16. Lee, K.-Y.; Lee, B.-J. Structure, Biology, and Therapeutic Application of Toxin-Antitoxin Systems in Pathogenic Bacteria. *Toxins* **2016**, *8*, 305. [CrossRef]
17. Yamaguchi, Y.; Park, J.-H.; Inouye, M. Toxin-Antitoxin Systems in Bacteria and Archaea. *Annu. Rev. Genet.* **2011**, *45*, 61–79. [CrossRef]
18. Lobato-Márquez, D.; Díaz-Orejas, R.; García-del Portillo, F. Toxin-antitoxins and bacterial virulence. *FEMS Microbiol. Rev.* **2016**, *40*, 592–609. [CrossRef]
19. Wang, Y.; Wang, H.; Hay, A.J.; Zhong, Z.; Zhu, J.; Kan, B. Functional RelBE-Family Toxin-Antitoxin Pairs Affect Biofilm Maturation and Intestine Colonization in Vibrio cholerae. *PLoS ONE* **2015**, *10*, e0135696. [CrossRef]
20. Wood, T.L.; Wood, T.K. The HigB/HigA toxin/antitoxin system of Pseudomonas aeruginosa influences the virulence factors pyochelin, pyocyanin, and biofilm formation. *Microbiol. Open* **2016**, *5*, 499–511. [CrossRef]
21. Leplae, R.; Geeraerts, D.; Hallez, R.; Guglielmini, J.; Drèze, P.; Van Melderen, L. Diversity of bacterial type II toxin–antitoxin systems: A comprehensive search and functional analysis of novel families. *Nucleic Acids Res.* **2011**, *39*, 5513–5525. [CrossRef]
22. Georgiades, K.; Raoult, D. Genomes of the Most Dangerous Epidemic Bacteria Have a Virulence Repertoire Characterized by Fewer Genes but More Toxin-Antitoxin Modules. *PLoS ONE* **2011**, *6*, e17962. [CrossRef] [PubMed]
23. Sala, A.; Bordes, P.; Genevaux, P. Multiple Toxin-Antitoxin Systems in Mycobacterium tuberculosis. *Toxins* **2014**, *6*, 1002–1020. [CrossRef]
24. Xie, Y.; Wei, Y.; Shen, Y.; Li, X.; Zhou, H.; Tai, C.; Deng, Z.; Ou, H.-Y. TADB 2.0: An updated database of bacterial type II toxin–antitoxin loci. *Nucleic Acids Res.* **2018**, *46*, D749–D753. [CrossRef] [PubMed]
25. Sevin, E.W.; Barloy-Hubler, F. RASTA-Bacteria: A web-based tool for identifying toxin-antitoxin loci in prokaryotes. *Genome Biol.* **2007**, *8*, R155. [CrossRef] [PubMed]

26. Akarsu, H.; Bordes, P.; Mansour, M.; Bigot, D.-J.; Genevaux, P.; Falquet, L. TASmania: A bacterial Toxin-Antitoxin Systems database. *PLoS Comput. Biol.* **2019**, *15*, e1006946. [CrossRef] [PubMed]
27. Butt, A.; Müller, C.; Harmer, N.; Titball, R.W. Identification of type II toxin–antitoxin modules in Burkholderia pseudomallei. *FEMS Microbiol. Lett.* **2013**, *338*, 86–94. [CrossRef]
28. Xu, J.; Zhang, N.; Cao, M.; Ren, S.; Zeng, T.; Qin, M.; Zhao, X.; Yuan, F.; Chen, H.; Bei, W. Identification of Three Type II Toxin-Antitoxin Systems in Streptococcus suis Serotype 2. *Toxins* **2018**, *10*, 467. [CrossRef]
29. Coussens, N.P.; Daines, D.A. Wake me when it's over—Bacterial toxin-antitoxin proteins and induced dormancy. *Exp. Biol. Med.* **2016**, *241*, 1332–1342. [CrossRef]
30. Makarova, K.S.; Wolf, Y.I.; Koonin, E.V. Comprehensive comparative-genomic analysis of type 2 toxin-antitoxin systems and related mobile stress response systems in prokaryotes. *Biol. Direct* **2009**, *4*, 19. [CrossRef]
31. Sánchez, M.B. Antibiotic resistance in the opportunistic pathogen Stenotrophomonas maltophilia. *Front. Microbiol.* **2015**, *6*, 658. [CrossRef]
32. Bjarnsholt, T. The role of bacterial biofilms in chronic infections. *Apmis. Suppl.* **2013**, 1–51. [CrossRef] [PubMed]
33. Di Bonaventura, G.; Spedicato, I.; D'Antonio, D.; Robuffo, I.; Piccolomini, R. Biofilm formation by Stenotrophomonas maltophilia: Modulation by quinolones, trimethoprim-sulfamethoxazole, and ceftazidime. *Antimicrob. Agents Chemother.* **2004**, *48*, 151–160. [CrossRef] [PubMed]
34. Armitano, J.; Méjean, V.; Jourlin-Castelli, C. Gram-negative bacteria can also form pellicles. *Environ. Microbiol. Rep.* **2014**, *6*, 534–544. [CrossRef] [PubMed]
35. Cress, B.F.; Englaender, J.A.; He, W.; Kasper, D.; Linhardt, R.J.; Koffas, M.A.G. Masquerading microbial pathogens: Capsular polysaccharides mimic host-tissue molecules. *FEMS Microbiol. Rev.* **2014**, *38*, 660–697. [CrossRef]
36. Niu, T.; Guo, L.; Luo, Q.; Zhou, K.; Yu, W.; Chen, Y.; Huang, C.; Xiao, Y. Wza gene knockout decreases Acinetobacter baumannii virulence and affects Wzy-dependent capsular polysaccharide synthesis. *Virulence* **2020**, *11*, 1–13. [CrossRef]
37. Geisinger, E.; Isberg, R.R. Antibiotic modulation of capsular exopolysaccharide and virulence in Acinetobacter baumannii. *PLoS Pathog.* **2015**, *11*, e1004691. [CrossRef]
38. Skerniškytė, J.; Krasauskas, R.; Péchoux, C.; Kulakauskas, S.; Armalytė, J.; Sužiedėlienė, E. Surface-Related Features and Virulence Among Acinetobacter baumannii Clinical Isolates Belonging to International Clones I and II. *Front. Microbiol.* **2018**, *9*, 3116. [CrossRef]
39. Van Melderen, L.; De Bast, M.S. Bacterial Toxin–Antitoxin Systems: More Than Selfish Entities? *PLoS Genet.* **2009**, *5*, e1000437. [CrossRef]
40. Rowe-Magnus, D.A. Comparative Analysis of Superintegrons: Engineering Extensive Genetic Diversity in the Vibrionaceae. *Genome Res.* **2003**, *13*, 428–442. [CrossRef]
41. Szekeres, S.; Dauti, M.; Wilde, C.; Mazel, D.; Rowe-Magnus, D.A. Chromosomal toxin-antitoxin loci can diminish large-scale genome reductions in the absence of selection: Chromosomal addiction loci minimize gene loss. *Mol. Microbiol.* **2007**, *63*, 1588–1605. [CrossRef]
42. Gotfredsen, M.; Gerdes, K. The Escherichia coli relBE genes belong to a new toxin-antitoxin gene family. *Mol. Microbiol.* **1998**, *29*, 1065–1076. [CrossRef] [PubMed]
43. Schureck, M.A.; Repack, A.; Miles, S.J.; Marquez, J.; Dunham, C.M. Mechanism of endonuclease cleavage by the HigB toxin. *Nucleic Acids Res.* **2016**, *44*, 7944–7953. [CrossRef] [PubMed]
44. Makarova, K.S.; Grishin, N.V.; Koonin, E.V. The HicAB cassette, a putative novel, RNA-targeting toxin-antitoxin system in archaea and bacteria. *Bioinformatics* **2006**, *22*, 2581–2584. [CrossRef] [PubMed]
45. Zhang, Y.; Zhu, L.; Zhang, J.; Inouye, M. Characterization of ChpBK, an mRNA interferase from Escherichia coli. *J. Biol. Chem.* **2005**, *280*, 26080–26088. [CrossRef] [PubMed]
46. Winther, K.S.; Gerdes, K. Enteric virulence associated protein VapC inhibits translation by cleavage of initiator tRNA. *Proc. Natl. Acad. Sci. USA* **2011**, *108*, 7403–7407. [CrossRef]
47. Tiwari, P.; Arora, G.; Singh, M.; Kidwai, S.; Narayan, O.P.; Singh, R. MazF ribonucleases promote Mycobacterium tuberculosis drug tolerance and virulence in guinea pigs. *Nat. Commun.* **2015**, *6*, 1–13. [CrossRef]
48. Wen, W.; Liu, B.; Xue, L.; Zhu, Z.; Niu, L.; Sun, B. Autoregulation and Virulence Control by the Toxin-Antitoxin System SavRS in Staphylococcus aureus. *Infect. Immun.* **2018**, *86*. [CrossRef]

49. Ramage, H.R.; Connolly, L.E.; Cox, J.S. Comprehensive Functional Analysis of Mycobacterium tuberculosis Toxin-Antitoxin Systems: Implications for Pathogenesis, Stress Responses, and Evolution. *PLoS Genet.* **2009**, *5*. [CrossRef]
50. Frampton, R.; Aggio, R.B.M.; Villas-Bôas, S.G.; Arcus, V.L.; Cook, G.M. Toxin-Antitoxin Systems of Mycobacterium smegmatis Are Essential for Cell Survival. *J. Biol. Chem.* **2012**, *287*, 5340–5356. [CrossRef]
51. Lopes, A.P.Y.; Lopes, L.M.; Fraga, T.R.; Chura-Chambi, R.M.; Sanson, A.L.; Cheng, E.; Nakajima, E.; Morganti, L.; Martins, E.A.L. VapC from the Leptospiral VapBC Toxin-Antitoxin Module Displays Ribonuclease Activity on the Initiator tRNA. *PLoS ONE* **2014**, *9*, e101678. [CrossRef]
52. Chan, W.T.; Espinosa, M.; Yeo, C.C. Keeping the Wolves at Bay: Antitoxins of Prokaryotic Type II Toxin-Antitoxin Systems. *Front. Mol. Biosci.* **2016**, *3*. [CrossRef]
53. Strateva, T.; Trifonova, A.; Savov, E.; Dimov, S. An update on the antimicrobial susceptibility and molecular epidemiology of Stenotrophomonas maltophilia in Bulgaria: A 5-year study (2011–2016). *Infect. Dis.* **2019**, *51*, 387–391. [CrossRef] [PubMed]
54. Travassos, L.H.; Pinheiro, M.N.; Coelho, F.S.; Sampaio, J.L.M.; Merquior, V.L.C.; Marques, E.A. Phenotypic properties, drug susceptibility and genetic relatedness of Stenotrophomonas maltophilia clinical strains from seven hospitals in Rio de Janeiro, Brazil. *J. Appl. Microbiol.* **2004**, *96*, 1143–1150. [CrossRef] [PubMed]
55. Schneider, B.; Weigel, W.; Sztukowska, M.; Demuth, D.R. Identification and functional characterization of type II toxin/antitoxin systems in Aggregatibacter actinomycetemcomitans. *Mol. Oral Microbiol.* **2018**, *33*, 224–233. [CrossRef] [PubMed]
56. Bliven, K.A.; Maurelli, A.T. Antivirulence Genes: Insights into Pathogen Evolution through Gene Loss. *Infect. Immun.* **2012**, *80*, 4061–4070. [CrossRef] [PubMed]
57. Lima-Mendez, G.; Alvarenga, D.O.; Ross, K.; Hallet, B.; Melderen, L.V.; Varani, A.M.; Chandler, M. Toxin-Antitoxin Gene Pairs Found in Tn3 Family Transposons Appear To Be an Integral Part of the Transposition Module. *mBio* **2020**, *11*. [CrossRef]
58. Hu, L.-F.; Chang, X.; Ye, Y.; Wang, Z.-X.; Shao, Y.-B.; Shi, W.; Li, X.; Li, J.-B. Stenotrophomonas maltophilia resistance to trimethoprim/sulfamethoxazole mediated by acquisition of sul and dfrA genes in a plasmid-mediated class 1 integron. *Int. J. Antimicrob. Agents* **2011**, *37*, 230–234. [CrossRef]
59. Walsh, T.R.; Macgowan, A.P.; Bennett, P.M. Sequence Analysis and Enzyme Kinetics of the L2 Serine β-Lactamase from Stenotrophomonas maltophilia. *Antimicrob. Agents Chemother.* **1997**, *41*, 5. [CrossRef]
60. Yang, Z.; Liu, W.; Cui, Q.; Niu, W.; Li, H.; Zhao, X.; Wei, X.; Wang, X.; Huang, S.; Dong, D.; et al. Prevalence and detection of Stenotrophomonas maltophilia carrying metallo-Î²-lactamase blaL1 in Beijing, China. *Front. Microbiol.* **2014**, *5*. [CrossRef]
61. Youenou, B.; Favre-Bonté, S.; Bodilis, J.; Brothier, E.; Dubost, A.; Muller, D.; Nazaret, S. Comparative Genomics of Environmental and Clinical Stenotrophomonas maltophilia Strains with Different Antibiotic Resistance Profiles. *Genome Biol. Evol.* **2015**, *7*, 2484–2505. [CrossRef]
62. Deredjian, A.; Alliot, N.; Blanchard, L.; Brothier, E.; Anane, M.; Cambier, P.; Jolivet, C.; Khelil, M.N.; Nazaret, S.; Saby, N.; et al. Occurrence of Stenotrophomonas maltophilia in agricultural soils and antibiotic resistance properties. *Res. Microbiol.* **2016**, *167*, 313–324. [CrossRef] [PubMed]
63. Berg, G.; Eberl, L.; Hartmann, A. The rhizosphere as a reservoir for opportunistic human pathogenic bacteria. *Environ. Microbiol.* **2005**, *7*, 1673–1685. [CrossRef] [PubMed]
64. Muder, R.R.; Harris, A.P.; Muller, S.; Edmond, M.; Chow, J.W.; Papadakis, K.; Wagener, M.W.; Bodey, G.P.; Steckelberg, J.M. Bacteremia due to Stenotrophomonas (Xanthomonas) maltophilia: A prospective, multicenter study of 91 episodes. *Clin. Infect. Dis. Off. Publ. Infect. Dis. Soc. Am.* **1996**, *22*, 508–512. [CrossRef] [PubMed]
65. Khan, I.A.; Mehta, N.J. Stenotrophomonas maltophilia endocarditis: A systematic review. *Angiology* **2002**, *53*, 49–55. [CrossRef]
66. Horio, N.; Horiguchi, M.; Murakami, K.; Yamamoto, E.; Miyake, Y. Stenotrophomonas maltophilia endophthalmitis after intraocular lens implantation. *Graefes Arch. Clin. Exp. Ophthalmol. Albrecht Von Graefes Arch. Klin. Exp. Ophthalmol.* **2000**, *238*, 299–301. [CrossRef]
67. de Vidipó, L.A.; de Marques, E.A.; Puchelle, E.; Plotkowski, M.-C. Stenotrophomonas maltophilia Interaction with Human Epithelial Respiratory Cells In Vitro. *Microbiol. Immunol.* **2001**, *45*, 563–569. [CrossRef]

68. De Oliveira-Garcia, D.; Dall'Agnol, M.; Rosales, M.; Azzuz, A.C.G.S.; Alcántara, N.; Martinez, M.B.; Girón, J.A. Fimbriae and adherence of Stenotrophomonas maltophilia to epithelial cells and to abiotic surfaces. *Cell. Microbiol.* **2003**, *5*, 625–636. [CrossRef]
69. Zhuo, C.; Zhao, Q.; Xiao, S. The Impact of spgM, rpfF, rmlA Gene Distribution on Biofilm Formation in Stenotrophomonas maltophilia. *PLoS ONE* **2014**, *9*, e108409. [CrossRef]
70. Branda, S.S.; Vik, Å.; Friedman, L.; Kolter, R. Biofilms: The matrix revisited. *Trends Microbiol.* **2005**, *13*, 20–26. [CrossRef]
71. Yamamoto, K.; Arai, H.; Ishii, M.; Igarashi, Y. Trade-off between oxygen and iron acquisition in bacterial cells at the air–liquid interface. *FEMS Microbiol. Ecol.* **2011**, *77*, 83–94. [CrossRef]
72. Campa, M.; Bendinelli, M.; Friedman, H. *Pseudomonas Aeruginosa as an Opportunistic Pathogen*; Springer Science & Business Media: New York, NY, USA, 2012; ISBN 978-1-4615-3036-7.
73. Tipton, K.A.; Dimitrova, D.; Rather, P.N. Phase-Variable Control of Multiple Phenotypes in Acinetobacter baumannii Strain AB5075. *J. Bacteriol.* **2015**, *197*, 2593–2599. [CrossRef]
74. Chin, C.Y.; Tipton, K.A.; Farokhyfar, M.; Burd, E.M.; Weiss, D.S.; Rather, P.N. A high-frequency phenotypic switch links bacterial virulence and environmental survival in Acinetobacter baumannii. *Nat. Microbiol.* **2018**, *3*, 563–569. [CrossRef] [PubMed]
75. Boulnois, G.J.; Roberts, I.S. Genetics of Capsular Polysaccharide Production in Bacteria. In *Proceedings of the Bacterial Capsules*; Jann, K., Jann, B., Eds.; Springer: Berlin/Heidelberg, Germany, 1990; pp. 1–18.
76. McKay, G.A.; Woods, D.E.; MacDonald, K.L.; Poole, K. Role of phosphoglucomutase of Stenotrophomonas maltophilia in lipopolysaccharide biosynthesis, virulence, and antibiotic resistance. *Infect. Immun.* **2003**, *71*, 3068–3075. [CrossRef] [PubMed]
77. Altschul, S.F.; Gish, W.; Miller, W.; Myers, E.W.; Lipman, D.J. Basic local alignment search tool. *J. Mol. Biol.* **1990**, *215*, 403–410. [CrossRef]
78. Sievers, F.; Higgins, D.G. Clustal Omega, accurate alignment of very large numbers of sequences. *Methods Mol. Biol.* **2014**, *1079*, 105–116. [CrossRef] [PubMed]
79. Nishimura, A.; Morita, M.; Nishimura, Y.; Sugino, Y. A rapid and highly efficient method for preparation of competent Escherichia coli cells. *Nucleic Acids Res.* **1990**, *18*, 6169. [CrossRef]
80. Ye, X.; Dong, H.; Huang, Y.-P. Highly efficient transformation of Stenotrophomonas maltophilia S21, an environmental isolate from soil, by electroporation. *J. Microbiol. Methods* **2014**, *107*, 92–97. [CrossRef]
81. Yanisch-Perron, C.; Vieira, J.; Messing, J. Improved M13 phage cloning vectors and host strains: Nucleotide sequences of the M13mpl8 and pUC19 vectors. *Gene* **1985**, *33*, 103–119. [CrossRef]
82. Motiejūnaitė, R.; Armalytė, J.; Markuckas, A.; Sužiedėlienė, E. *Escherichia coli dinJ-yafQ* genes act as a toxin-antitoxin module. *FEMS Microbiol. Lett.* **2007**, *268*, 112–119. [CrossRef]
83. Guzman, L.M.; Belin, D.; Carson, M.J.; Beckwith, J. Tight regulation, modulation, and high-level expression by vectors containing the arabinose PBAD promoter. *J. Bacteriol.* **1995**, *177*, 4121–4130. [CrossRef]
84. Petkevičius, V.; Vaitekūnas, J.; Stankevičiūtė, J.; Gasparavičiūtė, R.; Meškys, R. Catabolism of 2-Hydroxypyridine by *Burkholderia* sp. Strain MAK1: A 2-Hydroxypyridine 5-Monooxygenase Encoded by *hpdABCDE* Catalyzes the First Step of Biodegradation. *Appl. Environ. Microbiol.* **2018**, *84*. [CrossRef] [PubMed]
85. Armalytė, J.; Skerniškytė, J.; Bakienė, E.; Krasauskas, R.; Šiugždinienė, R.; Kareivienė, V.; Kerzienė, S.; Klimienė, I.; Sužiedėlienė, E.; Ružauskas, M. Microbial Diversity and Antimicrobial Resistance Profile in Microbiota From Soils of Conventional and Organic Farming Systems. *Front. Microbiol.* **2019**, *10*, 892. [CrossRef] [PubMed]
86. Jurėnaitė, M.; Markuckas, A.; Sužiedėlienė, E. Identification and Characterization of Type II Toxin-Antitoxin Systems in the Opportunistic Pathogen Acinetobacter baumannii. *J. Bacteriol.* **2013**, *195*, 3165–3172. [CrossRef] [PubMed]
87. Krasauskas, R.; Skerniškytė, J.; Armalytė, J.; Sužiedėlienė, E. The role of Acinetobacter baumannii response regulator BfmR in pellicle formation and competitiveness via contact-dependent inhibition system. *BMC Microbiol.* **2019**, *19*, 241. [CrossRef]

Article

Antitoxin ε Reverses Toxin ζ-Facilitated Ampicillin Dormants

María Moreno-del Álamo, Chiara Marchisone and Juan C. Alonso *

Department of Microbial Biotechnology, Centro Nacional de Biotecnología, CNB-CSIC, 3 Darwin Str., 28049 Madrid, Spain; mmoreno@cnb.csic.es (M.M.-d.Á.); cmarchisone@cnb.csic.es (C.M.)

* Correspondence: jcalonso@cnb.csic.es

Received: 5 November 2020; Accepted: 9 December 2020; Published: 15 December 2020

Abstract: Toxin-antitoxin (TA) modules are ubiquitous in bacteria, but their biological importance in stress adaptation remains a matter of debate. The inactive ζ-ε_2-ζ TA complex is composed of one labile ε_2 antitoxin dimer flanked by two stable ζ toxin monomers. Free toxin ζ reduces the ATP and GTP levels, increases the (p)ppGpp and c-di-AMP pool, inactivates a fraction of uridine diphosphate-*N*-acetylglucosamine, and induces reversible dormancy. A small subpopulation, however, survives toxin action. Here, employing a genetic orthogonal control of ζ and ε levels, the fate of bacteriophage SPP1 infection was analyzed. Toxin ζ induces an active slow-growth state that halts SPP1 amplification, but it re-starts after antitoxin expression rather than promoting abortive infection. Toxin ζ-induced and toxin-facilitated ampicillin (Amp) dormants have been revisited. Transient toxin ζ expression causes a metabolic heterogeneity that induces toxin and Amp dormancy over a long window of time rather than cell persistence. Antitoxin ε expression, by reversing ζ activities, facilitates the exit of Amp-induced dormancy both in *rec*$^+$ and *recA* cells. Our findings argue that an unexploited target to fight against antibiotic persistence is to disrupt toxin-antitoxin interactions.

Keywords: toxin-antitoxin system; cell wall inhibition; persistence; nucleotide hydrolysis; uridine diphosphate-*N*-acetylglucosamine

Key Contribution: The understanding of bacterial persistence might enable us to devise treatments against such threat. Toxin ζ, which is a UNAG-dependent ATPase, interferes with purine nucleotide homeostasis, and this physiological heterogeneity induces reversible dormancy and a subpopulation of survivors. Transient ζ expression indirectly reduces Amp persistence without massive autolysis. Antitoxin ε expression reverses toxin action and induces the exit of cells from an Amp dormant state. A deeper understanding of the metabolic state to inhibit toxin action, which sensitizes bacterial cells to different antibiotics, and disruption of antitoxin-mediated reversion of toxic action, are crucial to overcome antibiotic persistence.

1. Introduction

Bacteria have evolved complex regulatory controls and a diverse repertoire of cell transition states in response to various environmental stresses. In order to survive, cells slow down their growth rate and redirect their metabolic resources, until conditions improve and growth can increase [1]. It was shown early in *Staphylococcus aureus* that a large bacterial fraction is unable to survive to lethal doses of penicillin but a subpopulation survives, showing a biphasic inactivation curve [2,3]. It was proposed that in the presence of phenotypic heterogeneity (noise at the level of transcription of key genes during the life cycle), stochastic and antibiotic-induced stress leads to a non-genetic phenotypic "antibiotic insensitive" state, which was termed as persistence [3–6].Persisters have been featured throughout due to their important role in bacterial infections. In the Proteobacteria Phylum

(*Escherichia coli* being the best-characterized representative) starvation-induced (p)ppGpp contributes to antibiotic persistence, whereas in bacteria of the Firmicutes phylum, the persistence effectors are more complex [7–10]. It has been proposed that in *S. aureus* the *bona fide* persistence effector is ATP depletion [11], but treatment with ATP synthesis inhibitors also induces (p)ppGpp accumulation and decreases GTP levels [12]. In *Bacillus subtilis,* which is the best-characterized representative of the Firmicutes phylum, the synthase-hydrolase RelA, and the two small alarmone synthases SasA (also termed RelP or YwaC) and SasB (RelQ or YjbM) downregulate ATP and GTP levels, increase (p)ppGpp, and indirectly the cyclic 3,5-diadenosine monophosphate (c-di-AMP) pools [13,14]. Antibiotic persistence might be controlled by deregulating (p)ppGpp and c-di-AMP pools [10,15]. In response to antibiotics that target cell wall biosynthesis, as the bactericidal β-lactam ampicillin (Amp), (p)ppGpp and c-di-AMP synthesis are induced in *B. subtilis* cells [16]. Furthermore, the *sasA* transcripts are increased by certain classes of cell-wall-active antibiotics (Amp among them) that trigger a membrane stress, and increase (p)ppGpp and c-di-AMP levels [13,17].

Bacterial toxin-antitoxin (TA) modules, which are ubiquitous across a broad range of extrachromosomal elements and are present in bacterial and archaeal chromosomes, are operons encoding a toxin that interferes with vital cell processes and an antitoxin that counteracts toxin action [18–21]. The type I to VI TA modules, which are classified according to the nature of the antitoxin and to their mode of toxin inhibition, are implicated in the fine tuning of multiple cellular functions and are associated with cell survival under different stress conditions [20,21]. The largest group is formed by type II TAs, which were first described as loci that enforce the maintenance of plasmids via post-segregational killing [22], and their ability to stabilize their mobile elements is considered as a selfish property [21,23]. In a type II TA system, which comprises a pair of genes coding for two proteins, the antitoxin forms a tight complex with the toxin to avoid its action [18–21,23]. Under certain stress conditions, the unstable antitoxin is rapidly degraded by host ATP-dependent proteases, freeing the proteolytically stable toxin [18,20]. The free toxin, which targets essential cellular processes such as RNA, DNA and protein synthesis, cell division, general stress response, etc., triggers a biphasic cell inactivation curve [24].

A variant of the HipAB type II TA system, rather than deletion of *hipBA* operon, has been shown to contribute to the formation of antibiotic persistence in *E. coli* cells. The *hipA7* mutation, which maps in the HipA toxin, produces a high fraction of persisters [7,25]. HipA activates the stringent response, and by increasing (p)ppGpp levels, leads to persister formation [26–28]. However, a strong link between the induction of TA systems and antibiotic persistence is missing, and further work is needed to establish whether physiological levels of a toxin contribute to antibiotic persistence [23,29–31]. Deletion of *E. coli*- and *B. subtilis*-encoded chromosomal type II TA systems have no effect on antibiotic persistence [12,31].

The type II TAs of the ζ-ε superfamily, which is among one of the most broadly distributed TA modules in nature, are found in major human and plant pathogens. TAs of this superfamily are composed of two toxin monomers (with ζ and PezT as representatives), that interact with their cognate dimeric ε_2/PezA_2 antitoxin forming an inactive $\zeta\varepsilon_2\zeta$ or PezT-PezA_2-PezT complex [32,33]. ζ/PezT interacts with its activator/substrate, the peptidoglycan precursor uridine diphosphate-*N*-acetylglucosamine (UNAG) [34]. UNAG is an essential precursor of bacterial cell wall biosynthesis [34]. The structures of the inactive $\zeta\varepsilon_2\zeta$ or PezT-PezA_2-PezT complex, alone or UNAG bound, have been reported [32–35]. The interaction of the antitoxin with the toxin sterically blocks ATP-Mg^{2+} binding by the ζ/PezT toxin, but not with UNAG [32–35]. Therefore, the interactions with ATP and ε_2/PezA_2 are mutually exclusive [32,33]. The structure of the inactive $\zeta\varepsilon_2\zeta$ or PezT-PezA_2-PezT complex, both alone or UNAG bound, has been reported [32–35]. Toxin ζ hydrolyses ATP in the presence of UNAG and phosphorylates a small fraction of UNAG, rendering it irreversibly inactive [15,34,36,37].

Transient toxin ζ expression, at or near physiological concentrations, leads to several cellular responses in *B. subtilis* cells growing in S7 minimal medium, at 37 °C with aeration. At early stages of transient toxin ζ expression (up to 15 min), the ζ UNAG-dependent ATPase decreases the ATP pool, and indirectly it alters the intracellular ATP/GTP ratios [37,38]. Such perturbation affects the concentration of the transcription initiation nucleoside triphosphates (ATP, GTP), and indirectly regulates RNA polymerase transcription initiation [39]. This mechanism, which is crucial for rapid adjustment of gene expression in response to environmental changes [40], alters the expression of ~2% of total *B. subtilis* genes (among those, it induces *relA* and *comGA* gene expression) [38]. However, an increased expression of SOS response genes or proteases that selectively induce cell-cycle arrest is not observed [38]. At 15–30 min times, the concerted action of *relA* and *comGA* increases the (p)ppGpp pool and indirectly the levels of the essential c-di-AMP second messenger [15,38]. Toxin ζ expression induces a cellular dormant state, but a reduced subpopulation, which shows no genetic modification, survives toxin action [10,15,38]. At 30–90 min, the synthesis of macromolecules (DNA, RNA, proteins) is inhibited and the membrane potential is impaired, with ~35% of total cells transiently stained with propidium iodide (an indicator of membrane compromised cells) [38]. At 90–120 min, the cell wall biosynthesis is reduced by ζ-mediated phosphorylation of a small UNAG fraction, leading to the accumulation of unreactive UNAG-3P [34,38]. In addition, (p)ppGpp mediates the inhibition of peptidoglycan metabolism [41]. Antitoxin ε expression blocks ATP binding by toxin ζ, thereby inhibiting its ATPase and phosphotransferase activities. Antitoxin ε reverses the toxic effect exerted by ζ in cell proliferation, and only ~10% of total cells remained stained with propidium iodide [38,42,43]. This is consistent with the fact that the toxin ζ action is reversible by nature even in distantly related bacteria, such as *E. coli* and *B. subtilis* cells [38,43]. Transient toxin ζ expression sensitizes cells to Amp treatment, and antitoxin ε expression reverses the ζ-induced dormant state, and the steady state condition is re-established [15,38,43]. The reversibility of ζ action upon longer periods of time in the presence of Amp is unknown. Therefore, we have re-examined the mode of action of toxin ζ, at or near physiological concentrations in *B. subtilis* cells growing in the minimal S7 medium, to understand how toxin ζ contributes to reduce Amp persistence.

In this study, we show that toxin ζ induces a dormant although still metabolically active state. The study of SPP1 in cells where toxin ζ is expressed reveals that ζ induces an active slow-growth rather than a growth-arrested condition and abortive infection. Amp addition at twice the minimum inhibitory concentration (2× MIC) halts *B. subtilis* proliferation, leading to a biphasic time-inactivation curve, which is maintained by a long period of time. Transient toxin ζ expression sensitizes cells to Amp action during a long time period. Antitoxin ε expression switches back ζ-induced dormancy, and promotes the exit of toxin-facilitated Amp dormants without observing a point of no return. Inactivation of *recA*, which suppresses the SOS response, marginally reduced the relative rate of antibiotic persistence. Transient ε expression reverses the negative effect of the toxin and indirectly awakes Amp dormants to levels observed in the *rec*$^{+}$ control. This toxin-mediated sensitization of bacterial cells to Amp and the understanding of the conditions that inhibit antitoxin action will facilitate targeted engineering of their activity towards the development of anti-persistence agents.

2. Results and Discussion

2.1. Experimental Rationale

Resistance, tolerance, and persistence are independent solutions used by bacteria to survive an antibiotic action (Supplementary Figure S1) [1]. A bacterial population resistant to antimicrobial, which grows under drug pressure, uses mutation-associated defense mechanisms (such as antibiotic inactivation/modification, increased efflux, or target modification) to confer a resistant phenotype with or without a fitness cost (Figure S1) [19,44,45]. Persistence is a special case of tolerance, and often the two terms are interchangeable [45]. Tolerance is the general ability of a population to survive a longer antibiotic treatment without a significant change in the MIC (Figure S1, dotted line), whereas persistence represents a subpopulation of cells that can survive the antibiotic treatment (Figure S1, dotted vs. orange broken dotted line) [1].

The growth of susceptible bacteria is challenged by the addition of an antibiotic (e.g., Amp), but a subpopulation of stochastic variants, by undergoing a period of slow- or non-growth, escapes the antibiotic action, leading to a bimodal time-inactivation curve of cell persistence (Figure S1, orange broken dotted line) [1]. In *B. subtilis*, the mechanisms responsible for antibiotic persistence have not been delineated, and the fate of persistents after a prolonged exposure to the antibiotic before returning to permissive conditions is poorly understood. Here, Amp-induced cell-wall stress, by inducing *sasA* expression (one of the two small alarmone synthases), reduces the GTP pool and increases c-di-AMP and ppGpp levels [10,13,17]. This state might be transient and all cells might "awake" after removal of the antibiotic.

In previous studies, the number of toxin ζ and antitoxin ε molecules from a *S. pyogenes* plasmid-borne ε-ζ TA cassette was determined in the *B. subtilis* host grown in the minimal S7 medium [43,46]. An orthogonal control of ζ and ε was constructed by placing the ζ gene under the control of an isopropyl-β-D-thiogalactoside (IPTG) inducible promoter (*lacI*, P_{hsp}ζ, *spc* cassette), was ectopically integrated into the *B. subtilis amy* locus (Table 1) [38]. The antitoxin ε gene, which reads from a xylose (Xyl)-inducible promoter (*xylR*, P_{xylA}ε, *cat* cassette), was cloned into the middle-low-copy number pCB799 plasmid (~8 copies/cell) [38]. In *B. subtilis*, BG1125 pCB799 cells grown in S7 are prone to genome rearrangements, by transcriptional escape of the ζ gene from the P_{hsp} promoter. The Xyl concentration required to reduce genome rearrangements was estimated to be 0.005% [10,38]. BG1125 (pCB799) grows in the S7 medium with 0.005% Xyl at 37 °C with a doubling time of (58 ± 2 min) (Figure S2) [38].

Table 1. Bacterial strains used.

Strain	Relevant Genotype	Reference
BG214	*trpCE metA5 amyE1 ytsJ1 rsbV37 xre1 xkdA1* $att^{SP\beta}$, att^{ICEBs1}	Lab. Collection
BG1127	+ *lacI*, P_{hsp} [a], *spc*, [pCB799-borne *xylR*, P_{xylA}ε [b], *ermC*, *cat*]	[38]
BG1125	+ *lacI*, P_{hsp}ζ [a], *spc*, [pCB799-borne *xylR*, P_{xylA}ε [b], *ermC*, *cat*]	[38]
BG1125	+ *lacI*, P_{hsp}ζ [a], *spc*, [pCB1226-borne *xylR*, P_{xylA}ε [b], *cat*]	This work
BG1889	+ *lacI*, P_{hsp}ζ [a], *spc*, Δ*recA* (*recA:erm*) [pCB1226-borne *xylR*, P_{xylA}ε [b], *cat*]	This work

All *B. subtilis* strains are isogenic with the BG214 strain. [a] The P_{hsp} promoter is under the control of the LacI repressor. [b] The P_{xylA} promoter is under the control of the XylR repressor.

Then, the IPTG concentration necessary to express similar levels of toxin from the native system, and the Xyl concentration necessary to express sufficient $ε_2$ to inactivate the toxin ζ action was estimated [38]. Transient toxin ζ expression, at or near physiological concentrations, induces reversible cell dormancy. Here, a large cell fraction enters into a toxin-induced dormant state that cannot form colonies since they minimize their metabolic activity. By contrast, a small subpopulation escapes the toxin ζ action without diminishing their metabolic activity. When the inducer of toxin expression was removed, the cells exited the transient dormant state and formed colonies, leading to bimodal toxin survivors (Figure S1, blue broken dotted line) rather than toxin tolerants (Figure S1) [15]. These cells that escape the toxin ζ action show no genetic changes [38].

Transient toxin ζ expression sensitizes cells to the Amp treatment [10,15]. In the BG1125 [pCB799] strain, the presence of IPTG or Xyl does not affect the MIC for Amp [10]. The fate of the Amp persisters, which are sensitized by toxin ζ is unknown. At least two types of mechanisms can be envisioned. First, toxin-induced dormancy sensitizes cells to Amp action. Second, ζ-induced dormancy and stochastic slow growth cells trigger a viable but not-culturable state. In the latter condition, normal culturable cells stochastically lost their ability to grow on media, but remain viable [47]. If the first hypothesis is correct, antitoxin ε expression inactivates ζ, and indirectly reverses Amp dormants, as depicted in Figure S1 (red dotted curve). However, if the latter hypothesis is correct, transient $ε_2$ expression should reverse toxin-induced dormancy, but it should play no role on the exit from stochastic viable but the not-culturable state. These experiments performed here should contribute to understand the mechanism of genesis and exit of Amp persistence.

2.2. Toxin ζ Induces a Slow-Growth Active State

In γ-Proteobacteria, deletion of chromosomally-encoded type II TA loci neither confers a fitness benefit nor influences stress tolerance [48,49]. Similarly, deletion of *B. subtilis*-encoded type II TA systems has no effect on antibiotic persistence [12]. To gain insight on the molecular and physiological bases of toxin-induced dormancy, dormant cells were infected with a lytic bacteriophage (phage) and the infective cycle was followed.

Except TA systems that abort infection of a specific phage in selected hosts [50], and those phages that encode *bona fide* TA systems [23], it has been observed that overexpression of type II toxins contributes to abortive infection and causes exclusion of lytic phages from the bacterial population [26,51]. It has been also shown that in the arms race of a bacterium and its phage, a λ lytic variant was able to infect and kill survival bacteria, despite their dormancy state [52], and that T4 infection inactivates MazF toxin activity [53]. In contrast, it has been also shown that phage λ cannot overcome a toxin treat, since toxin overexpression, acting as a defense mechanism, protects the bacterial culture by increasing (p)ppGpp levels, and the alarmone might antagonize λ phage development [54]. Moreover, toxin RnlA overexpression cleaves phage T4 mRNAs to antagonize T4 amplification [55]. Therefore, phage infection of toxin-induced dormant cells can be a powerful tool to understand the physiology of toxin-induced dormancy and to address how the toxin ζ expression sensitizes cells to toxin ζ expression.

A set of isogenic *B. subtilis* strains, lacking prophages (SPβ and PBSX), conjugative genetic elements, CRISPR-Cas and restriction systems, and having a deficiency in the *rsbV* gene, whose function is crucial for the activation of RNA polymerase σ^B-mediated general stress response [56], were used to characterize the impact of toxin ζ expression on cell dormancy (Table 1). The BG1127 (pCB799) strain (in the presence or absence of IPTG, doubling time 59 ± 3 min) and BG1125 (pCB799) strain (in the absence of IPTG, doubling time 58 ± 2 min) grown in the minimal S7 medium supplemented with traces of Xyl (0.005%) at 37 °C with shaking had a similar doubling time (Figure S2). Both cell cultures reached a plateau at ~3.5×10^9 cells mL^{-1} (Figure S2). By contrast, if IPTG (2 mM) was added at OD_{560} = 0.2, the growth curve of the BG1125 (pCB799) strain was halted and after 720 min the OD_{560} was not significantly decreased (Figure 1 and Figure S2), but the plating efficiency was significantly reduced (Figure 2A) [38].

BG1125 (pCB799) cells were grown in the minimal S7 medium supplemented with 0.005% Xyl, to OD_{560} = 0.2 at 37 °C with shaking. At this time (-20 min), the culture was divided, 2 mM IPTG (time zero) was added to a half, and both cultures were incubated for 20 min to allow toxin-induced dormancy. The total number of cells were measured as colony-forming units (CFUs) by plating them on LB plates lacking IPTG (Figure 1A and Figure S3). At -5 min, both cell cultures, containing or not IPTG, were infected with phage SPP1 at a multiplicity of infection (*moi*) of 1 (to minimize the noise of free phages), and incubated for further 5 min. Then, free phages were removed by centrifugation, and this time was considered as 0 min in our experiment. At 0 min, the number of cells was measured by plating on LB agar plates and plaque-forming units (PFUs), to determine the number of infected centers, cells were measured by plating appropriate dilutions of the culture in a lawn of exponentially growing BG214 cells (indicator strain) onto LB-Mg (supplemented with 10 mM $MgCl_2$) plates (Figure 1A,B). In parallel, free phages from the supernatant of centrifugation were determined, to test whether centrifugation was sufficient to remove them or polyclonal anti-SPP1 antibodies were necessary to the inactive free phages. Since in the minus IPTG condition the number of total cells minus free phages was not significantly different from the number of infected centers, we have omitted the use of polyclonal anti-SPP1 antibodies.

After 20 min of toxin induction IPTG, CFUs were reduced by ~1000-fold at a time when compared to the parallel culture in the absence of IPTG (Figure 1A and Figure S3, black vs. grey bar). After infection with SPP1, the proportion of infected centers measured as PFUs was similar in the minus and plus IPTG conditions (Figure 1B, black and grey bars), suggesting that a large fraction of dormant cells were infected by SPP1. In the absence (grey bar) and presence of IPTG (black bar), the phage titer, which correlates to the cell population infected by SPP1, was similar and in good agreement with the expected for a one-hit kinetic. From the input phages, ~71% and ~67% infected the cells, respectively, defining the infected centers (Figure 1B) and ~30% were recovered from the supernatant after centrifugation. In parallel, the total number of cells at -20, 0, 60, and 120 min in the absence of phage SPP1 was also measured as CFUs by plating them on LB plates lacking IPTG (Figure S3).

The SPP1 infection cycle takes 35–45 min in cells growing in the minimal medium, thus the infected cells were incubated for 60 min, and then the relative number of progeny phage per infected cell (burst size) was deduced for the minus and plus IPTG conditions. At 60 min and in the presence of IPTG, cells remained in the dormant state (Figure 1A vs. Figure S3), suggesting that SPP1 does not encode (a) gene(s) whose function can neutralize toxin ζ action. At 60 min, the phage SPP1 titer from cultures where the toxin had been induced was similar to the number of infected cells at 0 min. However, in the absence of IPTG, the PFUs greatly increased. The SPP1 relative burst size (the ratio of the total number of SPP1 progeny to the number of input phage) from three independent experiments was quantified by counting cells and phages taking in to account the supernatant of the cultures (Figure 1A,B). The SPP1 burst size significantly increased (~80 SPP1 phages/infected cell). These results suggest that SPP1, upon infecting toxin-dormant cells neither enters in the amplification cycle nor promotes the exit of the dormant state.

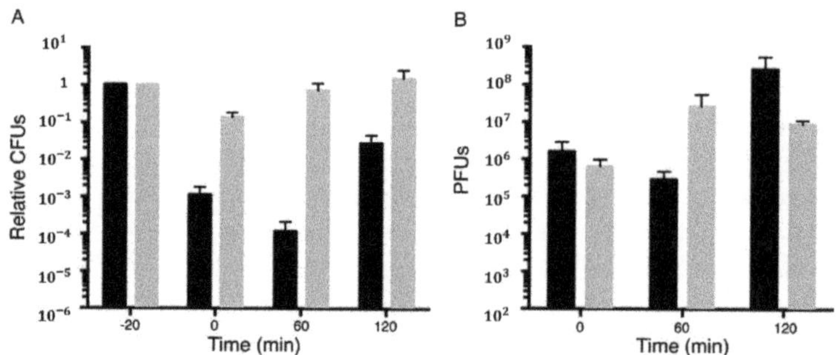

Figure 1. Toxin dormants infected with the lytic phage SPP1 are metabolically active. Survival rate of BG1125 (pCB799) cells infected by the SPP1 phage after ζ toxin induction. BG1125 (pCB799) cells were grown in the minimal S7 medium, containing traces of Xyl (0.005%) and 10 mM $MgCl_2$ to OD_{560} = 0.2 at 37 °C. At this time (-20 min) the culture was divided in two. To one half, IPTG (2 mM) was added (black bars) to induce ζ expression (-20 min) and incubated for 15 min. After 15 min, phage SPP1 at *moi* = 1 was added to both cultures and cells were incubated for further 5 min. At 0 min, cells were centrifuged to remove unadsorbed phages and resuspended in S7 medium, containing 0.005% Xyl and 10 mM $MgCl_2$. Colony-forming units (CFUs) were counted plating on a LB agar (time 0) and both infected cultures were incubated for further 60 min (60 min). At 60 min, 0.5% Xyl was added to induce antitoxin ε expression to neutralize toxin action, and both infected cultures were incubated for further 60 min (120 min). At 0, 60, and 120 min, samples were withdrawn and plated in LB agar plates lacking IPTG to count CFUs (**A**), or in a lawn of exponentially growing BG214 cells onto LB plates supplemented with 10 mM $MgCl_2$ to measure PFUs (**B**). Data are shown as mean ± standard error of the mean (SEM), from >3 independent experiments.

To test whether dormant cells antagonize or cannot support SPP1 amplification or if dormant cells can support SPP1 amplification upon toxin neutralization at 60 min post-infection, Xyl (0.5%) was added to induce antitoxin ε expression, and the cultures were incubated for further 60 min. At 120 min, the number of total cells and the SPP1 burst-size were quantified as described above. In the presence of IPTG and Xyl, the yield of SPP1 phage in cells expressing both ζ and ε genes (between 60 to 120 min) was significantly increased (~110 SPP1 phages/predicted infected cell), calculated from the PFU obtained and the number of infected centers at 60 min (Figure 1B). This suggests that only when infected cells exit toxin-induced dormancy, by the anti-toxin action, SPP1 was amplified. In the only Xyl condition (i.e., no toxin induced), the number of total cells increased (58 ± 2 min doubling time), but the phage titer did not, suggesting that these cells enter in a state that might not be further infected by the SPP1 phage.

These data altogether suggest that: (i) SPP1 infected toxin-induced dormant cells with a similar efficiency as normally growing cells (Figure 1B, time zero); (ii) ζ-induced dormancy does not impair phage absorption, but delays the SPP1 lytic cycle; (iii) toxin ζ-induced dormancy is a metabolically active state rather than a growth-arrested state leading to abortive infection; (iv) SPP1 does not undergo an abortive infection; and v) upon toxin ζ inactivation, by antitoxin ε expression, the exit of ζ-induced dormancy facilitates a viral infective cycle.

2.3. Discrete Subpopulations of Toxin ζ Survivors and Amp Persisters

The subpopulation of cells that survive an antibiotic treatment, are termed persisters [1,45]. The rate of killing by Amp is strictly proportional to the rate of bacterial growth, and Amp persistence is associated with a pre-existing phenotypic switching to a stochastic non-growing state (or leftover cells from high-stress stationary-phase inoculums) [57,58]. To analyze the interconnection between toxin survivors and Amp persisters and whether a prolonged growth arrest causes an irreversible cell cycle arrest (point of no return), we have used BG1125 (pCB799) cells (Table 1).

BG1125 (pCB799) cells were grown in the minimal S7 medium, supplemented with 0.005% Xyl, to OD_{560} = 0.2 at 37 °C with shaking, and then 2 mM IPTG (0 min) was added to induce toxin expression, and the reaction incubated for 900 min (Figure 2A). Toxin ζ expression quickly reduced the cell plating efficiency showing a biphasic time-inactivation curve (Figure 2A, blue empty rhombs), suggesting that toxin ζ expression reaches a steady-state in a short time and it remains for at least 900 min. It is likely that toxin ζ expression induces a slow-growth active state, but a rare fraction ($3–6 \times 10^{-5}$) exits this dormant state and survives, or a preexisting population of non-growing cells are insensitive to toxin action (type I-like persisters) (see [59]). Alternatively, expression of the ζ gene triggers the expression of any of the type I and/or type II TA modules present in the background, and the loss of cell viability and the observed surviving cells arise via the overlapping activation of multiple toxins, even with different activities.

To test whether the frequency of survivors may increase in response to the spontaneous induction of other chromosomally encoded toxins present in the background, IPTG was removed, by washing the culture with the pre-warmed S7 medium, containing 0.5% Xyl (to induce antitoxin ε expression), and cells were preincubated for 15 min before plating on LB plates containing 0.5% Xyl, but lacking IPTG. Antitoxin ε expression reversed ζ-induced dormancy, by significantly increasing cell viability nearly to full values almost similar to the time of IPTG addition (Figure 2A, blue filled rhombs). Similar results are observed when IPTG is not removed from the medium [15]. It is likely that: (i) Toxin ζ-induced dormancy rather than cell lysis, as judged by the non-significant variation in the OD_{560} = 0.2 during a 720-min interval (Figure S2); (ii) toxin ζ is not triggering a spontaneous induction of toxins from TA systems present in the background; and (iii) toxin ζ transiently halts DNA replication and triggers a phenotypic heterogeneity [38]. In addition, the expression of the antitoxin ε expression reversed ζ-induced growth arrest even after a prolonged dormancy (equivalent to ~15 mass doubling periods) (Figure 2A, blue filled rhombs).

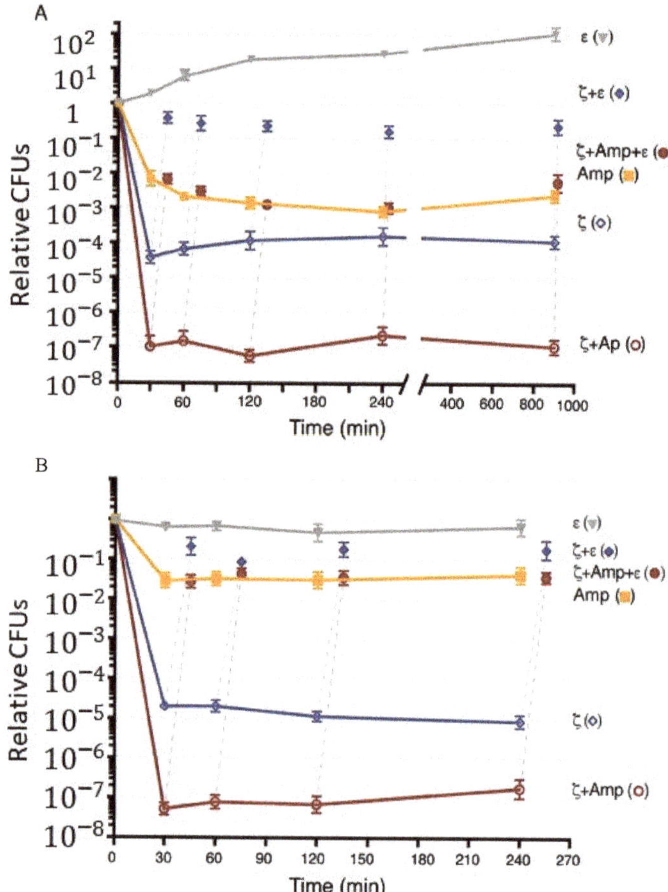

Figure 2. Toxin ζ induces reversible dormancy, and facilitates Amp dormancy. (**A**), BG1125 (pCB799) cells were grown in the S7 medium, containing traces of Xyl (0.005%) at 37 °C and the cultures were divided. Then, IPTG (2 mM) to induce ζ expression, Amp (at 2× MIC, 3 μg mL^{-1}), or both IPTG and Amp were added (0 min), and the cultures were incubated for 900 min. At various times, samples were withdrawn, centrifuged, and plated on LB agar plates to count ζ (empty blue rhomb), Amp (filled orange square), or both ζ and Amp survivals (empty purple circles). At 30, 60, 120, 240, and 900 min, aliquots were taken and 0.5% Xyl was added to induce antitoxin ε expression, and the cultures were incubated for 15 min before plating in LB agar plates containing Xyl, but lacking IPTG (filled blue rhomb) or both IPTG and Amp (filled purple circle). The vertical broken lines join the original point (ζ or ζ + Amp) with the reversed condition after ε expression. (**B**) An overnight culture of BG1125 (pCB799) cells grown in the S7 medium, containing traces of Xyl (0.005%) was normalized to ~1 × 10^{9} cells mL^{-1} (37 °C) and the cultures were divided. Xyl (0.5%) to induce ε expression as the control was added. IPTG to induce toxin, Amp, or both IPTG and Amp was added (0 min), and the cultures were incubated for 240 min. At various times, samples were withdrawn and plated in LB agar plates lacking IPTG (empty blue rhomb), Amp (filled orange square), or both IPTG and Amp (empty purple circles). At various times, 0.5% Xyl was added to induce antitoxin expression, and the cultures were incubated for 15 min before plating. Samples were withdrawn and plated in LB agar plates containing Xyl, but lacking IPTG (filled blue rhomb) or both IPTG and Amp (filled purple circle). The vertical broken lines join the original point (ζ or ζ + Amp) with the reversed condition after ε expression. Data are shown as mean ± standard error of the mean (SEM), from >4 independent experiments.

To address whether antibiotic persisters are produced due to a stochastic entrance into the stationary phase, the overnight culture was extensively diluted and BG1125 (pCB799) cells were grown in the minimal S7 medium, supplemented with 0.005% Xyl, to OD_{560} = 0.2 at 37 °C with shaking. It is expected that <0.3% of the total cells in the actively growing inoculum are old cells from the stationary-phase inoculum. Amp (at 2× MIC, 3 μg mL^{-1}) was added, and the culture was maintained up to 900 min. At indicated times, the cells were washed to remove the antibiotic, and plated in LB lacking Amp. The removal of Amp was sufficient to reveal that a large fraction of cells (>99% of total cells) cannot form colonies upon exposure to bacteriolytic Amp (Figure 2A, filled orange squares). As stated, the rate of Amp killing is proportional to the rate of bacterial growth [57], with >80% of total Amp-treated cells being stained with propidium iodide [10], which stains the membrane compromised cells. A small cell fraction (1–0.5 × 10^{-2} survivals), however, regained the ability to form colonies upon antibiotic removal (Figure 2A, filled orange squares). Similar results were observed when the overnight culture was less extensively diluted (~6% old cells), suggesting that old cells from the stationary-phase inoculum poorly contribute to the Amp persistence fraction.

Upon exposure to Amp for at least 900 min of incubation, cells survive following antibiotic removal (Figure 2A) rather than becoming metabolically inactive by entering a path of no return as described upon toxin MazF overexpression [60]. The culturable cells were still sensitive to Amp action, thus we have to believe that after prolonged incubation to the antibiotic there were no genetic changes [10].

2.4. Transient Toxin ζ Expression and Amp Addition Reduce the Subpopulation of Persisters

Transient toxin ζ expression sensitizes cells to the presence of different antibiotics [10,37]. At least two mechanisms can be envisioned: (i) Toxin dormants are sensitive to antibiotic action and stochastically enter in a viable but not-culturable state with resuscitation independent of antitoxin expression; or (ii) antibiotic persistence is the sum of stochastic and transient antibiotic-induced dormants with resuscitation dependent of antitoxin expression. To test the above hypotheses, BG1125 (pCB799) cells were grown in S7 medium, supplemented with 0.005% Xyl, to OD_{560} = 0.2 at 37 °C with shaking. Then, 2 mM IPTG and 3 μg mL^{-1} Amp were added, and at indicated times cells were washed and plated in the absence of both IPTG and Amp. A typical biphasic decline curve was observed (Figure 2A, purple empty circles). Amp and toxin action significantly decreased the proportion of survivals (1–2 × 10^{-7}) when compared to only ζ (3–6 × 10^{-5} survivors) or only Amp (1–0.5 × 10^{-2} persisters) (Figure 2A, purple empty circles vs. blue empty rhombs and orange filled squares). The persistent cells stochastically switch back to a growing state, upon plating in the absence of IPTG and Amp, with the fraction of persisters remaining nearly constant for ~900 min (Figure 2A, purple empty circles). It is likely that: (i) Amp persisters are in an active state that allows toxin expression; (ii) toxin ζ expression does not increase antibiotic persistence; (iii) toxin expression triggers a network of intracellular stress responses that significantly increases cell dormancy; and (iv) toxin-induced dormancy facilitates Amp dormants that resuscitate upon antitoxin expression.

2.5. Transient Antitoxin ε Expression Reverses ζ-Facilitated Exit of Amp Dormants

To test whether transient ζ expression enhances Amp efficacy with the cells entering in a viable but not-culturable state or if ζ expression induces dormancy of Amp persisters, BG1125 (pCB799) cells were grown in the minimal S7 medium, supplemented with 0.005% Xyl, to OD_{560} = 0.2 at 37 °C with shaking. Then, 2 mM IPTG and Amp (3 μg mL^{-1}) were added and the cells incubated up to 900 min (Figure 2A, purple empty circles). At indicated times, IPTG and Amp were removed by washing the culture with the pre-warmed S7 medium, containing 0.5% Xyl to induce antitoxin ε expression, and cells were preincubated (15 min) before plating (Figure 2A, purple filled circles). The antitoxin ε recovered cell proliferation to levels equivalent to that of Amp persisters, even after 900 min of toxin ζ and Amp action (Figure 2A, purple filled circles). These data altogether suggest that: (i) Toxin ζ expression and Amp addition induce prolonged dormancy (at least for 15 h) rather than programmed-cell death, as reported for the *E. coli* MazEF and RelBE locus (reviewed in [61]); (ii) ζ-induced dormancy is insensitive to Amp action; (iii) toxin-induced dormant cells awake and form colonies even after 15 h of toxin and Amp action, rather than triggering cell lysis or inducing a viable but not-culturable state (Figure 2A, purple filled circles); and (iv) the orthogonal control of ζ and ε_2 levels revealed that a TA system can be an important element in coping with Amp stress; and (v) toxin survivors and Amp persistence are not epistatic.

2.6. Toxin ζ Expression Induces Amp Dormancy in Non-Growing Cells

To re-evaluate whether Amp persisters are produced due to a stochastic entrance into the stationary phase or upon transient toxin ζ expression, and if toxin expression or Amp addition to stationary phase cells affects the persistence level and its subsequent exit, high-density slow- or non-growing stationary phase *B. subtilis* BG1125 (pCB799) cells, which grew overnight in the S7 medium supplemented with 0.005% Xyl, were used. The overnight culture was normalized to ~1×10^9 cells mL^{-1} with the pre-warmed S7 medium, supplemented with 0.005% Xyl and split. Amp (3 μg mL^{-1}), IPTG (2 mM), or both Amp and IPTG were added and the cells incubated up to 240 min. At the indicated times, IPTG, Amp, or both were removed from the medium and the culture was plated on LB plates. Since in the absence of both, Amp and IPTG, the number of cells did not significantly increase at 240 min (Figure 2B, filled grey triangles) and BG1125 (pCB799) cells grown in S7 supplemented with 0.005% Xyl usually had an usual lag-time of 30–45 min (Figure S2), we assume that the high-density cells remain in the stationary phase during the experimental period.

The MIC for Amp was estimated using low-density cells (1–3 × 10^6 cells mL^{-1} for 16 h, at 37 °C), and no correction by per-cell in the Amp concentration was performed. At a variable time, and upon removal of the antibiotic, a significant fraction could not form colonies, but a subpopulation of cells persisted to Amp action (3–5 × 10^{-2} survivals). As expected, the proportion of persisters was significantly higher than in exponentially growing cells (Figure 2A vs. Figure 2B, filled orange squares). This is consistent with the observation that the rate of killing of Amp is strictly proportional to the rate of bacterial growth [57]. Similar results were observed in the presence of 6 μg mL^{-1} Amp (data not shown). It is likely that the lower proportion of cell death might correlate with the less proportion of metabolically active cells in the stationary phase, but in the only Xyl condition the cells are not proliferating.

In the presence of IPTG, the subpopulation of toxin ζ (3–6 × 10^{-5} survivals) did not significantly increase during time and was similar in both, exponential growth and stationary phase (Figure 2A,B, empty blue rhombs). Similar results were observed when a low-lived ζ variant (ζY83C) is used [10], suggesting that when toxin ζ is expressed, it triggers dormancy in high-density stationary phase cells. When Amp and IPTG were added, a low proportion of high-density stationary phase cells survive toxin ζ and Amp action (4–7 × 10^{-8} survivals) (Figure 2B, empty purple circles).

To test whether the expression of ε antitoxin also reverses the effect of ζ-mediated dormancy in high-density non-growing cells, Xyl was added at the indicated times. The expression of ε antitoxin reversed the effect of the ζ toxin (3–5 × 10^{-1} survivals) and partially reversed toxin and Amp (4–5 × 10^{-2} survivals) (Figure 2B, filled blue rhombs and purple circles). It is likely that: (i) Toxin ζ induces a reversible dormant state in high-density, non-growing cells rather than entering in a stochastic viable but not-culturable state; (ii) there are two different subpopulations of Amp persisters: Stochastic (3–5 × 10^{-2} survivals) and toxin-induced (4–7 × 10^{-8} survivals); and iii) antitoxin expression reverses the apparent sensitization of toxin dormants to Amp action.

2.7. RecA Inactivation Does not Alter the Antitoxin Awake of Toxin and Amp Dormants

Toxin-induced dormants are metabolically active slow-growing cells, and this may enhance the development of heritable genetic changes, since many of the stress-response programs involved in the survival of persisters can also accelerate genome-wide mutagenesis [58]. In Proteobacteria, both stochastic and stress-induced slow-growth persisters have been shown to depend on the SOS responses [58,62]. Furthermore, the SOS response induces the expression of certain type I and II TA modules in a subpopulation of *E. coli* [5,63].

In Proteobacteria and Firmicutes, bactericidal antibiotics (e.g., Amp), regardless of the drug-target interaction, induce changes in cellular metabolism that promote the production of highly deleterious hydroxyl radicals, leading to cell killing [64]. Amp-induced hydroxyl radicals, may damage template bases and introduce single strand nicks. These nicks, through DNA replication, can be converted into one-ended double-strand breaks (DSBs), that cause a significant induction of the SOS response [64,65] and maintenance of Amp persistence in *E. coli* cells [66]. However, DNA damage caused by sublethal doses of a fluoroquinolone in *E. coli* promotes persistence, but not Amp [5]. The recombinase *recA*, which is a central player in the SOS response, contributes to the repair of the damaged template bases that escape specialized repair, and of the nicks that are converted in one-ended DSBs during DNA replication [67,68]. The contribution of the SOS response to the DSB repair, however, differs substantially between *E. coli* and *B. subtilis* cells. In *B. subtilis*, the inability to induce the SOS response does not compromise the DSB repair [69].

In *E. coli* cells, *recA* plays a critical role in the appearance of apoptotic phenotypes and in pushing the cell towards its death [65], and inactivation of *recA* potentiates killing by different antibiotics, Amp among them [64]. In *B. subtilis* cells, transient toxin ζ expression [38,43] or extended cell cycle arrest [70] does not induce the SOS response. To test whether *recA* inactivation sensitizes cells to toxin or Amp, a Δ*recA* mutation was mobilized onto the *B. subtilis* BG1125 (pCB1226) strain by SPP1-mediated generalized transduction, rendering the BG1889 (pCB1226) strain (Table 1). Plasmid pCB1226 is a derivative of pCB799, but the *ermC* gene was selectively inactivated so that the Δ*recA* mutation could be introduced in *B. subtilis* BG1125 (pCB1226) cells by Erm^R selection (Table 1).

BG1889 (pCB1226) cells were grown in S7 medium, supplemented with 0.005% Xyl, to $OD_{560} = 0.2$ at 37 °C with shaking. CFUs at this OD_{560} are 3–6 $\times 10^6$ cells mL^{-1}, in agreement with previous reports showing that in *B. subtilis* only 10–20% of Δ*recA* cells can form colonies in plates in the absence of DNA damage [71]. Then, IPTG (2 mM), Amp (3 µg mL^{-1}), or both IPTG and Amp were added (0 min), and the culture was maintained for 900 min (Figure 3). In the absence of ITPG and presence of traces of Xyl (0.005%), the BG1889 (pCB1226) cells had a poor fitness, formed small-size colonies, and the plating efficiency was reduced ~100-fold after 900 min of incubation (Figure 3, filled grey triangles). This result indicates that a significant proportion of Δ*recA* cells may have a prolonged defect in DNA repair and chromosomal segregation, which reduces cell viability during prolonged growth.

The number of Amp persisters and toxin survivors during the first 120 min was lower ~2×10^{-3} and ~4×10^{-6}, respectively, but persistence to both Amp and toxin (~4×10^{-7} frequency of persisters) in the Δ*recA* context (Figure 3, orange filled squares and blue empty rhomb vs. grey filled triangles) was similar to that of the *rec*$^+$ control (see Figure 2A). The decrease of toxin survivors and Amp persisters was observed at 900 min (~7×10^{-9}) after correction of the poor viability of the BG1889 (pCB1226) strain in the absence of IPTG and Amp (Figure 3, purple empty circles vs. grey filled triangles). What is the significance of such large variations in persisters to both toxin and Amp action in the Δ*recA* context? First, to measure persisters, the culture was concentrated ~100-fold. Second, under this experimental condition, the relative persistence frequency is similar to the spontaneous mutation rate. Finally, *B. subtilis* has multiple forms of differentiation and development, thus we cannot rule out that any uncontrolled condition might affect the outcome.

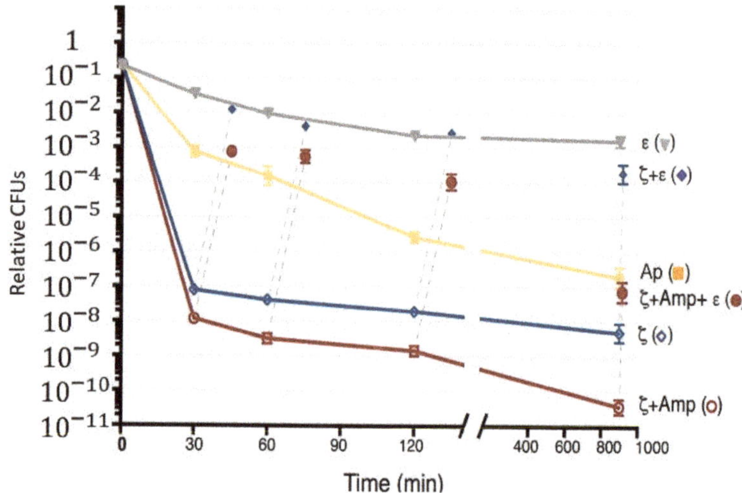

Figure 3. Inactivation of *recA* does not affect the frequency of ζ survivors and Amp persisters. BG1889 (pCB799) cells were grown in the minimal S7 medium, containing traces of xylose (Xyl; 0.005%) to OD_{560} = 0.2 (with CFUs of 1–3 × 10^6 cells mL^{-1}) at 37 °C. Then, the cultures were divided. To the indicated cultures, Xyl (0.5%) to induce ε expression as the control, IPTG (2 mM) to induce ζ expression, Amp (3 µg mL^{-1}) or both IPTG and Amp were added (0 min), and the cultures were incubated for 900 min. At various times, samples were withdrawn and plated in LB agar plates. At various times, aliquots were taken and 0.5% Xyl was added to induce antitoxin expression, and the cultures were incubated for 15 min before being plated in LB agar plates containing Xyl, but lacking IPTG (filled blue rhomb), or both IPTG and Amp (filled purple circle). Data are shown as mean ± standard error of the mean (SEM), from >4 independent experiments.

Transient antitoxin ε expression, by the addition of Xyl, reversed toxin action and the plating efficiency was increased to levels similar to the control strain in the absence of any treatment (Figure 3, blue filled rhombs vs. grey filled triangles). When Xyl was added to cells treated with both Amp and IPTG, the expression of ε antitoxin reversed the effect of ζ toxin and facilitated the exit of Amp dormants (Figure 3, purple filled circles). It is likely, therefore, that: (i) *RecA* inactivation, which impairs the SOS response, does not significantly increase the number of Amp or toxin survivors; (ii) *recA* inactivation reduces toxin or Amp persistence, suggesting that Amp-induced hydroxyl radicals accumulation is not significantly deleterious when compared to the *recA* control without Amp and IPTG (Figure 3, orange filled squares and blue empty rhombs vs. grey filled triangles); and (iii) *recA* does not induce an apoptotic phenotype upon Amp treatment, and *recA* inactivation does not sensitize cells to both toxin ζ and Amp action.

3. Conclusions

This study provides new insights that allow us to unravel the mechanisms underlying the mode of action of the toxin ζ dormancy and indirectly of toxin and Amp dormants. Transient toxin ζ expression, at or near physiological concentrations, induces a slow-growth dormant state, and a small fraction of cells (persisters) is insensitive to the toxin action in rec^+ and *recA* backgrounds. Toxin ζ-induced dormant cells can be infected with the lytic phage SPP1 with a similar efficiency as exponentially growing cells, but the amplification cycle is halted. Neutralization of the toxin, by inducing antitoxin expression, allows dormant cells to exit this state and this allows phage amplification. It is likely that toxin dormants are metabolically active rather than in a growth-arrested state.

Amp action kills ~99% of susceptible cells, but stochastic Amp persisters (frequency of 1–0.5 × 10^{-2}) can form colonies upon removal of the antibiotic. Toxin survivors are significantly rare (~3 × 10^{-5}), and in the presence of Amp the proportion of bacterial Amp and toxin survivors was further reduced (by 100–200-fold), suggesting a non-epistatic effect. To further evaluate the viable but non-culturable state hypothesis, we inactivated *recA*. In *E. coli* cells, *recA* contributes to apoptotic cell death in Amp treated cells [64], but the proportion of Amp persisters was not affected in the absence of *recA* [5]. In *B. subtilis* cells, toxin ζ and Amp persistence are not significantly affected in *recA* cells when compared to *rec*$^+$ control, and transient antitoxin ε expression is necessary and sufficient to switch off toxin-induced dormancy, and to awake the fraction of toxin-facilitated Amp-induced persisters both in the *rec*$^+$ and *recA* cells. We propose that in exponentially growing cells, two different subpopulations of Amp persisters are found, stochastic and toxin-induced dormancy, that also induce long term Amp dormancy.

4. Materials and Methods

4.1. Bacterial Strains and Plasmids

The bacterial strains and plasmids used in this study are listed in Table 1. All *B. subtilis* strains are isogenic with BG214. The strain BG1125 bearing *lacI-*P_{hsp} *wt*ζ and pCB799-borne *xylR-*P_{xylA} *wt*ε *ermC cat* cassette were previously reported (Table 1) [38]. The BG1885 strain was constructed in three steps. First, by site-directed mutagenesis (QuickChange Kit, Stratagene) using pCB799 as a template, the *ermC* gene was inactivated to render pCB1226. Second, the *lacI-*P_{hsp} *wt*ζ cassette was mobilized into BG214 cells bearing pCB1226-borne *xylR-*P_{xylA} *wt*ε *cat* cassette with selection for Cm^R. Third, the null *recA*::*Erm* gene was mobilized by chromosomal transformation into competent BG214 (pCB1226) cells with selection for Erm^R (Table 1). Integration by double crossover was analyzed by PCR.

Toxin ζ gene expression (transcribed from P_{hsp}) is regulated by IPTG (Calbiochem, Madrid, Spain) addition and the expression of ε gene (transcribed from P_{xylA}) is regulated by Xyl (Sigma-Aldrich, St. Louis, MO, USA) addition [38].

4.2. Growth Conditions

BG1125 (pCB799) and BG1125 (pCB1226) cells were grown to a mid-exponential phase (OD_{560} = 0.2) in S7 medium supplemented with methionine, tryptophan, and 0.005% Xyl at 37 °C with shaking [43]. Under this condition, cells grew in S7 medium, with a doubling time of 50–60 min. Transient toxin and/or antitoxin expression was induced by IPTG (2 mM) and/or Xyl (0.5%) addition. Before plating, cells were centrifuged and resuspended in the fresh LB medium to remove the inductor or the antibiotic, and dilutions were plated on LB agar plates. The survival rate was derived from the number of CFUs in a given condition relative to the CFU of the non-induced/non-antibiotic-treated control. All plates were incubated for 20 h at 37 °C.

The lytic SPP1 phage was used to test the toxin effect. BG1125 bearing *lacI-*P_{hsp} *wt*ζ and pCB799-borne *xylR-*P_{xylA} *wt*ε gene were grown to a mid-exponential phase at 37 °C in S7 medium, supplemented with methionine, tryptophan, 0.005% Xyl, and 10 mM $MgCl_2$. Toxin expression was induced by adding IPTG 2mM and after 15 min cells were infected with phage SPP1 at a multiplicity of infection (*moi*) of 1 in the presence or absence of IPTG. After 5 min, cells were centrifuged to remove unabsorbed phages and resuspended in S7 medium with 10 mM MgCl2 (0 min) and the culture was incubated for 60 or 120 min. The inducers were removed by washing, and the phage titer (PFUs) was calculated by plating the appropriate dilution deposited in a lawn of exponentially growing BG214 cells onto LB plates supplemented with 10 mM $MgCl_2$ (LB-Mg) to measure PFUs as described [72]. The plates were then incubated for 18–20 h at 37 °C.

The MIC for Amp (Sigma-Aldrich, St. Louis, MO, USA) was estimated by exposing 1–3 × 10^6 cells mL^{-1} (16 h, 37 °C) in the minimal S7 medium, with an increasing Amp concentration under shaking

(240 rpm). The minimal concentration that gave no growth overnight (1.5 μg mL^{-1}) was defined as MIC. The Amp concentration used was twice the MIC (2× MIC) or 3 μg mL^{-1}.

Supplementary Materials: The following data are available online at http://www.mdpi.com/2072-6651/12/12/801/s1, Figure S1: Graphic illustration of the distinct responses that may occur after Amp and toxin ζ action, Figure S2. Growth curves of BG1127 (pCB799) and BG1125 (pCB799) strains in the presence or absence of IPTG, Figure S3. BG1125 (pCB799) in the presence or the absence of IPTG.

Author Contributions: M.M.-d.Á. and J.C.A. designed the experiments; M.M.-d.Á. and C.M. performed the experiments, J.C.A. coordinated the research; M.M.-d.Á. and J.C.A. interpreted the data; M.M.-d.Á., C.M. and J.C.A. drafted the manuscript; and J.C.A. wrote the manuscript. All authors have read and agreed to the published version of the manuscript.

Funding: This research was funded by the Spanish Ministerio de Ciencia e Innovación/Agencia Estatal de Investigación (MCI/AEI)/FEDER, EU) PGC2018-097054-B-I00 to J.C.A.

Acknowledgments: We thank Silvia Ayora and Rubén Torres for critical reading and comments on the manuscript.

Conflicts of Interest: The authors declare no conflict of interest. The funders had no role in the design of the study; in the collection, analyses, or interpretation of data; in the writing of the manuscript, or in the decision to publish the results.

References

1. Balaban, N.Q.; Helaine, S.; Lewis, K.; Ackermann, M.; Aldridge, B.; Andersson, D.I.; Brynildsen, M.P.; Bumann, D.; Camilli, A.; Collins, J.J.; et al. Definitions and guidelines for research on antibiotic persistence. *Nat. Rev. Microbiol.* **2019**, *17*, 441–448. [CrossRef]
2. Hobby, G.L.; Meyer, K.; Chaffee, E. Observations on the mechanism of action of penicillin. *Proc. Soc. Exp. Biol. Med.* **1942**, *50*, 281–285. [CrossRef]
3. Bigger, J.W. Treatment of staphylococcal infections with penicillin. *Lancet* **1944**, *244*, 497–500. [CrossRef]
4. Johnson, P.J.; Levin, B.R. Pharmacodynamics, population dynamics, and the evolution of persistence in *Staphylococcus aureus*. *PLoS Genet.* **2013**, *9*, e1003123. [CrossRef]
5. Dörr, T.; Lewis, K.; Vulic, M. SOS response induces persistence to fluoroquinolones in *Escherichia coli*. *PLoS Genet.* **2009**, *5*, e1000760. [CrossRef]
6. Wu, Y.; Vulic, M.; Keren, I.; Lewis, K. Role of oxidative stress in persister tolerance. *Antimicrob. Agents Chemother.* **2012**, *56*, 4922–4926. [CrossRef]
7. Korch, S.B.; Henderson, T.A.; Hill, T.M. Characterization of the hipA7 allele of *Escherichia coli* and evidence that high persistence is governed by (p)ppGpp synthesis. *Mol. Microbiol.* **2003**, *50*, 1199–1213. [CrossRef]
8. Nguyen, D.; Joshi-Datar, A.; Lepine, F.; Bauerle, E.; Olakanmi, O.; Beer, K.; McKay, G.; Siehnel, R.; Schafhauser, J.; Wang, Y.; et al. Active starvation responses mediate antibiotic tolerance in biofilms and nutrient-limited bacteria. *Science* **2011**, *334*, 982–986. [CrossRef]
9. Amato, S.M.; Orman, M.A.; Brynildsen, M.P. Metabolic control of persister formation in *Escherichia coli*. *Mol. Cell* **2013**, *50*, 475–487. [CrossRef]
10. Tabone, M.; Lioy, V.S.; Ayora, S.; Machon, C.; Alonso, J.C. Role of toxin ζ and starvation responses in the sensitivity to antimicrobials. *PLoS ONE* **2014**, *9*, e86615. [CrossRef]
11. Conlon, B.P.; Rowe, S.E.; Gandt, A.B.; Nuxoll, A.S.; Donegan, N.P.; Zalis, E.A.; Clair, G.; Adkins, J.N.; Cheung, A.L.; Lewis, K. Persister formation in *Staphylococcus aureus* is associated with ATP depletion. *Nat. Microbiol.* **2016**, *1*, 16051. [CrossRef]
12. Fung, D.K.; Barra, J.T.; Schroeder, J.W.; Ying, D.; Wang, J.D. A shared alarmone-GTP switch underlies triggered and spontaneous persistence. *bioRxiv* **2020**. [CrossRef]
13. Nanamiya, H.; Kasai, K.; Nozawa, A.; Yun, C.S.; Narisawa, T.; Murakami, K.; Natori, Y.; Kawamura, F.; Tozawa, Y. Identification and functional analysis of novel (p)ppGpp synthetase genes in *Bacillus subtilis*. *Mol. Microbiol.* **2008**, *67*, 291–304. [CrossRef]
14. Kriel, A.; Brinsmade, S.R.; Tse, J.L.; Tehranchi, A.K.; Bittner, A.N.; Sonenshein, A.L.; Wang, J.D. GTP dysregulation in *Bacillus subtilis* cells lacking (p)ppGpp results in phenotypic amino acid auxotrophy and failure to adapt to nutrient downshift and regulate biosynthesis genes. *J. Bacteriol.* **2014**, *196*, 189–201. [CrossRef]

15. Moreno-Del Alamo, M.; Tabone, M.; Lioy, V.S.; Alonso, J.C. Toxin ζ triggers a survival response to cope with stress and persistence. *Front. Microbiol.* **2017**, *8*, 1130. [CrossRef]
16. Libby, E.A.; Reuveni, S.; Dworkin, J. Multisite phosphorylation drives phenotypic variation in (p)ppGpp synthetase-dependent antibiotic tolerance. *Nat. Commun.* **2019**, *10*, 5133. [CrossRef]
17. Eiamphungporn, W.; Helmann, J.D. The *Bacillus subtilis* σ^M regulon and its contribution to cell envelope stress responses. *Mol. Microbiol.* **2008**, *67*, 830–848. [CrossRef]
18. Gerdes, K.; Maisonneuve, E. Bacterial persistence and toxin-antitoxin loci. *Annu. Rev. Microbiol.* **2012**, *66*, 103–123. [CrossRef]
19. Page, R.; Peti, W. Toxin-antitoxin systems in bacterial growth arrest and persistence. *Nat. Chem. Biol.* **2016**, *12*, 208–214. [CrossRef]
20. Yamaguchi, Y.; Inouye, M. Regulation of growth and death in *Escherichia coli* by toxin-antitoxin systems. *Nat. Rev. Microbiol.* **2011**, *9*, 779–790. [CrossRef]
21. Diaz-Orejas, R.; Espinosa, M.; Yeo, C.C. The Importance of the expendable: Toxin-antitoxin genes in plasmids and chromosomes. *Front. Microbiol.* **2017**, *8*, 1479. [CrossRef]
22. Ogura, T.; Hiraga, S. Mini-F plasmid genes that couple host cell division to plasmid proliferation. *Proc. Natl. Acad. Sci. USA* **1983**, *80*, 4784–4788. [CrossRef]
23. Fraikin, N.; Goormaghtigh, F.; Van Melderen, L. Type II toxin-antitoxin systems: Evolution and revolutions. *J. Bacteriol.* **2020**, *202*, e00763-19. [CrossRef]
24. Alonso, J.C.; Balsa, D.; Cherny, I.; Christensen, S.K.; Espinosa, S.; Francuski, D.; Gazit, E.; Gerdes, K.; Hitchin, E.; Martín, M.T.; et al. *Bacterial Toxin-Antitoxin Systems as Targets for the Development of Novel Antibiotics*; ASM Press: Washington, DC, USA, 2007; pp. 313–329.
25. Moyed, H.S.; Bertrand, K.P. HipA, a newly recognized gene of *Escherichia coli* K-12 that affects frequency of persistence after inhibition of murein synthesis. *J. Bacteriol.* **1983**, *155*, 768–775. [CrossRef]
26. Bokinsky, G.; Baidoo, E.E.; Akella, S.; Burd, H.; Weaver, D.; Alonso-Gutierrez, J.; Garcia-Martin, H.; Lee, T.S.; Keasling, J.D. HipA-triggered growth arrest and β-Lactam tolerance in *Escherichia coli* are mediated by RelA-dependent ppGpp synthesis. *J. Bacteriol.* **2013**, *195*, 3173–3182. [CrossRef]
27. Germain, E.; Castro-Roa, D.; Zenkin, N.; Gerdes, K. Molecular mechanism of bacterial persistence by HipA. *Mol. Cell* **2013**, *52*, 248–254. [CrossRef]
28. Kaspy, I.; Rotem, E.; Weiss, N.; Ronin, I.; Balaban, N.Q.; Glaser, G. HipA-mediated antibiotic persistence via phosphorylation of the glutamyl-tRNA-synthetase. *Nat. Commun.* **2013**, *4*, 3001. [CrossRef]
29. Vogwill, T.; Comfort, A.C.; Furio, V.; MacLean, R.C. Persistence and resistance as complementary bacterial adaptations to antibiotics. *J. Evol. Biol.* **2016**, *29*, 1223–1233. [CrossRef]
30. Ramisetty, B.C.; Ghosh, D.; Roy Chowdhury, M.; Santhosh, R.S. what is the link between stringent response, endoribonuclease encoding type II toxin-antitoxin systems and persistence? *Front. Microbiol.* **2016**, *7*, 1882. [CrossRef]
31. Goormaghtigh, F.; Fraikin, N.; Putrins, M.; Hallaert, T.; Hauryliuk, V.; Garcia-Pino, A.; Sjodin, A.; Kasvandik, S.; Udekwu, K.; Tenson, T.; et al. Reassessing the role of type II toxin-antitoxin systems in formation of Escherichia coli type II persister cells. *MBio* **2018**, *9*, e00640-18. [CrossRef]
32. Khoo, S.K.; Loll, B.; Chan, W.T.; Shoeman, R.L.; Ngoo, L.; Yeo, C.C.; Meinhart, A. Molecular and structural characterization of the PezAT chromosomal toxin-antitoxin system of the human pathogen *Streptococcus pneumoniae*. *J. Biol. Chem.* **2007**, *282*, 19606–19618. [CrossRef]
33. Meinhart, A.; Alonso, J.C.; Strater, N.; Saenger, W. Crystal structure of the plasmid maintenance system ε/ζ: Functional mechanism of toxin ζ and inactivation by $\varepsilon_2\zeta_2$ complex formation. *Proc. Natl. Acad. Sci. USA* **2003**, *100*, 1661–1666. [CrossRef]
34. Mutschler, H.; Gebhardt, M.; Shoeman, R.L.; Meinhart, A. A novel mechanism of programmed cell death in bacteria by toxin-antitoxin systems corrupts peptidoglycan synthesis. *PLoS Biol.* **2011**, *9*, e1001033. [CrossRef]
35. Meinhart, A.; Alings, C.; Strater, N.; Camacho, A.G.; Alonso, J.C.; Saenger, W. Crystallization and preliminary X-ray diffraction studies of the εζ addiction system encoded by *Streptococcus pyogenes* plasmid pSM19035. *Acta Cryst. D Biol. Cryst.* **2001**, *57*, 745–747. [CrossRef]
36. Moreno-Del Alamo, M.; Tabone, M.; Munoz-Martinez, J.; Valverde, J.R.; Alonso, J.C. Toxin ζ reduces the ATP and modulates the uridine diphosphate-*N*-acetylglucosamine pool. *Toxins* **2019**, *11*, 29. [CrossRef]

37. Tabone, M.; Ayora, S.; Alonso, J.C. Toxin ζ reversible induces dormancy and reduces the UDP-N-acetylglucosamine pool as one of the protective responses to cope with stress. *Toxins* **2014**, *6*, 2787–2803. [CrossRef]
38. Lioy, V.S.; Machon, C.; Tabone, M.; Gonzalez-Pastor, J.E.; Daugelavicius, R.; Ayora, S.; Alonso, J.C. The ζ toxin induces a set of protective responses and dormancy. *PLoS ONE* **2012**, *7*, e30282. [CrossRef]
39. Krasny, L.; Gourse, R.L. An alternative strategy for bacterial ribosome synthesis: *Bacillus subtilis* rRNA transcription regulation. *EMBO J.* **2004**, *23*, 4473–4483. [CrossRef]
40. Kriel, A.; Bittner, A.N.; Kim, S.H.; Liu, K.; Tehranchi, A.K.; Zou, W.Y.; Rendon, S.; Chen, R.; Tu, B.P.; Wang, J.D. Direct regulation of GTP homeostasis by (p)ppGpp: A critical component of viability and stress resistance. *Mol. Cell* **2012**, *48*, 231–241. [CrossRef]
41. Eymann, C.; Homuth, G.; Scharf, C.; Hecker, M. *Bacillus subtilis* functional genomics: Global characterization of the stringent response by proteome and transcriptome analysis. *J. Bacteriol.* **2002**, *184*, 2500–2520. [CrossRef]
42. Lioy, V.S.; Rey, O.; Balsa, D.; Pellicer, T.; Alonso, J.C. A toxin-antitoxin module as a target for antimicrobial development. *Plasmid* **2010**, *63*, 31–39. [CrossRef]
43. Lioy, V.S.; Martin, M.T.; Camacho, A.G.; Lurz, R.; Antelmann, H.; Hecker, M.; Hitchin, E.; Ridge, Y.; Wells, J.M.; Alonso, J.C. pSM19035-encoded ζ toxin induces stasis followed by death in a subpopulation of cells. *Microbiology* **2006**, *152*, 2365–2379. [CrossRef]
44. McKeegan, K.S.; Borges-Walmsley, M.I.; Walmsley, A.R. Microbial and viral drug resistance mechanisms. *Trends Microbiol.* **2002**, *10*, S8–S14. [CrossRef]
45. Brauner, A.; Fridman, O.; Gefen, O.; Balaban, N.Q. Distinguishing between resistance, tolerance and persistence to antibiotic treatment. *Nat. Rev. Microbiol.* **2016**, *14*, 320–330. [CrossRef]
46. Camacho, A.G.; Misselwitz, R.; Behlke, J.; Ayora, S.; Welfle, K.; Meinhart, A.; Lara, B.; Saenger, W.; Welfle, H.; Alonso, J.C. *In vitro* and *in vivo* stability of the $\varepsilon_2\zeta_2$ protein complex of the broad host-range *Streptococcus pyogenes* pSM19035 addiction system. *Biol. Chem.* **2002**, *383*, 1701–1713. [CrossRef]
47. Ayrapetyan, M.; Williams, T.C.; Oliver, J.D. Bridging the gap between viable but non-culturable and antibiotic persistent bacteria. *Trends Microbiol.* **2015**, *23*, 7–13. [CrossRef]
48. LeRoux, M.; Culviner, P.H.; Liu, Y.J.; Littlehale, M.L.; Laub, M.T. Stress can induce transcription of toxin-antitoxin systems without activating toxin. *Mol. Cell* **2020**, *79*, 280–292e288. [CrossRef]
49. Rosendahl, S.; Tammana, H.; Brauera, A.; Remma, M.; Hõraka, R. *Pseudomonas putida* chromosomal toxin-antitoxin systems carry neither clear fitness benefits nor big costs. *bioRxiv* **2020**. [CrossRef]
50. Dy, R.L.; Przybilski, R.; Semeijn, K.; Salmond, G.P.; Fineran, P.C. A widespread bacteriophage abortive infection system functions through a Type IV toxin-antitoxin mechanism. *Nucleic Acids Res.* **2014**, *42*, 4590–4605. [CrossRef]
51. Hazan, R.; Engelberg-Kulka, H. *Escherichia coli mazEF*-mediated cell death as a defense mechanism that inhibits the spread of phage P1. *Mol. Genet. Genom.* **2004**, *272*, 227–234. [CrossRef]
52. Pearl, S.; Gabay, C.; Kishony, R.; Oppenheim, A.; Balaban, N.Q. Nongenetic individuality in the host-phage interaction. *PLoS Biol.* **2008**, *6*, e120. [CrossRef]
53. Alawneh, A.M.; Qi, D.; Yonesaki, T.; Otsuka, Y. An ADP-ribosyltransferase Alt of bacteriophage T4 negatively regulates the *Escherichia coli* MazF toxin of a toxin-antitoxin module. *Mol. Microbiol.* **2016**, *99*, 188–198. [CrossRef]
54. Lyzen, R.; Kochanowska, M.; Wegrzyn, G.; Szalewska-Palasz, A. Transcription from bacteriophage lambda P_R promoter is regulated independently and antagonistically by DksA and ppGpp. *Nucleic Acids Res.* **2009**, *37*, 6655–6664. [CrossRef]
55. Koga, M.; Otsuka, Y.; Lemire, S.; Yonesaki, T. *Escherichia coli rnlA* and *rnlB* compose a novel toxin-antitoxin system. *Genetics* **2011**, *187*, 123–130. [CrossRef]
56. Pane-Farre, J.; Quin, M.B.; Lewis, R.J.; Marles-Wright, J. Structure and function of the stressosome signalling hub. *Subcell Biochem.* **2017**, *83*, 1–41. [CrossRef]
57. Tuomanen, E.; Cozens, R.; Tosch, W.; Zak, O.; Tomasz, A. The rate of killing of *Escherichia coli* by β-lactam antibiotics is strictly proportional to the rate of bacterial growth. *J. Gen. Microbiol.* **1986**, *132*, 1297–1304. [CrossRef]
58. Cohen, N.R.; Lobritz, M.A.; Collins, J.J. Microbial persistence and the road to drug resistance. *Cell Host Microbe* **2013**, *13*, 632–642. [CrossRef]

59. Balaban, N.Q.; Merrin, J.; Chait, R.; Kowalik, L.; Leibler, S. Bacterial persistence as a phenotypic switch. *Science* **2004**, *305*, 1622–1625. [CrossRef]
60. Amitai, S.; Yassin, Y.; Engelberg-Kulka, H. MazF-mediated cell death in *Escherichia coli*: A point of no return. *J. Bacteriol.* **2004**, *186*, 8295–8300. [CrossRef]
61. Franch, T.; Gerdes, K. Programmed cell death in bacteria: Translational repression by mRNA end-pairing. *Mol. Microbiol.* **1996**, *21*, 1049–1060. [CrossRef]
62. Levin, B.R.; Rozen, D.E. Non-inherited antibiotic resistance. *Nat. Rev. Microbiol.* **2006**, *4*, 556–562. [CrossRef]
63. Dörr, T.; Vulic, M.; Lewis, K. Ciprofloxacin causes persister formation by inducing the TisB toxin in *Escherichia coli*. *PLoS Biol.* **2010**, *8*, e1000317. [CrossRef]
64. Kohanski, M.A.; Dwyer, D.J.; Hayete, B.; Lawrence, C.A.; Collins, J.J. A common mechanism of cellular death induced by bactericidal antibiotics. *Cell* **2007**, *130*, 797–810. [CrossRef]
65. Dwyer, D.J.; Camacho, D.M.; Kohanski, M.A.; Callura, J.M.; Collins, J.J. Antibiotic-induced bacterial cell death exhibits physiological and biochemical hallmarks of apoptosis. *Mol. Cell* **2012**, *46*, 561–572. [CrossRef]
66. Cui, P.; Niu, H.; Shi, W.; Zhang, S.; Zhang, W.; Zhang, Y. Identification of genes involved in bacteriostatic antibiotic-induced persister formation. *Front. Microbiol.* **2018**, *9*, 413. [CrossRef]
67. Ayora, S.; Carrasco, B.; Cardenas, P.P.; Cesar, C.E.; Canas, C.; Yadav, T.; Marchisone, C.; Alonso, J.C. Double-strand break repair in bacteria: A view from *Bacillus subtilis*. *FEMS Microbiol. Rev.* **2011**, *35*, 1055–1081. [CrossRef]
68. Kowalczykowski, S.C. An overview of the molecular mechanisms of recombinational DNA repair. *Cold Spring Harb. Perspect. Biol.* **2015**, *7*, a016410. [CrossRef]
69. Simmons, L.A.; Goranov, A.I.; Kobayashi, H.; Davies, B.W.; Yuan, D.S.; Grossman, A.D.; Walker, G.C. Comparison of responses to double-strand breaks between *Escherichia coli* and *Bacillus subtilis* reveals different requirements for SOS induction. *J. Bacteriol.* **2009**, *191*, 1152–1161. [CrossRef]
70. Arjes, H.A.; Kriel, A.; Sorto, N.A.; Shaw, J.T.; Wang, J.D.; Levin, P.A. Failsafe mechanisms couple division and DNA replication in bacteria. *Curr. Biol.* **2014**, *24*, 2149–2155. [CrossRef]
71. Sanchez, H.; Kidane, D.; Reed, P.; Curtis, F.A.; Cozar, M.C.; Graumann, P.L.; Sharples, G.J.; Alonso, J.C. The RuvAB branch migration translocase and RecU Holliday junction resolvase are required for double-stranded DNA break repair in *Bacillus subtilis*. *Genetics* **2005**, *171*, 873–883. [CrossRef]
72. Valero-Rello, A.; Lopez-Sanz, M.; Quevedo-Olmos, A.; Sorokin, A.; Ayora, S. Molecular mechanisms that contribute to horizontal transfer of plasmids by the bacteriophage SPP1. *Front. Microbiol.* **2017**, *8*, 1816. [CrossRef]

Publisher's Note: MDPI stays neutral with regard to jurisdictional claims in published maps and institutional affiliations.

 *toxins*

Review

Targeting Type II Toxin–Antitoxin Systems as Antibacterial Strategies

Marcin Równicki [1,2,*], Robert Lasek [2], Joanna Trylska [1] and Dariusz Bartosik [2,*]

1 Centre of New Technologies, University of Warsaw, ul. S. Banacha 2c, 02-097 Warsaw, Poland; joanna@cent.uw.edu.pl

2 Department of Bacterial Genetics, Institute of Microbiology, Faculty of Biology, University of Warsaw, Miecznikowa 1, 02-096 Warsaw, Poland; lasek@biol.uw.edu.pl

* Correspondence: marcin.rownicki1@gmail.com (M.R.); bartosik@biol.uw.edu.pl (D.B.)

Received: 19 August 2020; Accepted: 31 August 2020; Published: 4 September 2020

Abstract: The identification of novel targets for antimicrobial agents is crucial for combating infectious diseases caused by evolving bacterial pathogens. Components of bacterial toxin–antitoxin (TA) systems have been recognized as promising therapeutic targets. These widespread genetic modules are usually composed of two genes that encode a toxic protein targeting an essential cellular process and an antitoxin that counteracts the activity of the toxin. Uncontrolled toxin expression may elicit a bactericidal effect, so they may be considered "intracellular molecular bombs" that can lead to elimination of their host cells. Based on the molecular nature of antitoxins and their mode of interaction with toxins, TA systems have been classified into six groups. The most prevalent are type II TA systems. Due to their ubiquity among clinical isolates of pathogenic bacteria and the essential processes targeted, they are promising candidates for the development of novel antimicrobial strategies. In this review, we describe the distribution of type II TA systems in clinically relevant human pathogens, examine how these systems could be developed as the targets for novel antibacterials, and discuss possible undesirable effects of such therapeutic intervention, such as the induction of persister cells, biofilm formation and toxicity to eukaryotic cells.

Keywords: toxin–antitoxin systems; toxin activation; antibacterial agents; bacterial persistence

Key Contribution: Bacterial toxin–antitoxin (TA) systems are promising targets for the development of novel antimicrobial strategies.

---

1. Introduction

The emergence of antibiotic-resistant bacteria is a critical issue in modern medicine [1]. Considerable efforts are being made to identify novel antimicrobials to combat infectious diseases caused by evolving multi-resistant pathogens. Bacterial toxin–antitoxin (TA) systems represent promising targets for such compounds. These small genetic modules encode two components: a stable toxin (always a protein) that recognizes a specific cellular target, and a labile antitoxin (protein or RNA molecule), produced in excess, that counteracts the activity of the toxin [2]. TAs have been classified into several groups based on the molecular nature of the antitoxin and their mode of interaction with their cognate toxin. The most prevalent and most extensively studied are type II TA systems, encoding proteic antitoxins which neutralize the toxin by forming TA complexes under optimal growth conditions [3,4]. These two-gene *loci* are organized in operons, whose expression is tightly regulated by both the antitoxin and toxin–antitoxin complexes [5,6]. Degradation of the antitoxin by cellular proteases liberates the toxin, which elicits a bacteriostatic or bactericidal effect [7–10].

TA *loci* were initially identified within bacterial plasmids, where they function as post-segregational cell killing systems (PSK), providing stable maintenance for their carrier replicon in a bacterial population due to the elimination of plasmid-less cells [11]. A surprising observation was that TA systems are also highly abundant and widespread in the chromosomes of free-living bacteria (both Gram-negative and Gram-positive), as well as in archaea [12]. More detailed studies have revealed that these TAs may be involved in important biological processes, such as (i) the stringent response, helping cells to survive stressful conditions by limiting various metabolic activities [13], (ii) programmed cell death [14] and (iii) biofilm formation [15]. In the light of these observations, it seems likely that chromosomal TA *loci* have been transferred to extrachromosomal replicons and adopted as plasmid stabilization systems.

The toxins of TA systems target essential cellular processes of bacteria; therefore, they may be considered intracellular "molecular time bombs", which are activated under certain environmental conditions [16]. Compounds that can artificially activate TA toxins may form a new class of antimicrobials that could represent an alternative to antibiotics [17]. Type II TA systems are ubiquitous in bacteria [18], and their mechanisms are fairly well characterized [18], so they would appear to be excellent candidates for testing the merits of this idea.

The potential use of TA systems in combatting bacterial infections has been examined in several valuable reviews [19–22]. Here, we describe the most promising antibacterial strategies employing type II TA systems and discuss possible undesirable effects of their application, such as the induction of persister cells, biofilm formation and toxicity to eukaryotic cells. In addition, the distribution of type II TA systems among clinically relevant human pathogens is described, which may help to identify appropriate TA candidates for fighting particular infections.

2. Type II TA Systems in Pathogenic Bacteria

Since 2011, the studies of TA systems have been assisted by the Toxin–Antitoxin Database (TADB) [23], updated to version 2.0 in 2017 [24]. Currently, TADB 2.0 lists 6194 type II TA *loci* [24] categorized, where possible, in two partially interdependent ways. The first classification scheme uses a system based on the structural and functional characteristics of the toxin proteins [25,26]. There are 11 families of two-component type II TA systems: *ccd, hicBA, hipBA, mazEF, parD(PemKI), parDE, phd-doc, relBE* (with 5 subfamilies—*relBE, higBA, yfeM-yoeB, ygiTU(mqsAR)* and *prlF-yhaV*), *vapBC, mosAT* and *yeeU*. The less numerous three-component type II TA systems are *ω-ε-ζ, pasABC* and *paaR-paaA-parE*. The second classification scheme is based on a set of 44 conserved toxin–antitoxin protein domain pairings and better reflects the versatility and modularity of TA systems [27]. Figure 1 summarizes the variety of type II TA systems identified in bacterial species commonly associated with pathogenesis in humans. Among type II TA system (sub)families, four are most frequently encountered: *vapBC, relBE, mazEF* and *higBA*, and these account for ~80% of the listed *loci*. (An expanded list of type II TA systems of bacterial pathogens, organized into toxin–antitoxin domain pair groupings, is given in Supplementary Figure S1.)

Due to their widespread occurrence within the accessory genomes of human pathogens, their probable role in pathogenicity and potential as therapeutic targets, type II TA systems have been a subject of growing interest in recent years. Several systematic reviews have focused on the distribution and roles of the TA systems in clinically relevant pathogenic bacteria, including *Escherichia coli, Mycobacterium tuberculosis, Neisseria gonorrheae, Streptococcus* spp., *Burkholderia* spp. and species of the ESKAPE group (*Enterococcus faecium, Staphylococcus aureus, Klebsiella pneumoniae, Acinetobacter baumannii, Pseudomonas aeruginosa* and *Enterobacter* spp.) [19,28–32]. An increasing number of studies are being undertaken to describe the repertoire of TA systems in pathogenic bacteria at the level of a given strain or of a whole taxon. Not only do such approaches shed light on the importance of overall TA system networks with respect to the physiology, virulence and evolution of pathogens, but thanks to the large-scale genome comparisons involved, they also lead to the discovery of novel TA *loci*.

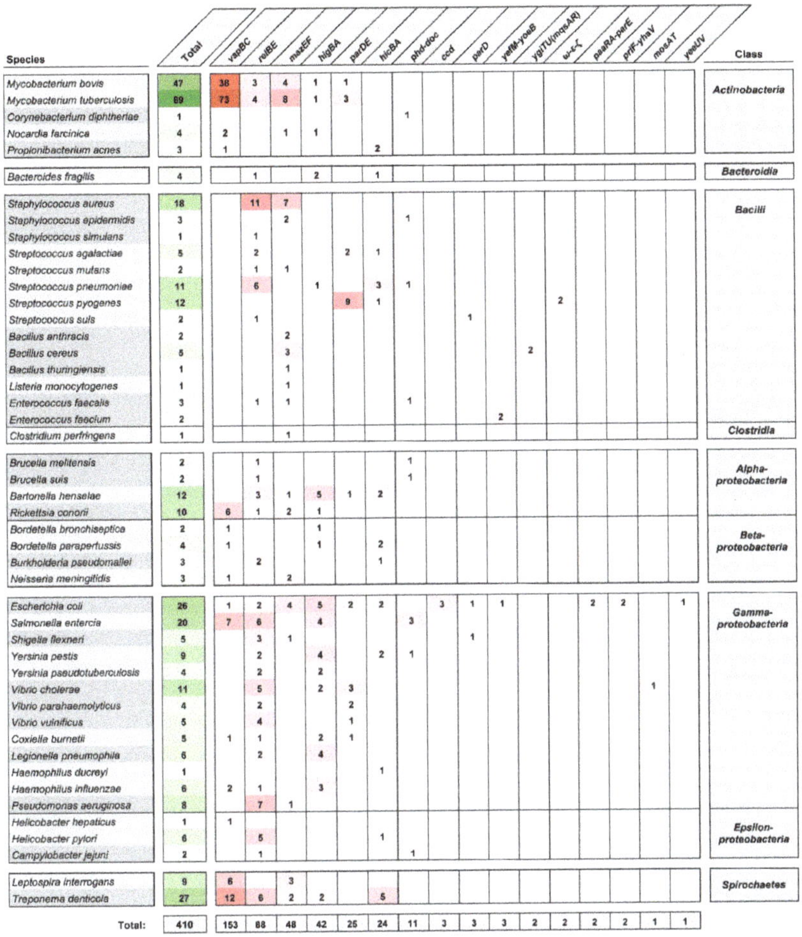

Species	Total	*vapBC*	*relBE*	*mazEF*	*higBA*	*parDE*	*hicBA*	*phd-doc*	*ccd*	*parD*	*yefM-yoeB*	*ygiTU(mqsAR)*	*ω-ε-ζ*	*paaRA-parE*	*prlF-yhaV*	*mosAT*	*yeeUV*	Class
Mycobacterium bovis	47	38	3	4	1	1												*Actinobacteria*
Mycobacterium tuberculosis	89	73	4	8	1	3												
Corynebacterium diphtheriae	1							1										
Nocardia farcinica	4	2		1	1													
Propionibacterium acnes	3	1					2											
Bacteroides fragilis	4		1		2		1											*Bacteroidia*
Staphylococcus aureus	18		11	7														*Bacilli*
Staphylococcus epidermidis	3			2				1										
Staphylococcus simulans	1		1															
Streptococcus agalactiae	5		2			2	1											
Streptococcus mutans	2		1	1														
Streptococcus pneumoniae	11		6		1		3	1										
Streptococcus pyogenes	12					9	1						2					
Streptococcus suis	2		1							1								
Bacillus anthracis	2			2														
Bacillus cereus	5			3								2						
Bacillus thuringiensis	1			1														
Listeria monocytogenes	1			1														
Enterococcus faecalis	3		1	1				1										
Enterococcus faecium	2										2							
Clostridium perfringens	1			1														*Clostridia*
Brucella melitensis	2		1					1										*Alpha-proteobacteria*
Brucella suis	2		1					1										
Bartonella henselae	12		3	1	5	1	2											
Rickettsia conorii	10	6	1	2	1													
Bordetella bronchiseptica	2	1			1													*Beta-proteobacteria*
Bordetella parapertussis	4	1			1		2											
Burkholderia pseudomallei	3		2				1											
Neisseria meningitidis	3	1		2														
Escherichia coli	26	1	2	4	5	2	2		3	1	1			2	2		1	*Gamma-proteobacteria*
Salmonella entercia	20	7	6		4			3										
Shigella flexneri	5		3	1						1								
Yersinia pestis	9		2		4		2	1										
Yersinia pseudotuberculosis	4		2		2													
Vibrio cholerae	11		5		2	3										1		
Vibrio parahaemolyticus	4		2			2												
Vibrio vulnificus	5		4			1												
Coxiella burnetii	5	1	1		2	1												
Legionella pneumophila	6		2		4													
Haemophilus ducreyi	1						1											
Haemophilus influenzae	6	2	1		3													
Pseudomonas aeruginosa	8		7	1														
Helicobacter hepaticus	1	1																*Epsilon-proteobacteria*
Helicobacter pylori	6		5				1											
Campylobacter jejuni	2		1					1										
Leptospira interrogans	9	6		3														*Spirochaetes*
Treponema denticola	27	12	6	2	2		5											
Total:	410	153	88	48	42	25	24	11	3	3	3	2	2	2	2	1	1	

Figure 1. Type II toxin–antitoxin (TA) systems identified in human pathogenic bacteria according to the toxin family classification system [25,26] as collected in the Toxin–Antitoxin Database (TADB) 2.0 [24]. The number given for an individual species is the sum of the values for all strains of a given taxon. This list was manually curated to represent strains and/or species frequently associated with infections in humans (including opportunistic pathogens).

M. tuberculosis is notable for its abundance of TA systems, especially of the *vapBC*, *mazEF* and *relBE* families (Figure 1), and their activity has been linked to the regulation of adaptive responses to stress caused by interaction with the host and drug treatment [33]. Predictably, *M. tuberculosis* has been the object of studies on the common expression patterns of TA *loci* [34–36], as well as cross-activation between homologous systems [37,38], and putative interaction between non-cognate toxin–antitoxin pairs, which remains controversial [39–41]. Furthermore, comparative studies of the distribution of TA systems in mycobacteria have demonstrated the variability of TA *loci* between different *M. tuberculosis* lineages, and led to the discovery of putative novel TA systems [42,43]. Similarly, the body of knowledge about the TA systems of *S. aureus* [44] was recently expanded by several studies providing insights into the association between the TA systems landscape and strain phenotypes [45–48]. A recent in-depth analysis of the diversity and distribution of type II and type IV TA systems in the genomes of

K. pneumoniae species complex identified several novel toxins and demonstrated the co-occurrence of TA *loci* and clinically relevant genes [49].

Comprehensive strategies for the identification of novel TA *loci* are required in order to gain a greater understanding of the role of TA networks in the biology of bacterial cells. In response to this need, Akarsu and colleagues created the "discovery-oriented" database TASmania in 2019 [50]. Using this newly developed pipeline, they annotated a set of over 41,000 assemblies from the EnsemblBacteria database, resulting in the identification of >2 × 10^6 candidate TA *loci* [50]. The greater flexibility of TASmania compared to the TAFinder search tool in TADB 2.0 allowed for the identification of a higher number of putative TA *loci*, thus providing a starting point for experimental analyses [50]. Moreover, the "guilt-by-association" strategy used throughout the annotation process, i.e., targeting *loci* directly neighboring orphan toxin or antitoxin genes, facilitated the discovery of new TA protein families. In the case of *Listeria monocytogenes*, TASmania-assisted large-scale genomic comparisons led to the identification of 14 putative TA genes [51].

3. Strategies for the Artificial Activation of Toxin–Antitoxin Systems

The growth-inhibitory and lethal consequences of the activity of TA systems led to the proposal that artificial activation of the toxins could provide an effective antibacterial strategy [17,52]. Type II TA systems seem to be most convenient for developing such strategies because many toxins of these systems have been thoroughly characterized and their cellular targets are known [53–55]. Moreover, as shown in Figure 1, the presence and conservation of type II TAs have been confirmed in major human-associated bacterial pathogens. Importantly, TA systems have no human homologs, and no pre-existing resistance against TA toxins has been observed. To date, several TA-based antibacterial strategies have been proposed [19,20], but in most cases they are not supported by significant experimental data.

In this review, we focus our attention on the most promising approaches: (i) the direct activation of TA systems by the use of specific molecules that interfere with TAs and imbalance the delicate stoichiometry of active toxin and antitoxin in bacterial cells, and (ii) the indirect activation of TAs by triggering other cellular components whose functions are interconnected with the TA system. Another novel and interesting approach that we consider involves the use of engineered species-specific toxins that can selectively kill selected strains of pathogenic bacteria.

3.1. Direct Activation of TA Systems

3.1.1. Disruption or Preventing the Formation of TA Complexes

The most straightforward strategy for direct activation of the antibacterial effect of type II TA systems requires disruption of the protein toxin and antitoxin complexes, leading to toxin liberation (Figure 2a). This strategy has been validated by several research groups who focused their efforts on identifying high affinity peptide inhibitors that can efficiently displace toxins from their cognate antitoxins.

Lioy and co-workers selected the ε-ζ TA system from *Streptococcus pyogenes* as their research model because the free ζ toxin was shown to trigger a loss of cell proliferation similar to that caused by known antimicrobials [56]. Using high-throughput methods, they screened several oligopeptide libraries for the ability to impair the assembly of ε–ζ complexes. This led to the identification of a library containing a mixture of 17-amino acid-long oligopeptides that interfered with the TA interaction. However, further subfractionation of this library resulted in a diminished effect. The authors suggested that the disruption of ε–ζ complexes that was originally observed might have been a consequence of the concerted action of several weak-binding oligopeptides. Nevertheless, this study provided a proof-of-concept for the antimicrobial potential of this strategy [56].

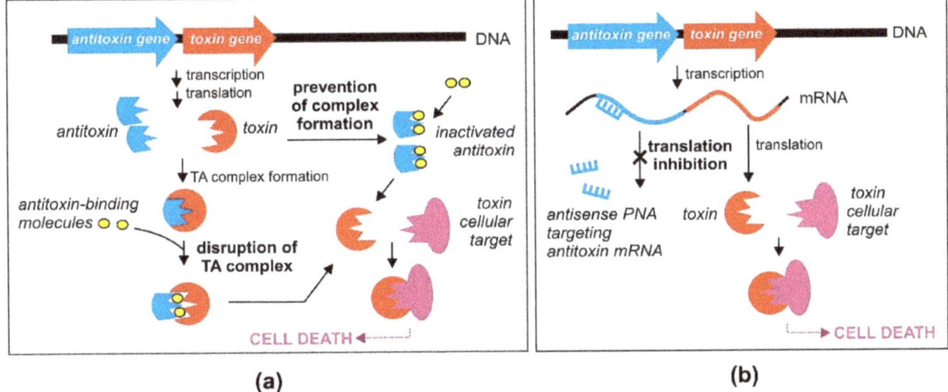

Figure 2. Proposed antibacterial strategies based on the direct activation of toxins of TA systems; (**a**) disruption of the protein toxin and antitoxin complex and/or prevention of protein complex formation; (**b**) inhibition of antitoxin translation (see text for details).

A similar approach was applied in the case of two related TA systems—*pemIK* and *moxXT* of *Bacillus anthracis*. The former module encodes the toxin PemK, a ribonuclease whose overexpression exerts a toxic effect in *B. anthracis* cells characterized by the drastic inhibition of protein synthesis [57]. Based on in silico protein structural modeling, several peptides were designed to mimic the C-terminal domain of the antitoxin PemI, which is involved in toxin binding. In vitro experiments indicated the effectiveness of the designed molecules and demonstrated the feasibility of disrupting the TA interaction using octapeptides [58]. Similar results were obtained in studies on the related TA system *moxXT*. Based on the crystal structure of the MoxX–MoxT complex, Verma and colleagues [59] designed a series of peptides that effectively disturbed the interaction of MoxT with antitoxin MoxX and stimulated MoxT ribonuclease activity [60].

Analogously, following a detailed analysis of *vapBC* TA systems of *Mycobacterium tuberculosis*, Lee and colleagues [61] designed VapB- and VapC-based peptides that effectively disrupted the TA complexes, causing activation of the VapC toxin ribonuclease [61].

3.1.2. Inhibition of Antitoxin Translation

Toxin release can also be caused by a reduction in the amount of antitoxin molecules in a bacterial cell. One way to achieve such an effect is by blocking antitoxin translation (Figure 2b) [20]. The toxin and antitoxin genes, co-transcribed as a single mRNA, possess separate Shine–Dalgarno sequences [62]. Therefore, inhibition of antitoxin translation should not influence the translation of the toxin protein. This strategy was tested by Równicki and co-workers who designed a peptide nucleic acid (PNA)-based treatment to inhibit translation of the antitoxins of the *mazEF* and *hipBA* TA systems of *E. coli* [63].

The sequence-specific antisense PNAs targeted either *mazE* or *hipB* antitoxin mRNAs and as a result lowered the cellular levels of these transcripts, which caused an effective inhibition of *E. coli* growth [63]. Importantly, the PNA treatment did not change the relative levels of the *mazF* or *hipA* toxin mRNAs. The crucial role of the MazF and HipA toxins in the observed growth inhibition was confirmed by showing that *E. coli* mutants lacking the genes encoding these proteins were "resistant" to treatment with the antitoxin-specific PNAs.

Another PNA-based strategy, described in the same report, used PNA oligomers directed at the cellular target of the toxin, thus bypassing the involvement of the TA system. It was demonstrated that PNAs can mimic the action of the HipA toxin by silencing the *gltX* gene encoding its cellular target, glutamyl-tRNA synthase [63]. For these experiments, the PNA oligomers were conjugated with a cell penetrating peptide—$(KFF)_3K$, which indicated the importance of employing an efficient carrier to introduce these antisense oligonucleotides into bacterial cells [63].

The above results confirmed experimentally that TA systems are susceptible to sequence-specific antisense agents and provided a basis for their further exploitation in antimicrobial strategies. Importantly, PNA oligomers exhibit nuclease and protease resistance, high binding affinity to natural nucleic acids, and negligible toxicity to eukaryotic cells [64,65].

3.2. Indirect Activation of TA Systems

3.2.1. Enhanced Expression of Proteases Degrading Antitoxins

As previously mentioned, the antitoxins of type II TA systems are more susceptible to degradation by host cytoplasmic proteases than their cognate toxins. In most cases, bacterial Lon protease is involved in antitoxin degradation; however, the two-component protease ClpP, acting in cooperation with the chaperones ClpA or ClpX, is used by some TA systems [66–68].

The depletion of antitoxin molecules results in the liberation of the toxin proteins from TA complexes and relieves transcriptional repression of TA operons, causing increased production of TA transcripts. These events alter the stoichiometry of TA molecules in a cell and ultimately have a lethal effect [69]. Therefore, increasing the level of cellular proteases or designing specific molecules that activate these proteases could be a promising strategy for the indirect activation of toxins (Figure 3a).

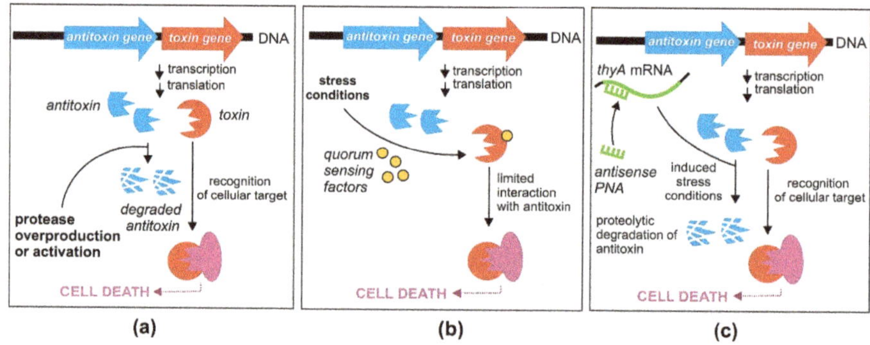

Figure 3. Proposed antibacterial strategies based on the indirect activation of toxins of TA systems: (**a**) activation of the Lon or ClpP proteases that degrade antitoxins; (**b**) triggering TA systems by quorum sensing factors; (**c**) triggering TAs by artificial induction of the stringent response (see text for details).

Increased protease expression can be achieved by introducing a plasmid carrying a cloned protease gene. Overproduction of the Lon protease is known to be lethal for *E. coli* cells [70]. However, by employing an inducible Lon overproduction system, Christensen and colleagues overcame this problem to demonstrate that the lethality is partially dependent on the *yefM-yoeB* TA system [66]. They showed that overproduction of Lon triggered YoeB-dependent mRNA cleavage, leading to translation inhibition. This, in turn, activated the YoeB toxin by preventing the synthesis of its unstable YefM antidote, which was eventually lethal to the host cells [66].

Proteolysis is a tightly controlled process which can be significantly influenced by specific molecules targeting proteases. For example, acyldepsipeptides (ADEPs) are compounds with antibiotic properties that specifically activate the bacterial protease ClpP. Uncontrolled proteolysis induced in this way inhibits bacterial cell division and results in cell death, possibly with the participation of activated toxins [71].

3.2.2. Triggering of TA Systems by Quorum-Sensing Factors

Another interesting antimicrobial strategy was developed by Kumar and Engelberg-Kulka [72,73]. Their approach targeted *mazEF* TA systems, which are among the most abundant bacterial TA *loci* (Figure 1). Toxin MazF is a sequence-specific endoribonuclease that initiates a programmed cell death pathway in response to environmental stress conditions [74], while MazE is a labile antitoxin that is preferentially degraded by the serine protease ClpAP [14].

The proposed strategy involves the use of a newly discovered group of pentapeptides secreted by bacteria, called extracellular death factors (EDFs), which act in quorum sensing and enhance MazF activity under stressful conditions (Figure 3b) [72]. Interestingly, it was shown that EDFs bind directly to MazF in a sequence-specific manner, and this binding is likely to limit interaction of the toxin with its cognate antitoxin MazE [73,75]. EDFs can also stimulate activation of *mazEF* in heterologous hosts, which might broaden the potential application of this strategy [73,75]. Although an early study on MazF classified its toxicity as lethal to cells [14], this statement has since been revisited, suggesting that MazF is involved in growth arrest rather than cell death [76,77].

3.2.3. Induction of the Stringent Response

Another antibacterial strategy involving the indirect activation of toxins is based on induction of the stringent response, a conserved mechanism that allows bacteria to adapt their metabolism in response to stressful environmental conditions, e.g., nutrient deprivation. Many chromosomal type II TA systems, including the aforementioned *mazEF loci*, are transcriptionally upregulated under stressful conditions [78], and their activity leads to remodeling of cellular metabolism and/or programmed cell death, affecting part of the bacterial community [74].

The stringent response (mediated by the alarmone guanosine 3,5 bispyrophosphate, ppGpp) is activated by different natural starvation and stress signals [79], and this can also be achieved by the application of artificial factors. Równicki and colleagues used sequence-specific PNAs targeting the *thyA* gene of *E. coli*, conjugated with a $(KFF)_3K$ peptide as a carrier, to trigger MazF toxin production by inducing thymine starvation [63] (Figure 3c). The *thyA* gene encodes thymidylate synthase, an enzyme involved in folic acid metabolism, which normally interferes with *mazEF*-mediated growth inhibition [74]. As shown for *E. coli*, thymine starvation leads to accumulation of ppGpp in bacterial cells, which reduces global transcription [80]. As a consequence, the inhibition of transcription of *mazEF* leads to activation of the MazF toxin [81]. The significantly reduced level of *thyA* mRNA after treatment with a complementary anti-*thyA* PNA and the resulting growth inhibition confirmed the effectiveness of this silencing strategy [63].

3.3. *Engineered TA Systems in a Targeted Killing Strategy*

An innovative strategy for the targeted killing of selected pathogenic bacteria, without harming beneficial members of the microbiota inhabiting eukaryotic host organisms, was recently proposed by López-Igual et al. [82] (Figure 4). This strategy is based on the use of the *ccdA-ccdB* type II TA system, encoding the toxin CcdB that poisons DNA gyrase—an enzyme responsible for the negative supercoiling of bacterial DNA [83]. CcdB interferes with the activity of DNA gyrase, inducing it to form a covalent GyrA–DNA complex that cannot be resolved, thus promoting DNA breakage and cell death. This mechanism is closely related to the action of quinolone antibiotics, which also target DNA gyrase [84].

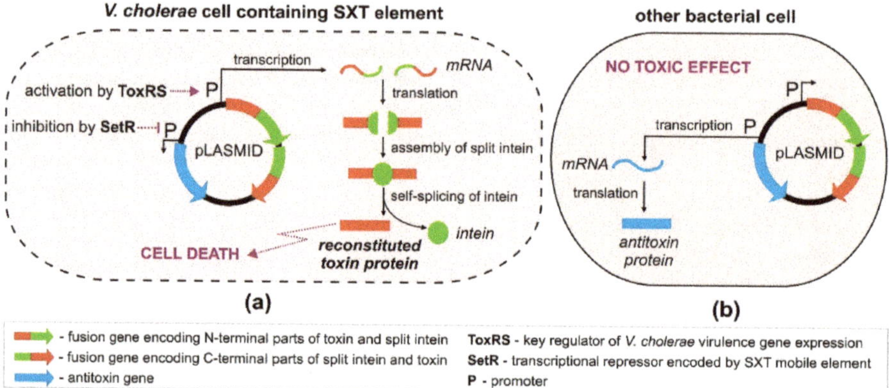

Figure 4. Engineered CcdB toxin as a *V. cholerae*-specific antimicrobial agent. pLASMID—carrier plasmid molecule containing engineered genes of the *ccdAB* TA system. (**a**) In *V. cholerae* cells, by producing ToxRS and SetR transcription factors, transcription of the antitoxin gene is repressed, and expression of engineered toxin genes is activated, leading to cell death; (**b**) in other bacterial cells (lacking both transcription factors), the antitoxin gene is preferentially expressed.

To better control CcdB production in vivo, the gene for this toxin was divided into two parts, and each was fused with DNA encoding split inteins. The generation of a functional toxin occurs in three stages: (i) expression of the two polypeptides, (ii) their association and ligation into a single fusion protein, and (iii) self-splicing of the intein (Figure 4). An important step in this strategy was the construction of a mobilizable plasmid containing genes encoding the antitoxin and the engineered toxin–intein, whose expression could be independently regulated in bacteria harboring specific transcription factors. Although the plasmid could be readily transferred by conjugation from *E. coli* to other bacteria, the toxic effect was observed exclusively in strains of *Vibrio cholerae*. This host specificity was achieved by cloning the engineered *ccdB* toxin–intein genes downstream of a promoter regulated by the transcriptional activator ToxRS, a cholera toxin-associated activator characteristic for *V. cholerae* [85] (Figure 4). Bacterial strains lacking ToxRS were unaffected, including the *E. coli* donor strain and non-pathogenic *Vibrio* spp.

The second component of the TA module, the antitoxin gene *ccdA*, was placed under the transcriptional control of the repressor SetR. The presence of *setR* is considered a hallmark of SXT—an integrative and conjugative mobile element (ICE) of *V. cholerae* that often includes various antibiotic resistance genes [86]. Therefore, expression of the antitoxin gene is repressed in pathogenic multidrug-resistant *V. cholerae* cells containing the *setR* gene, allowing the toxin to poison gyrase and cause a bactericidal effect (Figure 4). However, in cells that lack *setR*, the antitoxin CcdA is produced, neutralizing the effects of the toxin. Thus, the described approach allows species-specific killing of antibiotic-resistant *V. cholerae* strains without affecting the growth of other bacteria present in the mixed populations. The effectiveness of this strategy was confirmed in vivo, in the microbiota of zebrafish and crustacean larvae, where *Vibrio* spp. naturally occur [82].

4. Toxicity to Eukaryotic Cells

Although TA *loci* do not occur in eukaryotic genomes, most toxins of TA systems are endoribonucleases that cleave mRNAs irrespective of their origin, be that prokaryotic or eukaryotic. Therefore, before using these toxins in potential antibacterial strategies, it is important to consider possible cytotoxicity and side effects on human cells. Unfortunately, only a few studies have addressed this issue to date. Notably, toxin cytotoxicity was exploited in one study that employed an engineered version of the MazF toxin to reduce solid tumors in mice [87]. Similarly, the VapC toxin from a TA system of *M. tuberculosis* demonstrated pro-apoptotic activity in human cancer cells, regardless of the expression system used. In another study, Chono and colleagues found that the MazF toxin dosage is a critical factor in determining its activity and cytotoxicity in eukaryotic cells [88,89]. The above examples show that mammalian cells are sensitive to the ribonuclease activity of toxins. However, the ability of bacterial TA-system toxins to penetrate eukaryotic cells is currently unknown, and strategies that specifically target pathogens or utilize toxins that lack targets in human cells might prevent any deleterious effects.

5. Role of Type II TA Modules in Biofilm Formation and Bacterial Persistence

Two potentially undesirable effects of artificial activation of TAs are bacterial persistence and biofilm formation. Persister cells are a subpopulation of slow-growing or growth-arrested bacterial cells that have a decreased susceptibility to killing by bactericidal antibiotics within an otherwise susceptible clonal population [90]. It has been shown that increased tolerance of biofilms to antibiotics is due to the higher amounts of persister cells within the biofilm community [91,92]. While antibiotics kill the majority of biofilm cells, persisters remain viable and repopulate biofilms when the level of antibiotics drops [93]. Thus, persister cells appear to play a central role in the recalcitrance of chronic and biofilm-related infections [94].

One of the first items of evidence linking type II TA modules to biofilm formation comes from a well-characterized chromosomal TA system *mqsRA* in *E. coli* [95,96]. The *mqsRA* is a unique type II TA system where the toxin gene *msqR* precedes the antitoxin *msqA* [97]. It has been shown that a Tn*5* insertion mutant of the toxin *mqsR* formed less adherent biomass [96]. On the contrary, later reanalysis of the *msqRA* revealed that this TA module does not affect biofilm formation in nutrient-rich conditions [98]. The authors identified two new promoters located in the toxin coding sequence that allow the constitutive expression of *mqsA*, thereby allowing a constant and steady level of the MqsA antitoxin compared to the MqsR toxin. This work, quite understandably, has opened up the debate about the role of TA systems in persistence and biofilm formation [78,99]. This result disproves the role of *mqsRA* (and nine others) TA modules in spontaneous biofilm formation, but does not address the question of a role of TA systems in growth control and biofilm formation under stress conditions. The first persistence-related gene to be identified was *hipA*, encoding the toxin of the *E. coli hipBA* TA system—a serine/threonine kinase that inhibits cell growth by inactivating the glutamyl-tRNA synthetase GltX [100]. This discovery linked TA modules and bacterial persistence for the first time. Since then, correlations between antibiotic persistence and TA systems have been extensively studied. Evidence both for and against the participation of TA systems in persister cell formation has been accumulated, making their contribution to this phenomenon unclear [99,101–104]. For example, the deletion of 10 ribonuclease-encoding TA systems from the *E. coli* genome was found to decrease the number of persisters. However, it was subsequently shown that this result was influenced by the presence of φ80 bacteriophage contamination [105]. Moreover, reconstruction of this mutant strain demonstrated that deletion of the 10 TA systems did not affect susceptibility to ofloxacin or ampicillin [106]. These contradictory findings have given rise to considerable debate [78,107–109], and the relevance of TA systems to bacterial persistence remains unclear.

In contrast, the results of several other studies support the contribution of TA systems to bacterial persistence. For example, overexpression of the toxins RelE or MazF was shown to increase the survival of *E. coli* under antibiotic exposure [9,110]. In another study, the *dinJ/yafQ* TA module was found to be involved in tolerance to cephalosporin and aminoglycoside antibiotics [111]. Further evidence linking type II TA modules and bacterial persistence was obtained in a study on uropathogenic isolates of *E. coli* [112]. TA systems have also been linked to bacterial persistence in *Salmonella*, where the *shpB1* allele, carrying a mutation in the antitoxin of the *shpAB* TA module, was associated with the salmonella high persister phenotype [113]. In addition to the *shpAB* system, 13 other type II TA modules were shown to contribute to persistence in *Salmonella* triggered by stresses encountered during macrophage infection [114]. Furthermore, the overexpression of three acetyltransferase toxins (TacT, TacT2, TacT3—acetylating tRNA) in *Salmonella enterica* was found to increase the level of persisters in the population, and their deletion resulted in a decrease in the proportion of persister cells [115,116]. Since the general mechanism of persistence is still unclear, it is crucial to determine whether the activation of TAs influences persister or biofilm formation before they are considered for use as antimicrobial targets. Besides the type II TA modules, other TA systems have been shown to be involved in persistence in *E. coli*. For further information see important recent publications by Fisher et al. [117], Ronneau and Helaine [78], Wilmaerts et al. [118], Dorr et al. [119], and papers from the Wood group [120,121].

While persistence is among the possible side effects of using TAs as antibacterial targets, these systems can also serve in antipersister strategies [122]. Conlon and colleagues [123] reasoned that a compound capable of striking a target in dormant cells will kill persisters. They used an acyldepsipeptide antibiotic (ADEP4) to globally activate the protease ClpP and showed that this becomes fairly non-specific and kills persisters by degrading over 400 proteins, which forces cells to self-digest. Subsequently, this strategy was used to eradicate a biofilm in an animal model, which confirmed its potential as the basis of therapies to treat chronic infections. In a different approach, Li and co-workers [124] identified a novel inhibitor of the *E. coli* HipA toxin which interfered with persister formation in an antibiotic-independent manner. A comprehensive review describing current antipersister strategies was recently published by Defraine et al. [122].

6. Conclusions

Although TA-based antimicrobial strategies have great potential, it is still too early to assess the therapeutic value of toxin activation in clinical settings. Drug discovery and development are time consuming and costly processes, so selecting the right approach will be very important if efforts to produce new antimicrobials based on TA activators are to progress. It is also crucial to select the correct TA targets, and such decisions should be based on data concerning their clinical relevance (i.e., prevalence in antibiotic-resistant clinical isolates), effect exerted on bacterial cells (bactericidal, bacteriostatic), influence on persister cell and biofilm formation, cytotoxicity to human cells and lastly the necessary mode of drug delivery. It would appear advantageous to combine the use of TA activators with conventional antibiotics, since such a strategy could be more effective against a broader spectrum of multidrug-resistant strains [63]. The use of engineered toxins is another promising avenue of research [82,125]. According to the type and properties of such recombinant proteins, they might be applied in targeted antimicrobial strategies or directed against human cancer cells in novel anti-cancer therapies.

Supplementary Materials: The following is available online at http://www.mdpi.com/2072-6651/12/9/568/s1, Figure S1: The number of the type II TA systems identified in human pathogenic bacteria according to the toxin-antitoxin domain pair system [27] as collected in TADB 2.0 [24].

Funding: We acknowledge support from the National Science Centre, Poland (PRELUDIUM 2017/25/N/NZ1/01578 to M.R. and J.T.; UMO-2016/23/B/NZ1/03198 to J.T).

Conflicts of Interest: The authors declare no conflict of interest.

References

1. Ventola, C.L. The antibiotic resistance crisis: Causes and threats. *Pharm. Ther. J.* **2015**, *40*, 277–283.
2. Gerdes, K.; Nielsen, A.; Thorsted, P.; Wagner, E.G.H. Mechanism of killer gene activation. Antisense RNA-dependent RNase III cleavage ensures rapid turn-over of the stable Hok, SrnB and PndA effector messenger RNAs. *J. Mol. Biol.* **1992**, *226*, 637–649. [CrossRef]
3. Fraikin, N.; Goormaghtigh, F.; Van Melderen, L. Type II toxin-antitoxin systems: Evolution and revolutions. *J. Bacteriol.* **2020**, *202*, e00763-19. [CrossRef]
4. Page, R.; Peti, W. Toxin-antitoxin systems in bacterial growth arrest and persistence. *Nat. Chem. Biol.* **2016**, *12*, 208–214. [CrossRef]
5. Coussens, N.P.; Daines, D.A. Wake me when it's over-Bacterial toxin-antitoxin proteins and induced dormancy. *Exp. Biol. Med.* **2016**, *241*, 1332–1342. [CrossRef]

6. Yamaguchi, Y.; Inouye, M. Regulation of growth and death in *Escherichia coli* by toxin–antitoxin systems. *Nat. Rev. Microbiol.* **2011**, *9*, 779–790. [CrossRef]
7. Piscotta, F.J.; Jeffrey, P.D.; James Link, A. ParST is a widespread toxin–antitoxin module that targets nucleotide metabolism. *Proc. Natl. Acad. Sci. USA* **2019**, *116*, 826–834. [CrossRef]
8. Unterholzner, S.J.; Poppenberger, B.; Rozhon, W. Toxin–antitoxin systems. Biology, identification, and application. *Mob. Genet. Elem.* **2013**, *3*, 1–13. [CrossRef]
9. Keren, I.; Shah, D.; Spoering, A.; Kaldalu, N.; Lewis, K. Specialized persister cells and the mechanism of multidrug tolerance in *Escherichia coli*. *J. Bacteriol.* **2004**, *186*, 8172–8180. [CrossRef]
10. Robson, J.; McKenzie, J.L.; Cursons, R.; Cook, G.M.; Arcus, V.L. The vapBC operon from Mycobacterium smegmatis is an autoregulated toxin-antitoxin module that controls growth via inhibition of translation. *J. Mol. Biol.* **2009**, *390*, 353–367. [CrossRef]
11. Dziewit, L.; Jazurek, M.; Drewniak, L.; Baj, J.; Bartosik, D. The SXT conjugative element and linear prophage N15 encode toxin-antitoxin-stabilizing systems homologous to the tad-ata module of the Paracoccus aminophilus plasmid pAMI2. *J. Bacteriol.* **2007**, *189*, 1983–1997. [CrossRef]
12. Pandey, D.P.; Gerdes, K. Toxin-antitoxin loci are highly abundant in free-living but lost from host-associated prokaryotes. *Nucleic Acids Res.* **2005**, *33*, 966–976. [CrossRef]
13. Christensen, S.K.; Mikkelsen, M.; Pedersen, K.; Gerdes, K. RelE, a global inhibitor of translation, is activated during nutritional stress. *Proc. Natl. Acad. Sci. USA* **2001**, *98*, 14328–14333. [CrossRef]
14. Aizenman, E.; Engelberg-Kulka, H.; Glaser, G. An *Escherichia coli* chromosomal "addiction module" regulated by 3′,5′-bispyrophosphate: A model for programmed bacterial cell death. *Proc. Natl. Acad. Sci. USA* **1996**, *93*, 6059–6063. [CrossRef]
15. Wen, Y.; Behiels, E.; Devreese, B. Toxin-antitoxin systems: Their role in persistence, biofilm formation, and pathogenicity. *Pathog. Dis.* **2014**, *70*, 240–249. [CrossRef]
16. Kędzierska, B.; Lian, L.Y.; Hayes, F. Toxin-antitoxin regulation: Bimodal interaction of YefM-YoeB with paired DNA palindromes exerts transcriptional autorepression. *Nucleic Acids Res.* **2007**, *35*, 325–339. [CrossRef]
17. Williams, J.; Hergenrother, P. Artificial activation of toxin-antitoxin systems as an antibacterial strategy. *Trends Microbiol.* **2014**, *20*, 291–298. [CrossRef]
18. Harms, A.; Brodersen, D.E.; Mitarai, N.; Gerdes, K. Toxins, targets, and triggers: An overview of toxin-antitoxin biology. *Mol. Cell* **2018**, *70*, 768–784. [CrossRef]
19. Lee, K.Y.; Lee, B.J. Structure, biology, and therapeutic application of toxin-antitoxin systems in pathogenic bacteria. *Toxins (Basel)* **2016**, *8*, 305. [CrossRef]
20. Chan, W.T.; Balsa, D.; Espinosa, M. One cannot rule them all: Are bacterial toxins-antitoxins druggable? *FEMS Microbiol. Rev.* **2015**, *39*, 522–540. [CrossRef]
21. Bravo, A.; Ruiz-Cruz, S.; Alkorta, I.; Espinosa, M. When humans met superbugs: Strategies to tackle bacterial resistances to antibiotics. *Biomol. Concepts* **2018**, *9*, 216–226. [CrossRef] [PubMed]
22. Carlos, L.; Barbosa, B.; Sânder, A.; Cangussu, R.; Garrido, S.S.; Marchetto, R. Toxin-antitoxin systems and its biotechnological applications. *Afr. J. Biotechnol.* **2014**, *13*, 11–17.
23. Shao, Y.; Harrison, E.M.; Bi, D.; Tai, C.; He, X.; Ou, H.Y.; Rajakumar, K.; Deng, Z. TADB: A web-based resource for type 2 toxin-antitoxin loci in bacteria and archaea. *Nucleic Acids Res.* **2011**, *39*, D606–D611. [CrossRef]
24. Xie, Y.; Wei, Y.; Shen, Y.; Li, X.; Zhou, H.; Tai, C.; Deng, Z.; Ou, H.Y. TADB 2.0: An updated database of bacterial type II toxin-antitoxin loci. *Nucleic Acids Res.* **2018**, *46*, D749–D753. [CrossRef] [PubMed]
25. Gerdes, K.; Christensen, S.K.; Løbner-Olesen, A. Prokaryotic toxin-antitoxin stress response loci. *Nat. Rev. Microbiol.* **2005**, *3*, 371–382. [CrossRef] [PubMed]
26. Van Melderen, L.; De Bast, M.S. Bacterial toxin-antitoxin systems: More than selfish entities? *PLoS Genet.* **2009**, *5*, e1000437. [CrossRef]
27. Makarova, K.S.; Wolf, Y.I.; Koonin, E.V. Comprehensive comparative-genomic analysis of Type 2 toxin-antitoxin systems and related mobile stress response systems in prokaryotes. *Biol. Direct* **2009**, *4*, 19. [CrossRef]
28. Fernandez-Garcia, L.; Blasco, L.; Lopez, M.; Bou, G.; Garcia-Contreras, R.; Wood, T.; Tomas, M. Toxin-antitoxin systems in clinical pathogens. *Toxins (Basel)* **2016**, *8*, 227. [CrossRef]
29. Lobato-Márquez, D.; Díaz-Orejas, R.; García-del Portillo, F. Toxin-antitoxins and bacterial virulence. *FEMS Microbiol. Rev.* **2016**, *40*, 592–609. [CrossRef]
30. Kędzierska, B.; Hayes, F. Emerging roles of toxin-antitoxin modules in bacterial pathogenesis. *Molecules* **2016**, *21*, 790. [CrossRef]

31. Yang, Q.E.; Walsh, T.R. Toxin-antitoxin systems and their role in disseminating and maintaining antimicrobial resistance. *FEMS Microbiol. Rev.* **2017**, *41*, 343–353. [CrossRef] [PubMed]
32. Kang, S.M.; Kim, D.H.; Jin, C.; Lee, B.J. A systematic overview of type II and III toxin-antitoxin systems with a focus on druggability. *Toxins (Basel)* **2018**, *10*, 515. [CrossRef] [PubMed]
33. Slayden, R.A.; Dawson, C.C.; Cummings, J.E. Toxin-antitoxin systems and regulatory mechanisms in *Mycobacterium tuberculosis*. *Pathog. Dis.* **2018**, *76*, fty039. [CrossRef] [PubMed]
34. Korch, S.B.; Malhotra, V.; Contreras, H.; Clark-Curtiss, J.E. The *Mycobacterium tuberculosis* relBE toxin:antitoxin genes are stress-responsive modules that regulate growth through translation inhibition. *J. Microbiol.* **2015**, *53*, 783–795. [CrossRef] [PubMed]
35. Gupta, A.; Venkataraman, B.; Vasudevan, M.; Gopinath Bankar, K. Co-expression network analysis of toxin-antitoxin loci in *Mycobacterium tuberculosis* reveals key modulators of cellular stress. *Sci. Rep.* **2017**, *7*, 5868. [CrossRef] [PubMed]
36. Thakur, Z.; Saini, V.; Arya, P.; Kumar, A.; Mehta, P.K. Computational insights into promoter architecture of toxin-antitoxin systems of *Mycobacterium tuberculosis*. *Gene* **2018**, *641*, 161–171. [CrossRef]
37. Yang, M.; Gao, C.; Wang, Y.; Zhang, H.; He, Z.G. Characterization of the interaction and cross-regulation of three *Mycobacterium tuberculosis* RelBE modules. *PLoS ONE* **2010**, *5*, e10672. [CrossRef]
38. Agarwal, S.; Tiwari, P.; Deep, A.; Kidwai, S.; Gupta, S.; Thakur, K.G.; Singh, R. System-wide analysis unravels the differential regulation and in vivo essentiality of virulence-associated proteins B and C toxin-antitoxin systems of *Mycobacterium tuberculosis*. *J. Infect. Dis.* **2018**, *217*, 1809–1820. [CrossRef]
39. Zhu, L.; Sharp, J.D.; Kobayashi, H.; Woychik, N.A.; Inouye, M. Noncognate *Mycobacterium tuberculosis* toxin-antitoxins can physically and functionally interact. *J. Biol. Chem.* **2010**, *285*, 39732–39738. [CrossRef]
40. Ramirez, M.V.; Dawson, C.C.; Crew, R.; England, K.; Slayden, R.A. MazF6 toxin of *Mycobacterium tuberculosis* demonstrates antitoxin specificity and is coupled to regulation of cell growth by a Soj-like protein. *BMC Microbiol.* **2013**, *13*, 240. [CrossRef]
41. Tandon, H.; Melarkode Vattekatte, A.; Srinivasan, N.; Sandhya, S. Molecular and structural basis of cross-reactivity in M. tuberculosis toxin–antitoxin systems. *Toxins (Basel)* **2020**, *12*, 481. [CrossRef] [PubMed]
42. Solano-Gutierrez, J.S.; Pino, C.; Robledo, J. Toxin-antitoxin systems shows variability among *Mycobacterium tuberculosis* lineages. *FEMS Microbiol. Lett.* **2019**, *366*, fny276. [CrossRef] [PubMed]
43. Tandon, H.; Sharma, A.; Wadhwa, S.; Varadarajan, R.; Singh, R.; Srinivasan, N.; Sandhya, S. Bioinformatic and mutational studies of related toxin–antitoxin pairs in *Mycobacterium tuberculosis* predict and identify key functional residues. *J. Biol. Chem.* **2019**, *294*, 9048–9063. [CrossRef]
44. Schuster, C.F.; Bertram, R. Toxin-antitoxin systems of Staphylococcus aureus. *Toxins (Basel)* **2016**, *8*, 140. [CrossRef] [PubMed]
45. Bukowski, M.; Hyz, K.; Janczak, M.; Hydzik, M.; Dubin, G.; Wladyka, B. Identification of novel mazEF/pemIK family toxin-antitoxin loci and their distribution in the Staphylococcus genus. *Sci. Rep.* **2017**, *7*, 13462. [CrossRef] [PubMed]
46. Habib, G.; Zhu, Q.; Sun, B. Bioinformatics and functional assessment of toxin-antitoxin systems in Staphylococcus aureus. *Toxins (Basel)* **2018**, *10*, 473. [CrossRef]
47. Kato, F.; Yoshizumi, S.; Yamaguchi, Y.; Inouye, M. Genome-wide screening for identification of novel toxin-antitoxin systems in Staphylococcus aureus. *Appl. Environ. Microbiol.* **2019**, *85*, e00915-19. [CrossRef]
48. Sierra, R.; Viollier, P.; Renzoni, A. Linking toxin-antitoxin systems with phenotypes: A Staphylococcus aureus viewpoint. *Biochim. Biophys. Acta Gene Regul. Mech.* **2019**, *1862*, 742–751. [CrossRef]
49. Horesh, G.; Fino, C.; Harms, A.; Dorman, M.J.; Parts, L.; Gerdes, K.; Heinz, E.; Thomson, N.R. Type II and type IV toxin-antitoxin systems show different evolutionary patterns in the global *Klebsiella pneumoniae* population. *Nucleic Acids Res.* **2020**, *48*, 4357–4370. [CrossRef]
50. Akarsu, H.; Bordes, P.; Mansour, M.; Bigot, D.J.; Genevaux, P.; Falquet, L. TASmania: A bacterial toxin-antitoxin systems database. *PLoS Comput. Biol.* **2019**, *15*, e1006946. [CrossRef]
51. Agüero, J.A.; Akarsu, H.; Aguilar-Bultet, L.; Oevermann, A.; Falquet, L. Large-scale comparison of toxin and antitoxins in *Listeria monocytogenes*. *Toxins (Basel)* **2020**, *12*, 29. [CrossRef] [PubMed]
52. Shapiro, S. Speculative strategies for new antibacterials: All roads should not lead to Rome. *J. Antibiot. (Tokyo)* **2013**, *66*, 371–386. [CrossRef] [PubMed]
53. Germain, E.; Castro-Roa, D.; Zenkin, N.; Gerdes, K. Molecular mechanism of bacterial persistence by HipA. *Mol. Cell* **2013**, *52*, 248–254. [CrossRef] [PubMed]

54. Zhang, S.-P.; Wang, Q.; Quan, S.-W.; Yu, X.-Q.; Wang, Y.; Guo, D.-D.; Peng, L.; Feng, H.-Y.; Yong-Xing, H. Type II toxin–antitoxin system in bacteria: Activation, function, and mode of action. *Biophys. Rep.* **2020**, *6*, 68–79. [CrossRef]
55. Ramisetty, B.C.M.; Santhosh, R.S. Endoribonuclease type II toxin-antitoxin systems: Functional or selfish? *Microbiology* **2017**, *163*, 931–939. [CrossRef]
56. Lioy, V.S.; Rey, O.; Balsa, D.; Pellicer, T.; Alonso, J.C. A toxin-antitoxin module as a target for antimicrobial development. *Plasmid* **2010**, *63*, 31–39. [CrossRef]
57. Agarwal, S.; Agarwal, S.; Bhatnagar, R. Identification and characterization of a novel toxin-antitoxin module from Bacillus anthracis. *FEBS Lett.* **2007**, *581*, 1727–1734. [CrossRef]
58. Agarwal, S.; Mishra, N.K.; Bhatnagar, S.; Bhatnagar, R. PemK toxin of Bacillus anthracis is a ribonuclease: An insight into its active site, structure, and function. *J. Biol. Chem.* **2010**, *285*, 7254–7270. [CrossRef]
59. Verma, S.; Kumar, S.; Gupta, V.P.; Gourinath, S.; Bhatnagar, S.; Bhatnagar, R. Structural basis of Bacillus anthracis MoxXT disruption and the modulation of MoxT ribonuclease activity by rationally designed peptides. *J. Biomol. Struct. Dyn.* **2015**, *33*, 606–624. [CrossRef]
60. Chopra, N.; Agarwal, S.; Verma, S.; Bhatnagar, S.; Bhatnagar, R. Modeling of the structure and interactions of the B. anthracis antitoxin, MoxX: Deletion mutant studies highlight its modular structure and repressor function. *J. Comput. Aided. Mol. Des.* **2011**, *25*, 275–291. [CrossRef]
61. Lee, I.G.; Lee, S.J.; Chae, S.; Lee, K.Y.; Kim, J.H.; Lee, B.J. Structural and functional studies of the *Mycobacterium tuberculosis* VapBC30 toxin-antitoxin system: Implications for the design of novel antimicrobial peptides. *Nucleic Acids Res.* **2015**, *43*, 7624–7637. [CrossRef] [PubMed]
62. Chan, W.T.; Moreno-Córdoba, I.; Yeo, C.C.; Espinosa, M. Toxin-antitoxin genes of the Gram-positive pathogen Streptococcus pneumoniae: So few and yet so many. *Microbiol. Mol. Biol. Rev.* **2012**, *76*, 773–791. [CrossRef] [PubMed]
63. Równicki, M.; Pieńko, T.; Czarnecki, J.; Kolanowska, M.; Bartosik, D.; Trylska, J. Artificial activation of Escherichia coli mazEF and hipBA toxin–antitoxin systems by antisense peptide nucleic acids as an antibacterial strategy. *Front. Microbiol.* **2018**, *9*, 2870. [CrossRef] [PubMed]
64. Good, L.; Awasthi, S.K.; Dryselius, R.; Larsson, O.; Nielsen, P.E. Bactericidal antisense effects of peptide-PNA conjugates. *Nat. Biotechnol.* **2001**, *19*, 360–364. [CrossRef] [PubMed]
65. Wojciechowska, M.; Równicki, M.; Mieczkowski, A.; Miszkiewicz, J.; Trylska, J. Antibacterial peptide nucleic acids—Facts and perspectives. *Molecules* **2020**, *25*, 559. [CrossRef] [PubMed]
66. Christensen, S.K.; Maenhaut-Michel, G.; Mine, N.; Gottesman, S.; Gerdes, K.; Van Melderen, L. Overproduction of the Lon protease triggers inhibition of translation in Escherichia coli: Involvement of the yefM-yoeB toxin-antitoxin system. *Mol. Microbiol.* **2004**, *51*, 1705–1717. [CrossRef]
67. Brzozowska, I.; Zielenkiewicz, U. The ClpXP protease is responsible for the degradation of the Epsilon antidote to the Zeta toxin of the streptococcal pSM19035 plasmid. *J. Biol. Chem.* **2014**, *289*, 7514–7523. [CrossRef]
68. Van Melderen, L.; Thi, M.H.D.; Lecchi, P.; Gottesman, S.; Couturier, M.; Maurizi, M.R. ATP-dependent degradation of CcdA by Lon protease. Effects of secondary structure and heterologous subunit interactions. *J. Biol. Chem.* **1996**, *271*, 27730–27738. [CrossRef]
69. Muthuramalingam, M.; White, J.C.; Bourne, C.R. Toxin-antitoxin modules are pliable switches activated by multiple protease pathways. *Toxins (Basel)* **2016**, *8*, 214. [CrossRef]
70. Goff, S.A.; Goldberg, A.L. An increased content of protease La, the lon gene product, increases protein degradation and blocks growth in Escherichia coli. *J. Biol. Chem.* **1987**, *262*, 4508–4515.
71. Brötz-Oesterhelt, H.; Beyer, D.; Kroll, H.P.; Endermann, R.; Ladel, C.; Schroeder, W.; Hinzen, B.; Raddatz, S.; Paulsen, H.; Henninger, K.; et al. Dysregulation of bacterial proteolytic machinery by a new class of antibiotics. *Nat. Med.* **2005**, *11*, 1082–1087. [CrossRef] [PubMed]
72. Kumar, S.; Engelberg-Kulka, H. Quorum sensing peptides mediating interspecies bacterial cell death as a novel class of antimicrobial agents. *Curr. Opin. Microbiol.* **2014**, *21*, 22–27. [CrossRef]
73. Kumar, S.; Kolodkin-Gal, I.; Engelberg-Kulka, H. Novel quorum-sensing peptides mediating interspecies bacterial cell death. *MBio* **2013**, *4*, e00314-13. [CrossRef] [PubMed]
74. Engelberg-Kulka, H.; Hazan, R.; Amitai, S. mazEF: A chromosomal toxin-antitoxin module that triggers programmed cell death in bacteria. *J. Cell Sci.* **2005**, *118*, 4327–4332. [CrossRef] [PubMed]

75. Nigam, A.; Kumar, S.; Engelberg-Kulka, H. Quorum sensing extracellular death peptides enhance the endoribonucleolytic activities of *Mycobacterium tuberculosis* MazF toxins. *MBio* **2018**, *9*, e00685-18. [CrossRef] [PubMed]
76. Ramisetty, B.C.M.; Raj, S.; Ghosh, D. Escherichia coli MazEF toxin-antitoxin system does not mediate programmed cell death. *J. Basic Microbiol.* **2016**, *56*, 1398–1402. [CrossRef] [PubMed]
77. Fu, Z.; Tamber, S.; Memmi, G.; Donegan, N.P.; Cheung, A.L. Overexpression of MazF in *Staphylococcus aureus* induces bacteriostasis by selectively targeting mRNAs for cleavage. *J. Bacteriol.* **2009**, *191*, 2051–2059. [CrossRef]
78. Ronneau, S.; Helaine, S. Clarifying the link between toxin–antitoxin modules and bacterial persistence. *J. Mol. Biol.* **2019**, *431*, 3462–3471. [CrossRef]
79. Jimmy, S.; Saha, C.K.; Kurata, T.; Stavropoulos, C.; Oliveira, S.R.A.; Koh, A.; Cepauskas, A.; Takada, H.; Rejman, D.; Tenson, T.; et al. A widespread toxin–antitoxin system exploiting growth control via alarmone signaling. *Proc. Natl. Acad. Sci. USA* **2020**, *117*, 10500–10510. [CrossRef]
80. Török, I.; Kari, C. Accumulation of ppGpp in a relA mutant of Escherichia coli during amino acid starvation. *J. Biol. Chem.* **1980**, *255*, 3838–3840.
81. Ramisetty, B.C.M.; Natarajan, B.; Santhosh, R.S. MazEF-mediated programmed cell death in bacteria: "What is this?". *Crit. Rev. Microbiol.* **2015**, *41*, 89–100. [CrossRef] [PubMed]
82. López-Igual, R.; Bernal-Bayard, J.; Rodríguez-Patón, A.; Ghigo, J.M.; Mazel, D. Engineered toxin–intein antimicrobials can selectively target and kill antibiotic-resistant bacteria in mixed populations. *Nat. Biotechnol.* **2019**, *37*, 755–760. [CrossRef] [PubMed]
83. Collin, F.; Karkare, S.; Maxwell, A. Exploiting bacterial DNA gyrase as a drug target: Current state and perspectives. *Appl. Microbiol. Biotechnol.* **2011**, *92*, 479–497. [CrossRef] [PubMed]
84. Bernard, P.; Kézdy, K.E.; Van Melderen, L.; Steyaert, J.; Wyns, L.; Pato, M.L.; Higgins, P.N.; Couturier, M. The F plasmid CcdB protein induces efficient ATP-dependent DNA cleavage by gyrase. *J. Mol. Biol.* **1993**, *234*, 534–541. [CrossRef]
85. Childers, B.M.; Klose, K.E. Regulation of virulence in Vibrio cholerae: The ToxR regulon. *Future Microbiol.* **2007**, *2*, 335–344. [CrossRef]
86. Beaber, J.W.; Hochhut, B.; Waldor, M.K. SOS response promotes horizontal dissemination of antibiotic resistance genes. *Nature* **2004**, *427*, 72–74. [CrossRef]
87. Shimazu, T.; Mirochnitchenko, O.; Phadtare, S.; Inouye, M. Regression of solid tumors by induction of MazF, a bacterial mRNA endoribonuclease. *J. Mol. Microbiol. Biotechnol.* **2014**, *24*, 228–233. [CrossRef]
88. Chono, H.; Saito, N.; Tsuda, H.; Shibata, H.; Ageyama, N.; Terao, K.; Yasutomi, Y.; Mineno, J.; Kato, I. In vivo safety and persistence of endoribonuclease gene-transduced CD4+ t cells in cynomolgus macaques for HIV-1 gene therapy model. *PLoS ONE* **2011**, *6*, e23585. [CrossRef]
89. Chono, H.; Matsumoto, K.; Tsuda, H.; Saito, N.; Lee, K.; Kim, S.; Shibata, H.; Ageyama, N.; Terao, K.; Yasutomi, Y.; et al. Acquisition of HIV-1 resistance in T lymphocytes using an ACA-specific *E. coli* mRNA interferase. *Hum. Gene Ther.* **2011**, *22*, 35–43. [CrossRef]
90. Lewis, K. Persister cells. *Annu. Rev. Microbiol.* **2010**, *64*, 357–372. [CrossRef]
91. Lewis, K. Persister cells and the riddle of biofilm survival. *Biochemistry* **2005**, *70*, 267–274. [CrossRef] [PubMed]
92. Lewis, K. Multidrug tolerance of biofilms and persister cells. *Curr. Top. Microbiol. Immunol.* **2008**, *322*, 107–131. [PubMed]
93. Wood, T.K.; Knabel, S.J.; Kwan, B.W. Bacterial persister cell formation and dormancy. *Appl. Environ. Microbiol.* **2013**, *79*, 7116–7121. [CrossRef] [PubMed]
94. Lebeaux, D.; Ghigo, J.-M.; Beloin, C. Biofilm-related infections: Bridging the gap between clinical management and fundamental aspects of recalcitrance toward antibiotics. *Microbiol. Mol. Biol. Rev.* **2014**, *78*, 510–543. [CrossRef]
95. Soo, V.W.C.; Wood, T.K. Antitoxin MqsA represses curli formation through the master biofilm regulator CsgD. *Sci. Rep.* **2013**, *3*, 3186. [CrossRef]
96. González Barrios, A.F.; Zuo, R.; Hashimoto, Y.; Yang, L.; Bentley, W.E.; Wood, T.K. Autoinducer 2 controls biofilm formation in *Escherichia coli* through a novel motility quorum-sensing regulator (MqsR, B3022). *J. Bacteriol.* **2006**, *188*, 305–316. [CrossRef]

97. Kasari, V.; Kurg, K.; Margus, T.; Tenson, T.; Kaldalu, N. The Escherichia coli mqsR and ygiT genes encode a new toxin-antitoxin pair. *J. Bacteriol.* **2010**, *192*, 2908–2919. [CrossRef]
98. Fraikin, N.; Rousseau, C.J.; Goeders, N.; Van Melderen, L. Reassessing the role of the type II MqsRA toxin-antitoxin system in stress response and biofilm formation: MqsA is transcriptionally uncoupled from mqsR. *MBio* **2019**, *10*, e02678-19. [CrossRef]
99. Ramisetty, B.C.M.; Ghosh, D.; Chowdhury, M.R.; Santhosh, R.S. Corrigendum: What is the link between stringent response, endoribonuclease encoding type II toxin-antitoxin systems and persistence? *Front. Microbiol.* **2017**, *8*, 458. [CrossRef]
100. Moyed, H.S.; Bertrand, K.P. hipA, a newly recognized gene of Escherichia coli K-12 that affects frequency of persistence after inhibition of murein synthesis. *J. Bacteriol.* **1983**, *155*, 768–775. [CrossRef]
101. Van Melderen, L.; Wood, T.K. Commentary: What is the link between stringent response, endoribonuclease encoding type II toxin-antitoxin systems and persistence? *Front. Microbiol.* **2017**, *14*, 191. [CrossRef] [PubMed]
102. Shan, Y.; Gandt, A.B.; Rowe, S.E.; Deisinger, J.P.; Conlon, B.P.; Lewis, K. ATP-Dependent persister formation in Escherichia coli. *MBio* **2017**, *8*, e02267-16. [CrossRef] [PubMed]
103. Ramisetty, B.C.M.; Ghosh, D.; Chowdhury, M.R.; Santhosh, R.S. What is the link between stringent response, endoribonuclease encoding type II toxin-antitoxin systems and persistence? *Front. Microbiol.* **2016**, *7*, 1882. [CrossRef] [PubMed]
104. Harms, A.; Maisonneuve, E.; Gerdes, K. Mechanisms of bacterial persistence during stress and antibiotic exposure. *Science* **2016**, *16*, aaf4268. [CrossRef] [PubMed]
105. Maisonneuve, E.; Castro-Camargo, M.; Gerdes, K. Retraction notice to: (p)ppGpp controls bacterial persistence by stochastic induction of toxin-antitoxin activity. *Cell* **2018**, *172*, 1135. [CrossRef]
106. Goormaghtigh, F.; Fraikin, N.; Putrinš, M.; Hallaert, T.; Hauryliuk, V.; Garcia-Pino, A.; Sjödin, A.; Kasvandik, S.; Udekwu, K.; Tenson, T.; et al. Reassessing the role of type II toxin-antitoxin systems in formation of escherichia coli type II persister cells. *MBio* **2018**, *9*, e00640-18. [CrossRef]
107. Goeders, N.; Van Melderen, L. Toxin-antitoxin systems as multilevel interaction systems. *Toxins (Basel)* **2013**, *6*, 304–324. [CrossRef]
108. Van Melderen, L. Toxin-antitoxin systems: Why so many, what for? *Curr. Opin. Microbiol.* **2010**, *13*, 781–785. [CrossRef]
109. Pacios, O.; Blasco, L.; Bleriot, I.; Fernandez-Garcia, L.; Ambroa, A.; López, M.; Bou, G.; Cantón, R.; Garcia-Contreras, R.; Wood, T.K.; et al. (p)ppGpp and its role in bacterial persistence: New challenges. *Antimicrob. Agents Chemother.* **2020**. [CrossRef]
110. Tripathi, A.; Dewan, P.C.; Siddique, S.A.; Varadarajan, R. MazF-induced growth inhibition and persister generation in Escherichia coli. *J. Biol. Chem.* **2014**, *289*, 4191–4205. [CrossRef]
111. Harrison, J.J.; Wade, W.D.; Akierman, S.; Vacchi-Suzzi, C.; Stremick, C.A.; Turner, R.J.; Ceri, H. The chromosomal toxin gene yafQ is a determinant of multidrug tolerance for *Escherichia coli* growing in a biofilm. *Antimicrob. Agents Chemother.* **2009**, *53*, 2253–2258. [CrossRef] [PubMed]
112. Norton, J.P.; Mulvey, M.A. Toxin-Antitoxin Systems Are Important for Niche-Specific Colonization and Stress Resistance of Uropathogenic *Escherichia coli*. *PLoS Pathog.* **2012**, *8*, e1002954. [CrossRef] [PubMed]
113. Slattery, A.; Victorsen, A.H.; Brown, A.; Hillman, K.; Phillips, G.J. Isolation of highly persistent mutants of *Salmonella enterica* serovar typhimurium reveals a new toxin-antitoxin module. *J. Bacteriol.* **2013**, *195*, 647–657. [CrossRef] [PubMed]
114. Helaine, S.; Cheverton, A.M.; Watson, K.G.; Faure, L.M.; Matthews, S.A.; Holden, D.W. Internalization of salmonella by macrophages induces formation of nonreplicating persisters. *Science* **2014**, *343*, 204–208. [CrossRef] [PubMed]
115. Cheverton, A.M.; Gollan, B.; Przydacz, M.; Wong, C.T.; Mylona, A.; Hare, S.A.; Helaine, S. A Salmonella toxin promotes persister formation through acetylation of tRNA. *Mol. Cell* **2016**, *63*, 86–96. [CrossRef] [PubMed]
116. Rycroft, J.A.; Gollan, B.; Grabe, G.J.; Hall, A.; Cheverton, A.M.; Larrouy-Maumus, G.; Hare, S.A.; Helaine, S. Activity of acetyltransferase toxins involved in Salmonella persister formation during macrophage infection. *Nat. Commun.* **2018**, *9*, 1993. [CrossRef]
117. Fisher, R.A.; Gollan, B.; Helaine, S. Persistent bacterial infections and persister cells. *Nat. Rev. Microbiol.* **2017**, *15*, 453–464. [CrossRef]

118. Wilmaerts, D.; Dewachter, L.; De Loose, P.-J.; Bollen, C.; Verstraeten, N.; Michiels, J. HokB Monomerization and Membrane Repolarization Control Persister Awakening. *Mol. Cell* **2019**, *75*, 1031–1042. [CrossRef]
119. Dörr, T.; Vulic, M.; Lewis, K. Ciprofloxacin causes persister formation by inducing the TisB toxin in *Escherichia coli*. *PLoS Biol.* **2010**, *8*, e1000317. [CrossRef]
120. Cheng, H.Y.; Soo, V.W.C.; Islam, S.; McAnulty, M.J.; Benedik, M.J.; Wood, T.K. Toxin GhoT of the GhoT/GhoS toxin/antitoxin system damages the cell membrane to reduce adenosine triphosphate and to reduce growth under stress. *Environ. Microbiol.* **2014**, *16*, 1741–1754. [CrossRef]
121. Wang, X.; Lord, D.M.; Hong, S.H.; Peti, W.; Benedik, M.J.; Page, R.; Wood, T.K. Type II toxin/antitoxin MqsR/MqsA controls type V toxin/antitoxin GhoT/GhoS. *Environ. Microbiol.* **2013**, *15*, 1734–1744. [CrossRef] [PubMed]
122. Defraine, V.; Fauvart, M.; Michiels, J. Fighting bacterial persistence: Current and emerging anti-persister strategies and therapeutics. *Drug Resist. Updates* **2018**, *38*, 12–26. [CrossRef] [PubMed]
123. Conlon, B.P.; Rowe, S.E.; Gandt, A.B.; Nuxoll, A.S.; Donegan, N.P.; Zalis, E.A.; Clair, G.; Adkins, J.N.; Cheung, A.L.; Lewis, K. Persister formation in *Staphylococcus aureus* is associated with ATP depletion. *Nat. Microbiol.* **2016**, *1*, 16051. [CrossRef] [PubMed]
124. Li, T.; Yin, N.; Liu, H.; Pei, J.; Lai, L. Novel inhibitors of toxin HipA reduce multidrug tolerant persisters. *ACS Med. Chem. Lett.* **2016**, *7*, 449–453. [CrossRef] [PubMed]
125. Solecki, O.; Mosbah, A.; Baudy Floc'h, M.; Felden, B. Converting a *Staphylococcus aureus* toxin into effective cyclic pseudopeptide antibiotics. *Chem. Biol.* **2015**, *22*, 329–335. [CrossRef]

MDPI
St. Alban-Anlage 66
4052 Basel
Switzerland
Tel. +41 61 683 77 34
Fax +41 61 302 89 18
www.mdpi.com

Toxins Editorial Office
E-mail: toxins@mdpi.com
www.mdpi.com/journal/toxins

www.ingramcontent.com/pod-product-compliance
Lightning Source LLC
LaVergne TN
LVHW070854170726
843515LV00005B/649

* 9 7 8 3 0 3 6 5 0 6 7 4 6 *